INEVITABLY TOXIC

INTERSECTIONS: ENVIRONMENT, SCIENCE, TECHNOLOGY

Sarah Elkind and Finn Arne Jørgensen, Editors

EDITED BY BRINDA SARATHY, VIVIEN HAMILTON, AND JANET FARRELL BRODIE

INEVITABLY TOXIC

HISTORICAL PERSPECTIVES ON CONTAMINATION, EXPOSURE, AND EXPERTISE

UNIVERSITY OF PITTSBURGH PRESS

Published by the University of Pittsburgh Press, Pittsburgh, Pa., 15260

This paperback edition, 2019
Manufactured in the United States of America
Printed on acid-free paper

Cataloging-in-Publication data is available from the Library of Congress

ISBN 13: 978-0-8229-6612-8

Cover art: federicofoto / 123RF
Cover design: Alex Wolfe

To our students: your curiosity and courage sustain our work and give us hope.

Contents

Acknowledgments

The initial spark for this project came out of a reading group on Science, Expertise, and Environment that we organized along with Pey-Yi Chu at Pomona College. Our first thanks go to the Hixon-Riggs Forum for Responsive Science and Engineering at Harvey Mudd College for funding this reading group, which provided such a fruitful space for interdisciplinary conversation with colleagues across the Claremont Colleges. Brinda and Vivien would especially like to thank Janet for bringing us together in the first place. Claremont isn't big, but making these kinds of connections across the colleges takes effort, and we are grateful for Janet's vision and initiative.

Early drafts of most chapters in this volume were presented at the Contested Expertise, Toxic Environments (CETE) workshop held in Claremont in fall 2015. Papers were precirculated and received written feedback from each participant prior to the meeting. We are grateful to workshop participants for their rich and engaged discussion of the essays. The chapters in this volume are absolutely stronger as a result of those conversations.

As part of this workshop, we invited the wider Claremont Colleges community to an evening screening of *Containment* by Peter Galison and Robert Moss. Afterward, we had the great pleasure of having Peter Galison Skype in for a conversation with the audience—a call he made from an abandoned parking lot in Maine well past midnight. Everyone on our end appreciated the extra insight and inspiration from this shared dialogue. Thank you, Peter! This public event was generously funded by the Harvey Mudd College Center for Environmental Studies.

The biggest source of funding for this project came from the Bradshaw Fund of the Claremont Graduate University (CGU) School of Arts and Humanities. We are deeply grateful for this financial support, which allowed us to bring together scholars for the 2015 CETE workshop and which has funded the costs associated with the publication of this volume. Additional funds for the index were provided by the Faculty Research and Awards

Fund at Pitzer College and the Department of Humanities, Social Sciences, and the Arts at Harvey Mudd College.

Thank you to Debbie Laird in the Department of Humanities, Social Sciences, and the Arts at Harvey Mudd College for assisting with catering and facilities logistics for the CETE workshop.

Finally, a huge thank-you to CGU graduate student Hillary Kirkham for her help keeping us organized, overseeing planning for the workshop, and staying in touch with authors throughout rounds of revision and editing. We truly could not have done this without you.

INEVITABLY TOXIC

Introduction

Toxicity, Uncertainty, and Expertise

Vivien Hamilton and Brinda Sarathy

Almost every year, students taking environmental justice at Pitzer College go on a toxics tour of their backyards in the Inland Empire of Southern California. This trip usually includes a visit to the Stringfellow Acid Pits, California's first site to be designated as a Superfund in 1983.[1] Millions of gallons of chemical waste were dumped there from the mid-1950s to the early 1970s, and authorities estimate it will take at least four hundred years to clean up this contamination. Having learned about the history of this site in class (see chapter 5 of this volume), students anxiously anticipate seeing for themselves what this toxic disaster zone actually looks like. When they get to the site, however, most students are taken aback. It is not what they expected. There are no open pits of discolored or smoldering liquids; there is no acrid smell in the air. There are not even any noticeable signs alerting them to the contaminated landscape upon which they stand. Some students note that they have driven past Stringfellow on various occasions but would never have guessed that this barren canyon on the side of the highway has a history of contamination by chemical wastes. At the end of the day, one of the most impactful lessons for the class is that toxic environments are often invisible or appear innocuous, and that such spaces are more prevalent in our day-to-day lives than we either know or care to admit.

The questions that haunt students after a visit to the Stringfellow site are the questions that motivate this book. To what extent do we know about the processes resulting in contaminated places like Stringfellow, and do we, in fact, even recognize such spaces for what they are? How is it that toxic environments have become so pervasive while at the same time remaining invisible, overlooked, or ignored? Why do conditions of normalized toxicity fail to rouse mass outcry? Numerous scholars from a broad range of academic disciplines—from environmental history to public health, from sociology to geography, and from science and technology studies to environmental philosophy—have tackled such questions in their rich and diverse writings on toxic environments. This substantive and growing body of interdisciplinary scholarship, however, tends to be written by and for the consumption of other academics, who are themselves experts in their fields and who engage around toxicity through shared theoretical concepts.[2] The extension of Michel Foucault's concept of biopower, for example, asking us to pay attention to the ways in which states count and control populations, has generated much productive thinking and writing about toxicity at multiple scales.[3] In science and technology studies, Sheila Jasanoff's seminal work on the coproduction of scientific knowledge and social norms has made room for scholars to more explicitly focus on the inextricable and complex ways in which scientific knowledge and public policy are shaped together, elucidating the proposition that how "we know and represent the world (both nature and society) are inseparable from the ways we chose to live in it."[4] As academics ourselves, we use these kinds of intellectual frameworks in our own research and analysis.

Yet, as professors of the liberal arts, we are also keenly aware that there is a broader audience to be engaged. In tackling the emergence of toxic environments in multiple sites across the twentieth century and into the twenty-first, we have therefore purposely written this book for a nonexpert audience. For us, this has meant limiting academic jargon, clarifying terms when they are used, and imbuing theory implicitly into the very telling of our stories. We have often found compelling storytelling to be the most effective means of capturing our students' imaginations and sparking critical conversations. We hope that by conveying histories of toxicity in this intellectually rooted and evocative manner, a broader audience will be similarly engaged.

The stories in this volume draw attention to a diverse set of toxic spaces in the United States, Canada, and Japan, spaces filled with x-rays, nuclear radiation, industrial waste, pesticides, and other chemical contaminants.

Given the often-imperceptible nature of these agents, our first goal with this collection is simply one of illumination. Taken together, these chapters show us the ways in which exposure to toxicity has become routine, as toxic spaces have become increasingly interwoven into the economic structures and fabric of everyday life. Even more, these stories demonstrate that the burden of exposure continues to fall disproportionately on those already marginalized by class, race, and structures of colonization. Illuminating this reality, however, is just the first step. Our ultimate hope is that uncovering the histories of these spaces will make complacency impossible.

Given the pervasive nature of toxic spaces and the urgent need for action, it might be surprising that most of the work in this book is historical rather than contemporary. Why examine x-ray rooms in the 1920s or pesticide use in the 1970s when we need to address water contamination from fracking right now? If we know the current state of contamination at the Stringfellow site, why delve into archives to pull out debates and decisions that are over half a century old? We suggest that this kind of historical analysis is powerful precisely because it disrupts the sense that our current predicament is inevitable. Understanding how it is that these spaces came into being can help us identify contemporary institutions as well as modes of thinking and acting that continue to allow environments of toxicity to persist. This seems especially urgent given the current political climate of deregulation in the United States, in which calls to "grow the economy" have become routine and are decoupled from any meaningful analysis of the ecologically unsustainable, socially exploitative, and violent processes through which capitalist relations operate. Careful historical analysis can illuminate these realities and inspire us to see how we can intervene to stem the tide of toxic spaces now and in the future.

In this introductory chapter, we first briefly outline the broader context in which environments of toxicity have been produced by modern imperatives of technological progress and economic growth. We then turn to consider the ways in which institutions of scientific expertise often work to hide—whether intentionally or not—the uncertain nature of knowledge about toxicity, excluding the experiences of those exposed to toxic agents. Despite deep and persistent uncertainties, scientific experts and other authority figures have often been called on to mitigate concerns about harmful substances, thus facilitating industrial and military expansion. We contend that this general pattern—articulated uniquely in different times and places—has resulted in conditions of environmental contamination and, often, disproportionate harm to already marginalized groups. Finally, in our roadmap to this volume, we highlight some common themes across and between chapters, and reflect on the larger context of contestation and struggles for environmental justice in response to toxic environments.

SITUATING TOXICITY

Toxic environments are a characteristic of our political-economic system and, more broadly, represent what sociologist Anthony Giddens has termed "manufactured risks": risks created by "the very progression of human development, especially by the progression of science and technology for which history provides us with very little previous experience."[5] Historian of science Michelle Murphy points to the emergence of "a chemical regime of living," in which toxics pervade environments at multiple scales, from individual bodies to geographic terrains and processes of production and consumption. The fact that toxics traverse so many kinds of boundaries, Murphy argues, requires "us to tie the history of technoscience with political economy."[6] Scholarship in environmental history, in particular, has shown the ways in which attitudes of technological hubris and manipulations of scientific uncertainty have resulted in the creation of toxic landscapes.[7] The case studies in this volume build upon this work and harness ways of thinking critically about toxicity at multiple scales, to more explicitly lay bare the political-economic foundations, modes of logic, and bases of knowledge upon which toxic spaces have been produced and obscured.

On the one hand, these cases could be read as proof of an increasing awareness of the toxic landscapes we inhabit, demonstrated by attempts to regulate and manage toxic substances, to study their circulation in the environment, and to create structures of safety to minimize exposure and keep bodies safe. Such actions are indicative of what sociologist Ulrich Beck has termed a "risk society," in which we anticipate, organize around, and respond to manufactured risks.[8] On the other hand, a closer look reveals just how inadequate and compromised these processes of regulation have been, almost from their very inception.[9] As far back as the early twentieth century, toxic experiments were real-time practices: new toxic agents were released into human environments and fragile ecosystems, and not studied first in isolation to assess possible negative impacts. Scholars elsewhere and in this volume show how this trend has continued, evident in the development and deployment of nuclear weapons and technologies during and after World War II, in the marketing of pesticides to consumers in newly concocted battles against insects and weeds, and in the decisions made about how best to dispose of these substances.[10]

In the case of nuclear technologies, their proliferation and the resulting environmental contamination has been uniquely hidden by a culture of Cold War secrecy. Scholars have only recently started to show us the widespread global impacts of this vast nuclear complex, uncovering, for instance, the ways in which scientists and politicians have made decisions about where to dump nuclear waste.[11] The experiences of individuals impacted by these decisions are increasingly coming into focus, including stories about growing up near secret weapons facilities, working in plutonium plants, and surviving nuclear disasters like Chernobyl.[12] The struggle for recognition and reparation is ongoing for multiple communities impacted by the nuclear industry globally.[13]

In all of these cases, individuals in positions of authority—among them scientists, military officials, and politicians—have been willing to take risks with new toxic technologies for the sake of growth and progress, waiting to deal with the consequences later. In writing about toxic substance regulation in the United States, Sheila Jasanoff once asserted that the issue "is not whether expertise detracts from political processes, but how it is harnessed and steered to serve some political interests over others."[14] The majority of stories in this volume attend to this question. Chapters highlight the inherently political ways in which scientific expertise has been wielded in contexts of uncertainty to facilitate dominant economic and military interests, often at the expense of environmental and human health.

EXPERT KNOWLEDGE AND UNCERTAINTY

Understanding the ways in which toxic landscapes have become unremarkable and ubiquitous requires an examination of the development of modern institutions of scientific and technical expertise. Most of us know very little about the chemical and physical properties of particular toxic agents, their physiological impacts, or the ways in which they circulate in different ecological systems. But we feel confident that chemists, physicists, toxicologists, epidemiologists, and safety engineers have that knowledge and have worked with legislators to put adequate regulations in place to keep bodies and spaces safe. This division of labor and deference to the special knowledge of experts, then, is a crucial component of societal complacency. Yet, institutions of expertise and patterns of science communication tend to mask the uncertain, tentative, or disputed nature of expert knowledge claims, while excluding the voices of those most impacted by toxicity.

Uncertainty is a central and disconcerting feature of histories of toxicity,[15] wielded differently depending on the interests of government and industry. Recent historical work has shown, for instance, that appeals to the uncertainty of experts have been extremely successful in nurturing social inaction, even in the face of the increasingly evident links between smoking and cancer, and carbon emissions and global warming.[16] Those stories reveal the conscious efforts of a small group of influential scientists to "manufacture doubt." Our stories, however, are rarely ones of deliberate deception. Many of the cases in this collection focus on the judgments of scientists, doctors, and engineers who have been called on to decide whether a health impact exists or whether a given space is safe. In the course of passing judgment, developing protocols, and shaping regulations, these experts often unintentionally obscured all that was still unknown about a particular toxic agent. Such actions led to an appearance of safety, certainty, and consensus even when none existed. In this way, many of the chapters in our collection study the production of ignorance as much as the production of knowledge, contributing to the project outlined in Robert Proctor and Londa Schiebinger's *Agnotology*.[17] The imperceptibility and also proliferation of different kinds of toxic substances, the difficulty of untangling causes and clusters of symptoms, and the inevitable messiness of scientific measurement outside of a lab have worked together to make simple statements about the impact of toxic exposure rare.[18]

Stabilizing any scientific phenomenon and creating scientific consensus is always messy, but knowledge about toxicity is particularly incoherent. In part, this has to do with structures of national and industrial secrecy that have restricted the free communication of information about new chemicals, radioactive isotopes, and industrial waste products. The nuclear weapons development of the Manhattan Project during World War II ushered in a new era of classified military research and regimes of secrecy that have continued to structure contemporary institutions, hampering the circulation of knowledge about toxicity, and nuclear technologies in particular.[19]

While classified knowledge and military-industrial secrecy are inherently exclusive, broader patterns of science education and communication have also worked to prevent most people from participating in the production of scientific knowledge. Over the course of the nineteenth century, science, medicine, and engineering became professionalized and increasingly specialized with new societies, journals, and university programs. At the same time, multiple forms of popular science emerged that erected new barriers between members of the public and their meaningful participation in science.[20] In England, in the late nineteenth century, for instance, popular science writing was often imbued with a natural theology that understood science as a moral and religious project. This was increasingly at odds with the deliberate secularism of professional science.[21] Over the course of the twentieth century, themes of natural theology faded but books, radio programs, and television shows continued to draw sharp boundaries around the world of professional science. These vehicles for popular science often cast audiences as passive consumers of science entertainment, rather than active participants in the creation of scientific knowledge.[22]

In order to fully participate in science and be recognized as an expert, then, one must pursue years of higher education, gain membership in particular professional societies, and participate in conversations with highly specialized jargon, at conferences and in journals. But barriers of income and education raise concerns that institutions of expertise simply continue to reproduce existing structures of social and racial inequity. The continued underrepresentation of women and minorities in STEM fields—disciplines that are integral to the regulation of toxic substances—gives at least one clear indication of how exclusionary these structures continue to be.[23] The numerous qualifications needed to become an expert also have direct consequences for those most impacted by toxic environments. As scholars of environmental justice elsewhere and in this volume show, low income and minority communities most often bear the brunt of toxic spaces.[24] Yet, these groups typically do not have the means to join expert ranks, and outside experts tend to discount these communities' lived experiences of toxicity.[25]

As a result, community activists fighting toxics in various "sacrifice zones"[26] must try to educate themselves about current regulations and research in toxicology, trying to become, in sociologist Steven Lerner's words, "amateur toxicologists and epidemiologists, as well as fluent in regulatory jargon."[27] But learning to read specialized literature is only one step in gaining expertise. As scholars Harry Collins and Robert Evans have argued, a crucial component of expertise consists of gaining a kind of tacit knowledge through experience, which in many cases is only possible through admission to an exclusive disciplinary community.[28] Community members reading published literature on toxicology, for example, will not be able to contribute to this body of knowledge without spending time in laboratories learning from practicing scientists.

Taken together, this paints a decidedly grim picture about the exclusionary nature of expertise, its use in contexts of uncertainty to create a veneer of safety, and challenges faced by communities exposed to toxic spaces. Given this structurally uneven terrain, should those contesting expertise and fighting toxic environments simply give up? No, they should not. As various scholars have argued, there is room to expand our understanding of expertise and recognize knowledge gained outside of these formal institutions.[29] Even the model offered by Collins and Evans acknowledges that individuals without access to the formal institutions of science may still have important experiential understanding of an environment, place, or illness. In some environments of toxic exposure, community members have successfully gathered their own health data through surveys, partnered with allied scientists, and gained recognition of illness, engaging in what sociologist Phil Brown has called "popular epidemiology."[30] Research on environmental justice organizing has further shown that collaborations between scientists and activists are not only possible but have also produced new ethical frameworks for collecting and reporting data on toxicity.[31]

Clearly, communities on the ground will continue to contest toxic spaces and, sometimes, change structures of expertise in the process. While the majority of stories in our collection demonstrate how seldom this kind of experiential expertise has been acknowledged historically, either by scientific experts or government bodies, more contemporary mobilizations against toxicity indicate that this situation may be gradually changing. By revealing how our current condition of pervasive toxicity has come to be, the work in this volume, then, may be considered an important "prequel" to understanding our toxic present.

ROADMAP AND THEMES

While any substance causing an adverse health effect on humans can be considered toxic, in this collection we focus in particular on inorganic toxins: chemical substances and radiation produced in x-ray tubes or emitted during fission or by radioactive decay. In almost all of our cases, the toxicity remains invisible, with health effects sometimes manifesting immediately, and sometimes after long periods of chronic exposure. Many of these toxic agents are the by-products of industrial or military operations—for instance, radioactive fission products circulating through the environment after nuclear weapons testing and chemicals leaking into groundwater as a result of industrial waste disposal or fracking. But not all of the toxics studied here are waste products. In some cases, toxic agents such as x-rays or pesticides were deployed with a particular goal in mind, and with their toxicity not fully understood.

Inevitably Toxic is divided into three parts. In Part One, "Radiation," we explore particular U.S. encounters with toxic radiation in hospital x-ray rooms, at test sites for nuclear weapons, and in experimental nuclear reactors in the period prior to and shortly after World War II. Here, authors examine the process by which different experts have made decisions about safe radiation exposure in the context of many kinds of uncertainties, highlighting the consequences of those decisions for those exposed. Janet Farrell Brodie's and William Palmer's chapters investigate the impact of secrecy on nuclear research while also challenging our ability as historians to reconstruct this history. This section also drives home the willingness of those in positions of power and authority to take enormous risks with people's health and safety, choosing to deal with the consequences of radiation exposure during or after the fact. Vivien Hamilton demonstrates in her piece, for example, how national safety standards for x-rays in the United States were developed even while doctors were already subjecting their own bodies and the bodies of their patients to radiation. Similarly, Lindsey Dillon explores the new biomedical problems that emerged from the U.S. Navy's atomic experiments during Operation Crossroads in 1946, showing us that scientific attention turned to questions of radiological safety and best practices only after people and places had already suffered irreparable damage.

In Part Two, "Industrial Toxins," authors examine how U.S. economic growth in the post–World War II era resulted in the creation of contaminated environments by many different kinds of industrial chemicals. These chapters explore contexts in which legal frameworks to regulate toxics were just emerging, revealing the shaky and compromised foundations of regulation itself. In so doing, these cases look closely at the ways in which public officials mobilized scientific expertise while drawing our attention to the entangled relationships of government agencies and private economic and military interests. Brinda Sarathy argues that expertise on water pollution control in Southern California was filtered through a lens that favored economic growth, showing how interpretations of scientific data were never simply objective but, rather, inherently political-economic in nature. Similarly, Bhavna Shamasunder reveals how oil industry lobbyists in early twentieth-century Los Angeles ultimately quashed community organizing by forging political alliances with state and federal officials. Such alliances have shaped both the physical and regulatory landscape of toxicity. Part Two also elucidates how the drumbeat of "growth at all costs" was sustained by an attitude of technological hubris and scientific utopianism. Sarah Stanford McIntyre, for example, highlights a version of this unyielding faith in technological progress, as she chronicles how an entire region in Texas held fast to the promise of petrochemicals while also, almost willfully, denying the potentially deleterious consequences of these new industries to human health and ecology. Similarly, James G. Lewis and Char Miller's chronicle of herbicide use by the U.S. Forest Service epitomizes both the technocratic confidence through which experts remade entire landscapes and their outright refusals to explore alternative means of pest control despite growing public concern and opposition.

In Part Three, "Community Contestation, Expanding Expertise," community resistance and contaminated bodies become a central theme and unit of analysis. The chapters in this section foreground the experiences of individuals in contact with toxic environments and contaminants; the roles of race, gender, and class in spurring activism; and the growing significance of the local expertise of "ordinary citizens." More specifically, these authors focus on groups that have often been depicted as victims rather than as agents of resistance in the wake of toxic contamination: Japanese American women who utilized folk traditions to treat radiation sickness following the bombings of Hiroshima and Nagasaki, and First Nations peoples in the Canadian North articulating alternate visions of community health and risk. This unit continues to highlight inconsistencies in the ways in which risk has been perceived and addressed for different kinds of bodies. Taken together, these chapters highlight both the structural forces that help keep in place the status quo but also point to possibilities of reconfigured ways of living that prioritize health and well-being.

Our afterword captures a conversation with historian of science Peter Galison, whose recent film with Robb Moss, *Containment* (2015), returns to the fraught problem of long-term nuclear waste storage. Galison's work connects with many of the themes in this volume concerning regulation, complacency, and the assessment and communication of risk. Examining sites of nuclear contamination and waste both in the United States and Japan, the film asks us to imagine how we might warn future generations of this pernicious danger.

In closing, we should acknowledge that most chapters in this volume examine the production of toxic spaces in or by the United States. As such, readers should keep in mind that narratives of "progress" in the U.S. context—and for Western modernity more generally—cannot be disentangled from legacies of conquest and genocide of Native American populations, land theft, slavery, and institutions of white supremacy, colonialism, and structural racism.[32] These formations have fundamentally shaped the creation of toxic environments in historically specific ways, both within the United States and beyond.[33]

Many of the works in this book, for instance, reveal ways in which the production and management of toxic environments are linked to projects of nation building and related imperial ambitions. In writing about Japanese American atomic bomb survivors, Naoko Wake interrogates how imperial dynamics played out in the formation of the Atomic Bomb Casualty Commission and its gendered portrayals of "victors" and "victims" during the postbomb period. More generally, radiation contamination from nuclear weapons developed by the United States—as outlined at different historical moments by Janet Farrell Brodie, William Palmer, Lindsey Dillon, and Naoko Wake—all bear testament to a larger arsenal that continues to shore up American neo-imperialism globally. Chemical contamination has similarly served to further western ambitions. To this point, James G. Lewis and Char Miller detail the involvement of U.S. Forest Service personnel in torching vast swaths of Vietnam jungle in an overseas war, while Alexander Zahara chronicles the Canadian government's militarized displays of defense to reinforce the belonging of northern territories within its larger national body. Such stories raise questions about how contaminated peoples and places are not merely incidental to western projects of nation building at home and abroad but instead are constitutive of the process itself.[34]

Finally, for cases of toxicity in the United States, we urge readers to again be mindful of the broader relations of force that have created toxic places in the name of "progress." Scholarship in environmental justice has been instrumental in showing how specific processes of white flight, racial segregation, political marginalization, and institutional racism have led to low-income communities of color in the United States being disproportionately impacted by toxic spaces. More recently, the high-profile cases of toxic contamination in Vernon, California, and Flint, Michigan, have further laid bare the ways in which capitalism works through race and difference, and via the state, to quite literally poison devalued nonwhite peoples and places.[35] To this end, Bhavna Shamasunder's chapter on Los Angeles–area activists fighting the oil industry and Sarah Stanford-McIntyre's references to the disproportionate exposure of Black and Latinx residents to petrochemicals must be read against a larger historic tapestry of racial dispossession and racialized violence in the United States. While our stories thus collectively paint a picture of contaminated environments that have become pervasive over time and space, resulting in few "safe places"[36] for anyone, toxic environments still impact some bodies and places more than others. This reality itself should be a basis for action against our condition of normalized toxicity.

NOTES

1. "Superfund" relates to the Comprehensive Environmental Response, Compensation, and Liability Act of 1980, passed by the U.S. Congress in response to toxic waste disasters such as that in Love Canal, Niagara Falls, New York, in 1978.

2. See, for example, compiled volumes such as Mitman, Murphy, and Sellers, eds., *Landscapes of Exposure*, and Jorgensen, Jorgensen, and Pritchard, eds., *New Natures.*

3. Rose, *Politics of Life Itself*; Braun, "Biopolitics and the Molecularization of Life"; Mitman, Murphy, and Sellers, "Introduction"; Murphy, "Chemical Regimes of Living"; Petryna, *Life Exposed*; Nash, *Inescapable Ecologies.*

4. Jasanoff, ed., *States of Knowledge*, 2.

5. Giddens, "Risk and Responsibility," 4.

6. Murphy, "Chemical Regimes of Living," 697.

7. Langston, *Toxic Bodies*; Walker, *Toxic Archipelago.*

8. Beck, *Risk Society.*

9. The volume *Powerless Science?*, for example, shows various instances of how "the production of scientific knowledge and expertise on toxicants and their effects evolved alongside the modes of toxicant regulation." Boudia and Jas, *Powerless Science?*, 3.

10. Carson, *Silent Spring*; Kuletz, *The Tainted Desert*; Steingraber, *Living Downstream.*

11. Hamblin, *Poison in the Well.*

12. Iversen, *Full Body Burden*; Brown, *Plutopia*; Petryna, *Life Exposed.*

13. Johnston, *Half-Lives and Half-Truths.*

14. Brickman, Jasanoff, and Ilgen, *Controlling Chemicals*, 174.

15. Mitman, Murphy, and Sellers, "Introduction."

16. Oreskes and Conway, *Merchants of Doubt*; Proctor, *Cancer Wars.*

17. Proctor and Schiebinger, *Agnotology.*

18. Michelle Murphy has developed the idea of "regimes of perceptibility," arguing that a particular effect can be made both perceptible and imperceptible by different practices of measurement and argument. Murphy, *Sick Building Syndrome.*

19. Galison, "Removing Knowledge"; Dennis, "Secrecy and Science Revisited"; Schwartz, *Atomic Audit*, chapter 8.

20. Any form of science communication pitched at a nonexpert audience, including books, newspaper articles, museum exhibits, science demonstrations, and lectures, can be considered a form of popular science. See Secord, *Victorian Sensation*; Lightman, *Victorian Popularizers of Science*; Bowler, *Science for All*; Lafollette, *Science on the Air*; Radar and Cain, *Life on Display*; Nelkin, *Selling Science.*

21. Lightman, "'The Voices of Nature.'"

22. While this dominant model has received criticism from scholars, such as Stephen Hilgartner, who have argued persuasively that it is impossible to clearly demarcate popular from "genuine" science, it is evident that modes of science communication often solidify boundaries between scientists and nonscientists. Felicity Mellor points out, for instance, that when popular physics books lay out an explicit goal of making their material accessible, they are implicitly constructing physics as inaccessible to most readers. Hilgartner, "The Dominant View of Popularization"; Mellor, "Between Fact and Fiction."

23. National Science Foundation, *Women, Minorities, and Persons with Disabilities in Science and Engineering: 2017.*

24. Bullard, *Unequal Protection*; Bullard et al., *Toxic Wastes and Race at Twenty 1987–2007*; Mohai, Pellow, and Roberts, "Environmental Justice"; Pulido, "Rethinking Environmental Racism"; Szasz, *Ecopopulism.*

25. Lerner, *Sacrifice Zones*, 3.

26. Sacrifice zones as redefined by sociologist Steven Lerner include a broader array of "fenceline communities or hot spots of chemical pollution where residents live immediately adjacent to heavily polluting industries or military bases." Lerner, *Sacrifice Zones*, 3.

27. Lerner, *Sacrifice Zones*, 8.

28. Collins and Evans, *Rethinking Expertise.*

29. Wynne, "May the Sheep Safely Graze?"; Epstein, "The Construction of Lay Expertise"; Collins and Evans, *Rethinking Expertise*; Brown, *Toxic Exposures*; Chambers, *Whose Reality Counts?*

30. Brown, *Toxic Exposures.*

31. Ottinger and Cohen, eds., *Technoscience and Environmental Justice.*

32. Limerick, *The Legacy of Conquest*; Cronon, "The Trouble with Wilderness; or Getting Back to the Wrong Nature"; Merchant, "Shades of Darkness"; Almaguer, *Racial Fault Lines*; Gilmore, "Growth"; Gilmore, "Fatal Couplings of Power and Difference"; Smith, "Heteropatriarchy and the Three Pillars of White Supremacy."

33. Pulido, "Rethinking Environmental Racism"; Dillon and Sze, "Police Powers and Particulate Matters"; Sze, *Noxious New York*; Pellow, "Environmental Inequality Formation"; Pellow, *Resisting Global Toxics*; Brown, *Toxic Exposures*; Lerner, *Sacrifice Zones*; Langston, *Toxic Bodies.*

34. For more, see Brown, *Plutopia*; Hecht, *Entangled Geographies.*

35. Pulido, "Geographies of Race and Ethnicity II"; Pulido, "Flint, Environmental Racism, and Racial Capitalism"; Pulido, Kohl, and Cotton, "State Regulation and Environmental Justice."

36. Brown and Mikkelsen, *No Safe Place.*

BIBLIOGRAPHY

Almaguer, Tomas. *Racial Fault Lines: The Historical Origins of White Supremacy in California.* Berkeley: University of California Press, 1994.

Beck, Ulrich. *Risk Society: Towards a New Modernity.* Translated by Mark Ritter. London: Sage Publications, 1992.

Boudia, Soraya, and Nathalie Jas, eds. *Powerless Science? Science and Politics in a Toxic World.* New York: Berghahn Books, 2014.

Bowler, Peter. *Science for All: The Popularization of Science in Early Twentieth-Century Britain.* Chicago: University of Chicago Press, 2009.

Braun, Bruce. "Biopolitics and the Molecularization of Life." *Cultural Geographies* 14, no. 1 (January 1, 2007): 6–28, doi:10.1177/1474474007072817.

Brickman, Ronald, Sheila Jasanoff, and Thomas Ilgen. *Controlling Chemicals: The Politics of Regulation in Europe and the United States.* Ithaca, NY: Cornell University Press, 1985.

Brown, Kate L. *Plutopia: Nuclear Families, Atomic Cities, and the Great Soviet and American Plutonium Disasters.* New York: Oxford University Press, 2013.

Brown, Phil. *Toxic Exposures: Contested Illnesses and the Environmental Health Movement.* New York: Columbia University Press, 2007.

Brown, Phil, and Edwin J. Mikkelsen. *No Safe Place: Toxic Waste, Leukemia, and Community Action.* Berkeley: University of California Press, 1990.

Bullard, Robert D. *Unequal Protection: Environmental Justice and Communities of Color.* New York: Random House, 1994.

Bullard, Robert D., Paul Mohai, Robin Saha, and Beverly Wright. *Toxic Wastes and Race at Twenty 1987–2007.* United Church of Christ Justice and Witness Ministries. March 2007. http://www.ucc.org/assets/pdfs/toxic20.pdf.

Carson, Rachel. *Silent Spring.* New York: Houghton Mifflin Harcourt, 2002.

Chambers, Robert. *Whose Reality Counts? Putting the First Last.* London: Intermediate Technology Publications, 1997.

Collins, Harry M., and Robert Evans. *Rethinking Expertise.* Chicago: University of Chicago Press, 2009.

Cronon, William. "The Trouble with Wilderness; or Getting Back to the Wrong Nature." In *Uncommon Ground: Rethinking the Human Place in Nature,* edited by William Cronon, 69–90. New York: W. W. Norton, 1996.

Dennis, Michael Aaron. "Secrecy and Science Revisited: From Politics to Historical Practice and Back." In *The Historiography of Contemporary Science, Technology and Medicine,* edited by Ronald E. Doel and Thomas Söderqvist, 172–84. London: Routledge, 2006.

Dillon, Lindsey, and Julie Sze. "Police Powers and Particulate Matters: Environmental Justice and the Spatialities of In/securities in U.S. Cities." *English Language Notes* 54, no. 2 (2016): 13–24.

Epstein, Steven. "The Construction of Lay Expertise: AIDS Activism and the Forging of Credibility in the Reform of Clinical Trials." *Science, Technology and Human Values* 20, no. 4 (1995): 408–37.

Galison, Peter. "Removing Knowledge: The Logic of Modern Censorship." In *Agnotology: The Making and Unmaking of Ignorance*, edited by Robert Proctor and Londa Schiebinger, 37–54. Stanford, CA: Stanford University Press, 2008.

Giddens, Anthony. "Risk and Responsibility." *The Modern Law Review* 62, no. 1 (1999): 1–10.

Gilmore, Ruth Wilson. "Growth: From Military Keynesianism to Post-Keynesian Militarism." *Race and Class* 40, no. 2/3 (1998–1999): 171–88.

Gilmore, Ruth Wilson. "Fatal Couplings of Power and Difference: Notes on Racism and Geography." *Professional Geographer* 54, no. 1 (2002): 15–24.

Hamblin, Jacob Darwin. *Poison in the Well: Radioactive Waste in the Oceans at the Dawn of the Nuclear Age*. New Brunswick, NJ: Rutgers University Press, 2008.

Hecht, Gabrielle. *Entangled Geographies: Empire and Technopolitics in the Global Cold War*. Cambridge, MA: MIT Press, 2011.

Hilgartner, Stephen. "The Dominant View of Popularization: Conceptual Problems, Political Uses." *Social Studies of Science* 20, no. 3 (1990): 519–39.

Iversen, Kristen. *Full Body Burden: Growing Up in the Nuclear Shadow of Rocky Flats*. New York: Crown Publishers, 2012.

Jasanoff, Sheila, ed. *States of Knowledge: The Co-production of Science and Social Order*. London: Routledge, 2010.

Johnston, Barbara Rose, ed. *Half-Lives and Half-Truths: Confronting the Radioactive Legacies of the Cold War*. Santa Fe, NM: School for Advanced Research Press, 2007.

Jorgensen, Dolly, Finn Arne Jorgensen, and Sara B. Pritchard, eds. *New Natures: Joining Environmental History with Science and Technology Studies*. Pittsburgh, PA: University of Pittsburgh Press, 2013.

Kuletz, Valerie. *The Tainted Desert: Environmental Ruin in the American West*. New York: Routledge, 1998.

Lafollette, Marcel Chotkowski. *Science on the Air: Popularizers and Personalities on Radio and Early Television*. Chicago: University of Chicago Press, 2008.

Langston, Nancy. *Toxic Bodies: Hormone Disruptors and the Legacy of DES*. New Haven, CT: Yale University Press, 2010.

Lerner, Steve. *Sacrifice Zones: The Front Lines of Toxic Chemical Exposure in the United States*. Cambridge, MA: MIT Press, 2012.

Lightman, Bernard. *Victorian Popularizers of Science: Designing Nature for New Audiences*. Chicago: University of Chicago Press, 2007.

Lightman, Bernard. "'The Voices of Nature': Popularizing Victorian Science." In *Victorian Science in Context*, edited by Bernard Lightman, 187–211. Chicago: University of Chicago Press, 1997.

Limerick, Patricia Nelson. *The Legacy of Conquest: The Unbroken Past of the American West*. New York: W. W. Norton, 1987.

Mellor, Felicity. "Between Fact and Fiction: Demarcating Science from Non-Science in Popular Physics Books." *Social Studies of Science* 33, no. 4 (2003): 509–38.

Merchant, Carolyn. "Shades of Darkness: Race and Environmental History." *Environmental History* 8 (2003): 380–94.

Mitman, Gregg, Michelle Murphy, and Christopher Sellers. "Introduction: A Cloud over History." *Osiris* 19 (2004): 1–17. doi:10.1086/649391.

Mitman, Gregg, Michelle Murphy, and Christopher Sellers, eds. *Landscapes of Exposure: Knowledge and Illness in Modern Environments. Osiris* 19 (2004).

Mohai, Paul, David Pellow, and J. Timmons Roberts. "Environmental Justice." *Annual Review of Environment and Resources* 34 (2009): 405–30.

Murphy, Michelle. "Chemical Regimes of Living." *Environmental History* 13, no. 4 (2008): 695–703.

Murphy, Michelle. *Sick Building Syndrome and the Problem of Uncertainty: Environment Politics, Technoscience, and Women Workers.* Durham, NC: Duke University Press, 2006.

Nash, Linda Lorraine. *Inescapable Ecologies: A History of Environment, Disease, and Knowledge.* Berkeley: University of California Press, 2006.

National Science Foundation, National Center for Science and Engineering Statistics. *Women, Minorities, and Persons with Disabilities in Science and Engineering: 2017.* Special Report NSF 17–310. Arlington, VA, 2017. www.nsf.gov/statistics/wmpd/.

Nelkin, Dorothy. *Selling Science: How the Press Covers Science and Technology.* New York: W. H. Freeman, 1995.

Oreskes, Naomi, and Erik M. Conway. *Merchants of Doubt: How a Handful of Scientists Obscured the Truth on Issues from Tobacco Smoke to Global Warming.* New York: Bloomsbury Press, 2011.

Ottinger, Gwen, and Benjamin R. Cohen, eds. *Technoscience and Environmental Justice: Expert Cultures in a Grassroots Movement.* Cambridge, MA: MIT Press, 2011.

Pellow, David N. "Environmental Inequality Formation: Toward a Theory of Environmental Injustice." *American Behavioral Scientist* 43, no. 4 (2000): 581–601.

Pellow, David N. *Resisting Global Toxics: Transnational Movements for Environmental Justice.* Cambridge, MA: MIT Press, 2007.

Petryna, Adriana. *Life Exposed: Biological Citizens after Chernobyl.* Princeton, NJ: Princeton University Press, 2013.

Proctor, Robert. *Cancer Wars: How Politics Shapes What We Know and Don't Know About Cancer.* New York: Basic Books, 1995.

Proctor, Robert N., and Londa Schiebinger. *Agnotology: The Making and Unmaking of Ignorance.* Stanford, CA: Stanford University Press, 2008.

Pulido, Laura. "Flint, Environmental Racism, and Racial Capitalism." *Capitalism Nature Socialism* 27, no. 3 (2015): 1–16.

Pulido, Laura. "Geographies of Race and Ethnicity II: Environmental Racism, Racial Capitalism and State-Sanctioned Violence." *Progress in Human Geography* 41, no. 4 (2017): 524–33.

Pulido, Laura. "Rethinking Environmental Racism: White Privilege and Urban Development in Southern California." *Annals of the Association of American Geographers* 90, no. 1 (2000): 12–40.

Pulido, Laura, Ellen Kohl, and Nicole-Marie Cotton. "State Regulation and Environmental Justice: The Need for Strategy Reassessment. *Capitalism Nature Socialism* 27, no. 2 (2016): 12–31.

Radar, Karen A., and Victoria E. M. Cain. *Life on Display: Revolutionizing U.S. Museums of Science and Natural History in the Twentieth Century.* Chicago: University of Chicago Press, 2014.

Rose, Nikolas S. *Politics of Life Itself: Biomedicine, Power, and Subjectivity in the Twenty-First Century.* Princeton, NJ: Princeton University Press, 2007.

Schwartz, Stephen I., ed. *Atomic Audit: The Costs and Consequences of U.S. Nuclear Weapons since 1940.* Washington, DC: Brookings Institution Press, 1998.

Secord, James. *Victorian Sensation: The Extraordinary Publication, Reception, and Secret Authorship of Vestiges of the Natural History of Creation.* Chicago: University of Chicago Press, 2003.

Smith, Andrea. "Heteropatriarchy and the Three Pillars of White Supremacy." In *Color of Violence: The INCITE! Anthology,* edited by Incite! Women of Color against Violence, 70–76. Cambridge, MA: South End Press, 2006.

Steingraber, Sandra. *Living Downstream: An Ecologist Looks at Cancer and the Environment.* London: Virago, 1999.

Szasz, Andrew. *Ecopopulism: Toxic Waste and the Movement for Environmental Justice.* Minneapolis: University of Minnesota Press, 1994.

Sze, Julie. *Noxious New York: The Racial Politics of Urban Health and Environmental Justice.* Cambridge, MA: MIT Press, 2007.

Walker, Brett L. *Toxic Archipelago: A History of Industrial Disease in Japan.* Seattle: University of Washington Press, 2010.

Wynne, Brian. "May the Sheep Safely Graze? A Reflexive View of the Expert-Lay Knowledge Divide." In *Risk, Environment and Modernity: Towards a New Ecology,* edited by Scott Lash, Bronislaw Szerszynski, and Brian Wynne, 44–83. London: Sage Publications, 1996.

PART ONE

Radiation

1

X-ray Protection in American Hospitals

Vivien Hamilton

The diagnostic and therapeutic promise of x-rays drew Americans into hospitals at an unprecedented rate in the early twentieth century, and yet the dangers associated with medical x-rays were significant and well known. Anyone walking into an x-ray room—patients, doctors, technicians—risked electrocution from high-voltage power sources, fire from tremendously flammable x-ray film, and even blunt trauma from falling apparatus.[1] When the National Bureau of Standards (NBS) published the first nationally recognized set of guidelines for x-ray protection in 1931,[2] a significant portion of the pamphlet offered recommendations intending to minimize these electrical and fire hazards. Overall, these recommendations for fire and electrical safety were straightforward and uncontroversial. Far more contentious, however, were the guidelines intending to provide protection against unwanted radiation exposure. The same x-rays that could produce images of the hidden body, and shrink cancerous tumors, could also cause horrific burns, blood changes, and even new cancers. In the words of NBS physicist Lauriston Taylor, this powerful medical technology was a "two-edged sword."[3] As soon as an x-ray tube was turned on, it transformed its surroundings into a toxic environment.

Unlike many of the other chapters in this collection that focus on environmental contamination by radioactive isotopes or chemical pollution from industry, the toxicity in hospital x-ray rooms is both contained and finite. Unwanted radiation lasts only as long as an x-ray tube is on. But while x-rays are present, emanating from the tube and scattering off of multiple surfaces in the room, many of the same challenges are present. X-rays are invisible—there is no way to sense whether a room is filled with radiation or not—and the effects of exposure might take weeks or months to manifest. Without the benefit of any immediate warning of danger, doctors were forced to find ways of ensuring their own safety as well as that of their patients.

In this chapter, I trace shifts in doctors' perceptions of the dangers of x-rays as well as negotiations between different constituents of the x-ray community concerning who was ultimately responsible for ensuring safety in these spaces. I focus in particular on conversations surrounding the publication of the 1931 NBS handbook, asking how the authors of these guidelines and the wider x-ray community envisioned the risks and responsibilities of different bodies in this toxic environment. In her study of recent changes to radiation protection standards in Belarus, Olga Kuchinskaya has argued that formal standards tend to "legitimize some actors and judgments and delegitimize others."[4] In this story, I have pulled out five categories of actors: patients, radiologists, and technicians whose bodies were in frequent and close contact with radiation from x-ray tubes, and physicists and x-ray manufacturers, whose bodies were not, but whose voices were prominent as the guidelines were drafted. I find that those bodies at highest risk, the patients and the technicians, had the least agency in this environment. The strongest voices are those of the doctors and the physicists. In particular, the clear, quantitative rules showing appropriate thicknesses of lead shielding appear to legitimize the physicists' vision of certainty and safety. I argue, however, that these reassuring numbers hid deep uncertainties about the effects of x-ray exposure, and doctors were ultimately unwilling to relinquish their responsibility. In the broader context of the history of medicine in the United States, this resistance from doctors makes sense. By the 1930s, the medical profession had achieved a position of strong cultural and social authority.[5] The x-ray room was a clinical space, and medical judgment was still paramount. Manufacturers, physicians, and even physicists agreed that it was not the lead shielding but the doctor in charge of the x-ray room who was ultimately responsible for the safety of each of the bodies who entered into that space.

EARLY FAITH IN PROTECTIVE EQUIPMENT

X-rays were discovered in 1895, and yet the first safety rules for doctors and x-ray operators in Britain were not published until 1916. In the United States, formal safety guidelines appeared even later. Historians of radiology have come back to this same question again and again, asking why the first generation of doctors was seemingly so unconcerned about the dangers of x-rays. One common explanation tells a story of the heroic pioneers of radiology who "wandered blindly,"[6] falling victim to hidden dangers.[7] But this telling fails to acknowledge the fact that the physiological dangers associated with x-rays were widely acknowledged within months of their discovery.

X-ray burns often appeared weeks after an initial exposure, making it difficult at first to trace their cause. Yet even though the effects of the new rays were "so occult and so long-delayed,"[8] the knowledge that x-rays could cause burns and unwanted epilation came almost immediately, and x-rays were linked to cancer within a few years.[9] Matthew Lavine has shown that fear of physical trauma developed quickly in the public sphere. By the turn of the century, newspaper accounts of x-rays routinely referred to the rays' ability to "burn" and cause "withering" of tissues.[10] In the early twentieth century, doctors most often worked with unshielded tubes and frequently used their own hands as test objects to gauge the output of the tube. These doctors developed burns, warts, ulcers, and tumors that forced the amputation of fingers, hands, and arms. So many of the early x-ray workers died due to radiation injuries that the German Röntgen Society erected a monument in Hamburg dedicated to the x-ray martyrs containing the names of almost two hundred individuals from Western Europe and North America.[11]

The monument commemorating these deaths symbolizes a second common telling of this story. Instead of unknown victims, the early radiologists who forged ahead with their practice became martyrs who willingly embraced the danger posed by x-rays and their own suffering. In 1936 the American radiologist Percy Brown compiled a small volume with the stories of twenty-seven x-ray workers whom he called "American martyrs to science." He asked why so many of his colleagues continued to work with the rays, often ignoring the available protection equipment, when the dangers were known almost immediately. To answer this question, Brown points to excitement over the new rays, to the obscurity of work on physiological dangers and confusion over the cause of the burns. He also implicates apparatus makers, arguing that they often advertised the simplicity of their machines, foregoing fussy shields and protective measures in favor of aesthetic minimalism.[12] Rebecca Herzig has noted the importance of the ethos of deliberate suffering to the professional ideology of radiology, arguing that the willingness of these early x-ray experimenters to suffer forged a shared sense of community. She notes that only certain sufferers were afforded martyrdom—the bodies of doctors being privileged in a way that technicians' bodies were not.[13]

However, the attitude that I find most common in the first decades of the twentieth century is not a conscious martyrdom or, in fact, any acceptance that x-ray work was necessarily dangerous but instead a widespread sense that the dangers of x-rays had been conquered. By 1910 doctors were aware of the harmful effects of x-rays and felt that they had these unwanted side effects under control, using basic lead shielding, taking care with maximum exposure times, and using remedies for burns when they did occur.

Many x-ray manufacturers began to offer protective equipment designed to shield against unnecessary radiation quite early in the twentieth century. The Kny-Scheerer Company declared in 1905 that "it has become a question of vital interest as to how to protect the operator . . . The loss of a hand or an arm is not a thing to be contemplated with equanimity."[14] The company offered protective cabinets for the operator to stand in with lead glass windows that would allow the doctor to observe the patient during an x-ray procedure. If the operator needed to approach the patient during an exam or therapeutic session, she or he could don a full bodysuit like the one shown in figure 1.1. Kny-Scheerer additionally offered lead glass and sheets of lead foil covered in rubber so that physicians could construct their own cabinets and aprons to their desired specifications. Many companies also offered lead shielding to cover x-ray tubes or x-ray tubes made almost entirely out of lead glass to ensure that the rays were not leaving the tube in all directions.[15]

These measures were designed to protect the operator who would receive multiple exposures over the course of weeks and years in a busy x-ray department. But doctors were equally worried about the possibility of needlessly harming a patient. Doctors were warned by manufacturing companies that "it is possible to inflict injury with the x-ray," but the message from those companies was that the machines could be operated safely if the right procedures were followed. In 1905 the Wagner Company judged that as long as the operator paid attention to the current in the x-ray tube and the distance of the tube to the patient, "there is not as much danger from the use of x-rays

Figure 1.1. Full protective suit, c. 1905. Rubber-covered lead foil including spectacles made of leaded glass. Kny-Scheerer Company, Illustrated and Descriptive Catalogue (1905).

as there is in prescribing strychnine and many of our most efficient remedies."[16] In the event that an x-ray burn did occur, the Kny-Scheerer Company kept a salve in stock and gave advice on how to treat the three levels of burn that might develop.[17]

Doctors echoed the sentiment prevalent among manufacturers that x-ray burns in patients were avoidable and treatable. In 1914 Dr. W. J. Dodd of Boston presented a paper on best practices for treating x-ray dermatitis but apologized "for mentioning such a subject before a gathering of Roentgenologists, some of whom have probably not seen such a condition, certainly not in their own practice." He felt justified returning to this topic because of the new, more powerful Coolidge tubes, even though "we all know that there is absolutely no danger when the trained Roentgenologist does this work."[18] Dodd was confident that even if a reaction did occur in a patient, it could easily be treated. He offered his own homemade salve as an "exceedingly simple and absolutely efficacious remedy."[19] Other doctors continued to acknowledge the possibility of harming a patient but were confident that these "accidents" could be avoided using a sheet of aluminum as a "protective ray filter."[20]

As for their own safety, doctors felt sure that the measures that they took to protect themselves were effective. In the pages of the *American Journal of Roentgenology,* the French doctor T. H. Nogier advocated complete adherence to safety procedures, using gloves, aprons, and goggles, and avoiding all direct radiation: "When combining all these procedures and using them in a systematic way, one may not avoid all the dangers, but one may reduce them to an infinitesimal proportion."[21]

In the first two decades of the twentieth century, doctors acknowledged that the x-ray room was a potentially dangerous environment for both themselves and their patients but placed their trust in existing protective equipment to combat that risk. It is not surprising, then, that there was no sustained push in the early twentieth century for a formal set of safety rules to guide radiologists. When those rules did appear, they were inspired first by physics research into the penetrability of actual protective materials and then by the deaths of a number of prominent radiologists who developed blood diseases after exposure to the more powerful x-rays generated by new Coolidge tubes.

PHYSICISTS SOUND THE ALARM

There were very few American physicists working on issues related to the medical use of x-rays and radium in the 1910s and 1920s—I can name eight—a tiny group compared to the many hundreds of American doctors using x-ray equipment at that time. Yet these very few physicists exerted a disproportionate influence on a number of crucial facets of radiology, including the development of safety guidelines. It was physicists who first sounded the alarm about inadequate protective materials in the late 1910s and a physicist, Lauriston Taylor, who convened the first nationally representative committee on x-ray safety.

W. S. Gorton was one of the first American physicists to investigate issues of x-ray safety. In 1917 he began working on a project at the NBS to test the effectiveness of commercially available x-ray protective devices. He purchased twelve pieces of lead glass from "representative dealers" and found that many of them afforded very little protection, and two of the samples were no more protective than ordinary plate glass. On behalf of the NBS, Gorton went on to correspond and collaborate with various manufacturers who were able to significantly improve their lead glass and lead-impregnated rubber materials.[22]

Physics research demonstrating the variability of protective materials caused renewed interest in the problem of protection, but it was the deaths

Table 1.1. Safety committees (c.1928)

Country	Organization
England	X-ray and Radium Protection Committee
United States	The Safety Committee of the American Roentgen Ray Society
Germany	The German X-ray Society
Sweden	The X-ray and Radium Protection Committee
Russia	The Radiological Congress of the Soviet Federation
Holland	The Protection Committee of the Board of Health

of a number of radiologists in the early 1920s that gave the problem of x-ray safety a sense of urgency. These deaths not only shook the radiological community in Britain and North America, but, as British physicist Sidney Russ later remembered: "There was a degree of public alarm which . . . threatened to restrict the use of x-rays among the public generally."[23] In particular, it was the death of Dr. W. Ironside Bruce in Britain that stirred the community of radiologists in London into action, prompting a group of doctors and physicists to form the British X-ray and Radium Protection Committee. Dr. Bruce's death was covered prominently in newspapers in both Britain and the United States, bringing attention once again to the problem of x-ray safety.[24] The medical correspondent to the *Times* blamed the power of the new x-ray tubes, telling readers that the powerful rays had caused the destruction of Dr. Bruce's blood. And this was not an isolated incident. Sounding the alarm, this author warned that several radium workers had recently died of this condition, as had an x-ray worker in Italy.[25] The death of Dr. Adolphe Leray, a French radiologist, was reported the same week.[26]

The first set of recommendations from this new British committee came out in July 1921[27] and was revised twice over the next few years. When the International Congress of Radiology met in 1928 in Stockholm, a new international committee on x-ray safety was formed. By this time, many countries had x-ray safety committees (table 1.1), and there was a sense that international cooperation was desirable in order to come up with a unified set of rules. The British X-ray and Radium Protection Committee offered their guidelines as a potential template and sent George Kaye, the National Physical Laboratory physicist, as their spokesperson, demonstrating the perceived expertise of physicists in questions of safety. The British guidelines were eventually adopted as the basis for the international guidelines, and these in turn informed the first national recommendations in the United States.[28]

Prior to the International Congress, the United States did not have a safety committee that was nationally representative. The American Roentgen Ray Society had its own safety committee, but the Radiological Society of North America did not. Nationally coordinated action on safety was organized by Lauriston Taylor, a young physicist who had been hired by the NBS in 1927. Taylor had recently graduated with an undergraduate degree in physics from Cornell and was working on his PhD requirements when he accepted the position. He was new to the world of medical x-rays, having been, in his recollection, "booby-trapped into it."[29] He was hired by the NBS to look into issues of x-ray measurement and did not receive much advice about protecting himself. He remembered being told not to stand in front of the beam, but little else. In order to try to estimate the radiation reaching him, he made a belt of photographic film to wrap around the tube that he was using for research. Every piece of film turned black, and he even measured radiation over at his desk, which sat quite a distance away from the tube. This was all the more disturbing given that the tube was lined with a protective lead shield and then covered with a big brass cylinder. Eventually, Taylor realized that the lead lining had been fastened to the outer brass cylinder with little screws and that x-rays were leaking out wherever there was a screw.[30]

At the NBS, he met the British physicist George Kaye, who was visiting the United States, and Kaye invited him to attend the International Congress of Radiology in Stockholm. Taylor was more than happy to accept and came back with recommendations from Kaye to start working on a unified set of American safety standards. Taylor contacted the presidents of the two radiological societies based in the United States, the American Roentgen Ray Society and the Radiological Society of North America, asking each of these bodies to nominate one physicist and one radiologist to sit on a new national safety committee. Two members were also nominated from the manufacturing community and one from the American Medical Association. Taylor was the final member of the committee, representing both the International Safety Committee and the NBS.

Half of the members of the new United States Advisory Committee on X-ray and Radium Protection were physicists—a percentage hugely disproportionate to their overall numbers in the wider radiological community. That physicists were given such a prominent place demonstrates once again how important their voices were deemed to be (table 1.2).[31] This committee published their initial set of recommendations in 1931. These appeared in American radiological journals and in an official NBS handbook.[32] This

Table 1.2. Inaugural membership of the United States Advisory Committee on X-ray and Radium Protection

Name	Occupation	Organization
Lauriston Taylor	Physicist	International Safety Commission; National Bureau of Standards
H. K. Pancoast	Doctor	American Roentgen Ray Society; University of Pennsylvania Hospital
J. L. Weatherwax	Physicist	American Roentgen Ray Society; Philadelphia General Hospital
R. R. Newell	Doctor	Radiological Society of North America; Stanford University Hospital
G. Failla	Physicist	Radiological Society of North America; Memorial Hospital
Francis Carter Wood	Doctor	American Medical Association; St. Luke's Hospital, New York
W. D. Coolidge	Physicist	Manufacturing; General Electric
W. S. Werner	Business	Manufacturing; Kelley-Koett

first handbook was updated in 1936 and was joined by a handbook on radium safety in 1934 and one covering rules for handling radioactive luminous compounds in 1941.[33]

THE 1931 GUIDELINES: PHYSICS VERSUS MEDICINE

The 1931 guidelines published by the NBS appear at first to solidify the authority of physics on questions of safety. The comfortingly precise quantitative recommendations for lead shielding seemed to rest firmly on physics research. But alongside the quantitative guidelines were a second set of recommended procedures that revealed all that was uncertain about the response of actual bodies exposed to x-rays.

At the heart of the 1931 guidelines was the chart shown in table 1.3, giving the minimum equivalent lead thicknesses that were deemed to be adequate protection for each corresponding exciting voltage. These numbers were identical to those set out by the International Safety Committee, with the only difference being that the American table continued up into higher voltages.

When tables like these were published in handbooks, textbooks, and journal articles, there was no justification given for these numbers, but these

Table 1.3. Guidelines showing the thickness of lead needed in protective materials according to the 1931 NBS handbook

X-rays generated by peak voltages not in excess of – (kV)	Minimum equivalent thickness of lead (mm)
75	1.0
100	1.5
125	2.0
150	2.5
175	3.0
200	4.0
225	5.0
300	9.0
400	15.0
500	22.0
600	34.0

guidelines were often published in conjunction with physics research in a way that suggested that their rationalization lay within the physics lab. The 2 mm of lead shielding deemed safe for voltages up to 125 kV had been a standard guideline for diagnostic work since the original 1916 British regulations. To scale this number up for higher voltages and to determine the equivalent thicknesses of different kinds of protective materials required research into the absorption and transmission of x-rays. This kind of work had been done by Kaye at the National Physical Laboratory and Taylor at the NBS. A summary of some of Kaye's research was published in the *British Journal of Radiology* in 1928 alongside the internationally agreed upon safety recommendations.[34] In the article, Kaye shows the reduction in the intensity of x-rays transmitted through different thicknesses of lead at different exciting voltages. If we correlate Kaye's measurements with the safety guidelines in table 1.3, it is possible to find the percentage of x-ray energy transmitted for some of the recommended thicknesses of lead. The guidelines stated, for instance, that an x-ray bulb excited to 200 kV required 4 mm of lead as protection. Kaye's research showed that at this voltage, 4 mm of lead transmitted 0.00196 percent of the incoming x-rays. This is a comfortingly precise figure. But how had 0.00196 percent of x-rays generated at 200 kV been deemed physiologically safe? That step was not articulated in any of the published guidelines in the United States or Britain. And it was not articulated because it was nothing more than a best guess.

In Kaye's recollection: "The best the [British] committee could do was try and translate into specific recommendation a sort of grand average of the protective measures which could be gleaned from the working conditions of a number of experienced radiologists who had escaped injury and still enjoyed normal health."[35]

In the United States, the work to establish a tolerance dose—the threshold under which exposure to radiation was deemed safe—was done by Dr. Arthur Mutscheller, who followed the same procedure. Mutscheller's rule was no more than 1/100th of an erythema dose in a year, with an erythema dose measuring the intensity of x-rays needed to produce reddening of the skin. He recalled arriving at that number by measuring the intensity of stray radiation in several x-ray laboratories in the country that for a number of years had "been proven to be entirely safe" due to "no detectable injurious effect during these several years."[36]

Physics research was, of course, helpful in determining the lead equivalency of different materials, but the practical difference in the health of an operator who stood behind 2 mm versus 3 mm of lead was unknown. Whether knowingly or not, by presenting the guidelines alongside his own research, Kaye ensured that the doctors uninvolved in the setting of these safety standards did not see the guess. These doctors saw a reassuring table of protective values backed by precise physical measurements.

For all the appearance of certainty provided by physicists' firm numbers, Taylor did acknowledge that this first set of guidelines was provisional. But, rather than explaining their provisional nature by appealing to all that was unknown at the time, he noted instead that future changes in the practice of radiology might require future revisions to the guidelines. When he first sent letters out explaining the decision at the International Congress in Stockholm to form a working committee in each country to consider issues of safety, he made it clear that the work of this committee would be ongoing: "With such a committee we can hope to keep abreast of any advances in the art and technique and alter the recommendations to suit."[37] When the guidelines came out in 1931, Taylor continued to talk about them in terms that showed that they would be constantly evolving. There was no legislation to back up the recommendations, but Taylor declared that the committee did not want legislation, since "legislative enacting tends to stunt development and prevent healthy changes."[38]

What Taylor failed to acknowledge was all that was currently unknown at the time of the publication of these safety rules. While admitting that it was "practically impossible to completely absorb x-rays or gamma rays," he still professed a belief that there was some threshold of intensity below which the x-rays were "relatively harmless."[39] It appears, however, that the rest of the committee was not comfortable promising to provide complete protection. In an early draft, the title of the guidelines reads: "Compiled for the Purpose of Avoiding Injuries to X-rays." In Taylor's handwriting on the draft, "avoiding injuries" was crossed out and replaced with a greater promise: "Affording Complete Protection against X-ray Exposure."[40] The final version, however, is simply titled "Protection from X-rays."

The doctors on the committee were much more willing than Taylor to acknowledge all that was unknown. One of the big problems, for instance, was that there was no clear sense of how x-ray energy was absorbed in different tissues. As Dr. R. R. Newell said, "We would like to record the x-ray energy absorbed in the patient's tissues (in orgs per cm^3) as the 'dose.' But I think we are not wise enough to do that with certainty."[41] The unit of measurement, the roentgen (R), measured the intensity of the rays, giving information about the x-rays being delivered to the skin but not information about how those x-rays would be absorbed by the skin. And, as Newell pointed out, the same application of x-ray energy would produce a larger biological effect in a larger volume because of scattering.

This uncertainty is evident in a second layer of recommendations within the published guidelines, a set of procedures and policies in addition to the lead shielding. While doctors and x-ray workers were instructed that "the omission of any protective device for the sake of expediency of operation is strictly forbidden," they were also warned that "protective gloves and aprons, even though in compliance with these recommendations, do not afford adequate protection. Consequently, care shall be taken at all times not to expose the body unnecessarily to radiation."[42]

Radioscopic work should be performed in the shortest possible time and with the lowest x-ray intensity possible. All x-ray workers were to test lead rubber for cracks because of the tendency of rubber to become brittle with age. The "Rules for the Physician in Charge" included an admonition that the efficacy of x-ray protective devices should be checked yearly with a fluorescent screen closely fitted over the eyes. "The detection of any fluorescence by the well, dark adapted eye shall be considered a hazard and remedied." But these precautions were still not sufficient. The doctor in charge of the x-ray department was to ensure that at least every four months, every worker would be supplied with dental x-ray film to be worn for fifteen working days. Any darkening of this film would be evidence that the protective measures were not adequate. And even that extra measure was not enough. Every worker must also be given four weeks' vacation and have complete blood counts done every two months with the results kept on permanent record in order to track any unusual changes.[43]

Evident in the 1931 document is a continued tension between physics and medicine, between a completely standardized, quantified, and quantifiable problem, and bodies that could react in unpredictable and idiosyncratic ways to radiation exposure. The lead gloves and lead aprons stamped with a particular lead equivalency provided comfort, but the blood tests and the dental film tests all spoke to doubt about the reliability of equipment, the efficacy of the rules, and the possibility of personal susceptibility. These uncertainties ultimately undermined the authority of physics, leaving responsibility for safety up to the radiologists.

PHYSICISTS

As the 1931 guidelines were developing, it was not yet clear who would be responsible for ensuring that these recommendations were actually followed. At first, physicists appeared to be the natural candidates to bear this responsibility, and they were increasingly asked for advice. This was encouraged by the physicists who saw themselves as safety experts. Talking about the work of committees on x-ray and radium protection, Taylor lamented "the existence of some impossible and contradictory recommendations," which meant that in many cases none of the recommendations were actually followed. He did not blame doctors, since, after all, "the average radiologist is neither engineer nor physicist and he could not be expected to choose accurately the salient points from among all of the less useful ones."[44] Taylor expected decisions on safety to rest on special knowledge of the physicist.

The manufacturing community shared this judgment, hoping that physicists would officially take on the role of safety inspectors. At the meeting of the Advisory Committee on X-ray and Radium Protection in Atlantic City in September 1931, the representatives of the manufacturing community, Dr. W. D. Coolidge and Dr. W. S. Werner, "asked that the Bureau of Standards undertake to examine and certify commercial x-ray equipment as to its satisfaction of the committee's recommendations."[45]

Despite this desire, the NBS simply did not have enough employees to take on this job and was only able to offer limited help in safety inspection. Taylor noted that "while we cannot undertake any sort of 'police work' we can make tests of apparatus to determine its satisfaction of any safety code. This of course falls in with our service of testing lead glass, gloves, aprons."[46] It seems likely that Taylor himself would have welcomed this responsibility. He fielded numerous letters from doctors about x-ray protection and replied promptly, even offering to meet and talk in person.[47] But he acknowledged that lack of funds and personnel meant that the staff was simply too busy to take on this extra work in any organized way. He did hope that the NBS would be able to provide a list of physicists who would be willing to do this work.[48]

MANUFACTURERS

Without legislation to back up the 1931 guidelines, and without the personnel resources necessary to send safety inspectors out into hospitals, Taylor hoped that he could rely on manufacturers to take on the primary responsibility for x-ray safety. He hoped in particular that he could rely on x-ray companies to persuade reluctant doctors to purchase safe equipment. He wrote to each of the major x-ray manufacturers wanting to "obtain some idea of the attitude which the manufacturer would take to the problem on x-ray protection when confronted by a prospective purchaser who does not wish to install protective devices . . . realiz[ing] that such situations must frequently occur."[49]

In particular, he "wished to learn what form of persuasion the manufacturer will use in confronting the doctor and to what extent he will try to insist on the purchase of safe installations."[50] The responses show that the x-ray companies were unwilling to insist, preferring instead to leave the responsibility for safety in the hands of the doctors.

While manufacturers had been providing lead screens, gloves, and goggles for over two decades, radiation safety was not their primary concern. W. S. Werner explained that, "in the past, electrical hazards, because they were easily visualized, have been given close attention by manufacturers."[51] Protection from stray radiation, however, had not been given as much attention. This is evident in the standard form that Picker's employees used to generate an inspection report for newly installed x-ray apparatus. All of the points are electrical and mechanical. There is not a single bullet point checking for adequate protection from radiation.[52]

When the 1931 safety guidelines were published, some manufacturers were clearly unhappy. A representative of Westinghouse sent a long list of concerns to Taylor, complaining that "a strict observance of these various points would be placing an enormous burden on a very large number of physicians and institutions who now possess x-ray installations which are not in accordance with the suggestions set forth in the handbook." It was not only the time and cost of installing new equipment that concerned him; he called into question the necessity of various provisions. Some, like the objection to portable screens, seemed to him to be unnecessarily cautious. He argued that if a room was lead-lined, there would be no chance of secondary radiation hitting the operator from behind, and so a portable screen would be perfectly adequate. In other cases, the recommendations seemed to pose new dangers. For instance, the recommendation that a foot switch be shielded seemed dangerous: "unless a foot switch with shield is carefully designed, it will be impracticable, since in the dark, an operator fumbling with his foot for the button may accidentally complete the x-ray circuit and cause more damage than if the shield were omitted."[53]

Other manufacturers were much more positive about the work of the committee. Mr. Edwin Goldfield wrote hoping for approval from Taylor and offering nothing but support and praise: "I would like to take this opportunity to say that I believe that a great deal of good will come from the work you have started and you may be sure that this is in exact accordance with our plans. We will therefore appreciate any suggestions that you may have to offer."[54]

A common thread through all of the responses—from Westinghouse, General Electric X-ray Corporation, Kelly-Koett, James Picker, and Standard X-ray—however, was the concern that cost would be the primary issue for doctors. With the country suffering from the Great Depression, manufacturers felt that cost was already a consideration that caused doctors to skimp on protection. "Too frequently the operator has no protection except a lead glass shield, through refusal of the customer to invest in the cost of a control booth or a lead protective screen."[55] As the vice president of General Electric X-ray Corporation explained to Taylor: "The manufacturer generally tries to make recommendations which will give his customer the maximum protection from x-radiation. Quite often his recommendations are not followed because of the cost involved."[56] Rather than drastic action, W. S. Werner recommended a "tapering period building up to the desired protection to avoid undue unfavorable comment due to cost."[57]

The manufacturers' sense was that doctors were weighing many factors, in addition to cost, when they made a purchasing decision. In the case of vertical and horizontal fluoroscopes, R. S. Landauer of Standard X-ray noted that despite the "unanimity of opinion" that a lead-lined box for the x-ray tube offers "far more protection" than a lead glass enclosure, "it is practically impossible to sell such a piece of equipment." He points to the fact that the lead-lined fluoroscope is "more bulky, slightly harder to handle and does not look quite so nice, and it is slightly more expensive. These are pitifully poor objections to place against maximum safety, yet they overbalance it."[58] Floor stand protectors were similarly preferred by doctors, even though they did not provide enough protection, because they did not take up as much space compared to an entirely lead-lined cubicle.

The issue of trust was vital for the manufacturers. Salespeople needed doctors to trust them. They argued that if a vendor insisted on a particular piece of protective equipment, the doctor might think that x-ray apparatus from that company was more dangerous or that the salesperson simply wanted the doctor to spend more money.[59] The situation was made more difficult because the manufacturers had no reason to trust each other. A representative from Westinghouse worried about "unscrupulous" manufacturers who would benefit compared to those who follow the new rules rigidly.[60] W. Steinkamp, the secretary and general manager of James Picker, agreed that the problem lay in not being able to control other companies' salespeople. Invoking the possibility of some kind of "binding code" agreed upon by all manufacturers, he did tell Taylor that his company would "be agreeable to any program outlined that [would] be agreed upon by the other manufacturers as well."[61]

Overall, most of the manufacturers did what they could to minimize their own responsibility, declaring that "the salesman is not a policeman, nor is he the doctor's guardian."[62] In the view of these companies, it was enough to provide the option of safe equipment meeting the standards set out by the NBS. Manufacturers preferred to leave the choice of purchasing this equipment up to the doctors. Partly this was motivated by the sense that after the installation, so much was out of their hands. The purchase of safe equipment was only the first step in keeping bodies in an x-ray room safe. The vice president of General Electric X-ray Corporation placed this burden firmly on doctors in charge of x-ray departments. In a large x-ray department with many people, the regulations "will often be disregarded although provided for if the roentgenologist in charge does not insist on a strict observance of the rules."[63] Even Taylor ultimately left the final judgment to the doctors: "I feel very strongly that the x-ray companies should instruct their men in the future to point out the need of protection from as impartial a view point as possible, leaving the decision as to purchase up to the doctor, without making him feel that anything is being forced upon him."[64]

In a meeting of the Advisory Committee with manufacturers in late 1931, the manufacturers once again noted how difficult it was to sell protective installations, airing their concerns about competitors who would take over a job if their salespeople refused to sell unsafe apparatus. It was recommended at the meeting that the manufacturers get written proof that unsafe installation had been pointed out to doctors. Abdicating responsibility, the manufacturers expressed hope that the professional radiological societies would be the ones to enforce the recommendations. They also hoped that insurance companies would adjust their rates for doctors with safe apparatus, feeling that this "would do much to promote the use of safe apparatus."[65]

TECHNICIANS

The radiologist in charge of a hospital x-ray room might be responsible for purchasing equipment, but by the early 1930s, most hospitals employed technicians or radiographers to actually run the apparatus. These individuals occupied a tenuous place in the 1931 guidelines. Working every day in an x-ray room, they put their bodies at the highest risk, and while they were given some responsibility for ensuring their own safety, it was ultimately up to the radiologist who supervised them to ensure their continued health. Dr. Mutscheller spoke of several technicians who had lost hair because of x-rays hitting wood in the operator's booth and producing secondary radiation.[66] The 1931 guidelines gave doctors the responsibility of ensuring the technicians wore dental film, had blood checks, and were given adequate vacation time. But the guidelines also emphasized the day-to-day responsibility of the technicians to check that the safety equipment was in good shape, ensuring, for instnace, that the lead rubber had not started to crack.[67] Technicians were told to make sure "to perform radioscopic work in the minimum time possible, using the lowest x-ray intensity and smallest aperture consistent with demands."[68] Just like the doctors, the technicians had to balance radiation safety with other practical concerns. For instance, when doing fluoroscopic work, the intensity could not be too low or the technician would risk eye strain.

While these guidelines position the technician as a responsible agent in this environment, one to be trusted to check for faulty equipment and to judge appropriate intensity levels for particular kinds of x-ray work, there are many hints that the technician was not always trusted by, or granted the same level of autonomy as, doctors.

The guidelines themselves assume a technician might cut corners in order to perform tasks more quickly, explicitly stating that "the omission of any protective devices for the sake of expediency of operation shall be strictly forbidden."[69] And doctors did not always trust those working for them. Dr. Mutscheller, for instance, clearly held technicians in very low opinion, talking about how easy it was for them to acquire bad habits such as standing in front of the tube. He recommended that someone in the x-ray department keep a list of every radiation injury to use to combat complacency. Besides his lack of faith in technicians' understanding of the gravity of the risk, he was also critical of the dental film guideline, imagining that it would be easy for a dishonest worker to expose his film in order to get a vacation.[70] Technicians were also seen as a source of potential error, putting the wrong filter in place for x-ray therapy and thereby endangering the patient.[71] Even the bodies of technicians could be untrustworthy. A low blood count could be due to radiation or it could just as easily be due to a completely unrelated infection of some kind.[72]

While the 1931 guidelines were intended to provide protection to these x-ray workers, their safety was ultimately in the hands of the doctor in charge of the unit, who had to oversee installation of safe equipment in the first place, and who then had to believe that the darkening of an x-ray film or an abnormal blood count meant that something was actually wrong.

DOCTORS

Just as the manufacturers put the final responsibility in the hands of the doctors, so too did the printed version of the 1931 guidelines. As we have already seen, these guidelines made the doctor in charge responsible for checking the efficacy of the protective devices. Using a fluoroscope to check for stray radiation, the doctor might catch "carelessness," such as nails driven into sheets of protective lead covering during the installation of an x-ray room.[73] The doctor in charge was likewise responsible for ensuring that workers were supplied with dental film to wear at regular intervals. Similarly, bimonthly blood counts, adequate vacation time, and a well-lit and ventilated work environment were all the responsibility of the physician in charge of the unit.

The doctors themselves expressed a wide range of opinions on the necessity of new safety regulations, and the merits of the NBS guidelines in particular. Dr. Newell, one of the members of the committee, felt that the guidelines were far too strict. Writing to Taylor with his notes on a draft of the regulations, he complained: "I think the present draft is so severe that it will be ignored. We are seeking freedom from harm, not necessarily freedom from x-rays."[74] Other doctors were much more likely to embrace what Edward Chamberlain called "the necessity of drastic protective measures."[75] His opinion as to the necessity of these measures had been formed because of his own personal experience. One of his technicians had developed anemia and had had to quit x-ray work. It turned out that the problem was the operator's room. It had lead walls and a lead glass window but no roof, so x-rays were still able to enter through the top of the booth.[76]

Chamberlain's caution appears to have been exceptional, however. His colleagues instinctively resisted any rule that might guide their behavior, adamant that they must follow their own judgment. Newell, for instance, was critical of the idea of a timer that would shut off the x-ray apparatus after a maximum dose had been given over one area of the patient. He argued that "the machine will shut off on almost every patient. This will only teach the radiologist to ignore the warning. This is a matter in which I do not see how one can make a mechanical substitute for care on the part of the radiologist."[77] A draft of the guidelines included a table that would have given doctors a quick reference to look up the surface dose in R/sec for different parts of the body and particular levels of voltage and current. This was omitted in the final document. The draft shows Taylor's notes in pencil: "Dangerous unless more specific."[78] Both the timer and the reference chart were rejected in favor of the clinical skill of the doctor.

One of the primary reasons that doctors were so loath to substitute their professional judgment for a mechanism or a shorthand rule was a continued belief in idiosyncratic bodies. In a discussion at the Radiological Society of North America following the publication of the NBS handbook, Newell offered as an example a patient who had taken much more than the recommended tolerance dose of x-rays without harm. The patient, who was being treated for Hodgkin's disease, had received double the ten-year quota in six months—"instead of killing her it cured her."[79] When Newell asked a colleague, Dr. Ross, what he considered his maximum safe dose, Dr. Ross replied: “It varies so with different individuals and different skins that, as you know, what will produce an erythema with one will not with the other.” Newell does, however, go on to say: “But it is not safe to figure from what one person can stand, or from what ninety-nine persons out of a hundred can stand. We must go on the basis of what might be dangerous for the exceptional person.”[80] For every body that could withstand much more than others, there was another body that was much more sensitive. Both gave reason for doctors to resist set guidelines in favor of the personal knowledge they gleaned about their own patients from multiple clinical interactions.

PATIENTS

Of all the bodies in the x-ray room, patients had the least agency and were wholly dependent on the clinical judgment of the doctor. Even their testimony might be discounted. Like technicians, patients were seen by doctors as potentially untrustworthy. Doctors worried that they might not be told about previous exposure to x-rays. In one case, a patient’s family sued a doctor after a family member had died following x-ray therapy. The suit was only dropped when the family and the doctor found out that this individual had seen two other doctors previously for x-ray treatment.[81]

Doctors used clinical signs to judge the reaction of a patient’s body to x-rays and weighed the risk to the body of the patient differently depending on whether the patient was receiving diagnostic or therapeutic x-rays. The 1931 guidelines identified three classes of x-ray apparatus: diagnostic, superficial therapy, and deep therapy. The only section with rules aimed at ensuring patient safety is the diagnostic section, where “it is recommended that for prolonged fluoroscopic work an accumulative timing device be used which will either indicate or turn off the apparatus when the total exposure exceeds a certain previously predetermined limit.”[82] In this chapter, we have already seen doctors reacting unfavorably to this provision, perceiving it as an attack on their judgment. But there was no attempt in the guidelines to address patient safety when the patient was receiving therapeutic x-rays, even though the beam in this case was more powerful and therefore more toxic. Here, the potential healing power appears to have outweighed the potential damage. Doctors wanted to educate the public so that they did not sue at the first sight of reddened skin. Newell argued that patients needed to understand, "first, that x-ray is potent, and second, that a reddened skin may be the necessary price of sufficient treatment."[83]

A patient entering an x-ray room, then, was wholly dependent on the doctor to keep him or her safe. The 1931 guidelines only briefly addressed the issue of patient safety, and where they did attempt to provide guidance they were severely criticized by doctors who resented this encroachment on their clinical autonomy. The only agency the patient had in this space was as a potential source of dangerous secondary radiation, evident in Taylor's handwritten note on a draft of the guidelines: "May need more for radiation from patient."[84]

CONCLUSION

Emerging out of the new concern about x-ray safety in the 1920s, the 1931 NBS guidelines reveal a strong hope that standardized equipment and strict procedures could minimize or even eliminate all danger. These guidelines were an attempt to renew trust in safety equipment, backed by the authority of physics. However, the desire for a set of rules and reassuring charts of adequate lead shielding clashed with doctors' desire for professional autonomy. The NBS guidelines were met with resistance from radiologists who remained the ultimate arbiters of safety in the x-ray room. I have offered a number of explanations for this locus of responsibility, including a physics community too small in number to supply safety inspectors and an economic climate in which manufacturers were unwilling to lose their competitive edge by insisting on the purchase of particular equipment. Most importantly, though, the continued responsibility of doctors was necessitated by deep uncertainties concerning the physiological effects of x-rays. The numbers in these guidelines looked certain but were based on what was merely a best guess, derived from informal observations of x-ray rooms where operators were still healthy. With this kind of uncertainty, the bodies of x-ray workers needed to be under constant medical surveillance. This is even more evident in the minimal guidelines for safe exposure for patients' bodies. The task of achieving a balance between the diagnostic or therapeutic benefit of x-rays and potential injury was left firmly in the hands of doctors.

Since these first national guidelines for x-ray safety, the history of radiation protection can best be characterized by repeating waves of optimism followed by concern. In 1934 the American radiological community felt confident enough to set a safe dose of x-rays at 0.1 R/day.[85] But by the late 1940s, increasing attention to possible genetic damage caused by radiation contributed to a lower recommendation of 0.05 R/day.[86] These discussions primarily stayed focused on occupational rather than patient safety. But it was Alice Stewart, a British epidemiologist, who shattered the perception that diagnostic x-rays could ever be perfectly safe. In the 1950s, she showed that a single diagnostic x-ray performed on pregnant women, a procedure delivering a dose much smaller than the accepted minimum safe dose, was enough to significantly increase the rates of pediatric leukemia in their children. Yet Stewart came up against huge opposition when she announced these findings.[87] The certainty of a minimum safe dose—the certainty provided by physicists' sharp numbers—has been, and continues to be, hard to abandon.

ACKNOWLEDGMENTS

Many thanks to the archivists at the Francis A. Countway Library of Medicine for help locating materials. Thanks also to colleagues who provided generous feedback at meetings for the History of Science Society and the Canadian Society for History and Philosophy of Science and to the Social Sciences and Humanities Research Council of Canada and Harvey Mudd College for funding this research.

NOTES

1. In a letter from Don S. Brown of Westinghouse Corporation to Lauriston Taylor, January 27, 1932, Brown refers to "several serious accidents" due to breaking of cables and dropping of heavy weight on a patient. Taylor Papers, box 8, folder 27.

2. U.S. Department of Commerce: National Bureau of Standards, "X-ray Protection" (1931).

3. Taylor, *X-ray Measurements and Protection, 1913–1964*, vii.

4. Kuchinskaya, "Twice Invisible," 80.

5. Starr, *The Social Transformation of American Medicine*.

6. Butcher, "The Measurement of X-rays," 70.

7. A small pamphlet telling the early story of x-rays published by the Victor X-ray Corporation speaks of the "pioneers" who had "unknowingly harmed themselves." See Victor X-ray Corporation, *A Little Journey into the Realms of the X-ray*, 23.

8. Butcher, "The Future of Electricity in Medicine," 8.

9. For a detailed examination of the first instances of reported burns and possible links between x-rays and cancer see Goldberg, "Suffering and Death among Early American Roentgenologists."

10. Lavine, "The Early Clinical X-ray in the United States," 9.

11. Burrows, *Pioneers and Early Years*, 237.

12. Brown, *American Martyrs to Science through the Roentgen Ray*, 18–21.

13. Herzig, *Suffering for Science*, 85–99.

14. Kny-Scheerer Company, *Illustrated and Descriptive Catalogue*, n.p.

15. See, for instance, Waite and Bartlett Manufacturing Company, *X-ray and High Frequency Apparatus*, 21.

16. R. V. Wagner Company, *Catalogue of Electrical Instruments for Physicians and Surgeons*, 61.

17. Kny-Scheerer Company, *Illustrated and Descriptive Catalogue*.

18. Dodd, "Treatment of Acute Roentgen Ray Dermititis," 430.

19. Dodd, "Treatment of Acute Roentgen Ray Dermititis," 431.

20. Caldwell, "Skiagraphy of the Accessory Sinuses of the Nose," 570.

21. Lund, "Procédés De Protection," 448.

22. Gorton, "Roentgen Ray Protective Materials." He thanks, in particular, the cooperation of Corning Glass Works, Pittsburgh Plate Glass Company, B. F. Goodrich Company, Manhattan Rubber Manufacturing Company, and Price Electric Company.

23. Russ, "A Personal Retrospect," 554.

24. "Dangers of X-ray," *New York Times*, May 15, 1921.

25. Medical Correspondent, "An X-ray Crisis," *Times (London)*, March 31, 1921, 10.

26. "The Danger of X-ray Therapeutics," 9.

27. "X-ray and Radium Protection," 17.

28. For easy comparison, both the 1927 (third revision) British guidelines and the first 1928 international guidelines are reproduced in appendix A and B of Kaye, "Protection and Working Conditions in X-ray Departments."

29. Interview of Lauriston Sale Taylor by Gilbert Whittemore on August 11, 1990, Niels Bohr Library and Archives, American Institute of Physics, College Park, MD, http://www.aip.org/history/ohilist/5153_1.html.

30. Interview of Lauriston Sale Taylor by Gilbert Whittemore.

31. U.S. Department of Commerce: National Bureau of Standards, "X-ray Protection" (1931).

32. U.S. Department of Commerce: National Bureau of Standards, "X-ray Protection" (1931). Reprinted in *Radiology* 17 (1931): 542 and in *American Journal of Roentgenology and Radium Therapy* 26 (1933): 436.

33. U.S. Department of Commerce: National Bureau of Standards, "X-ray Protection" (1931); U.S. Department of Commerce: National Bureau of Standards, "X-ray Protection" (1936); U.S. Department of Commerce: National Bureau of Standards, "Radium Protection for Amounts Up to 300 Milligrams"; U.S. Department of Commerce: National Bureau of Standards, "Radium Protection"; U.S. Department of Commerce: National Bureau of Standards, "Safe Handling of Radioactive Luminous Compound."

34. Kaye, "Protection and Working Conditions in X-ray Departments."

35. Kaye, "The Story of Protection," 299.

36. "Discussion," *Radiology* 19 (1932): 34.

37. Taylor to Dr. E. H. Skinner, St. Luke's Hospital, Kansas City (October 29, 1928). Taylor Papers, box 8, folder 21.

38. Taylor, "The Work of the National and International Committees on X-ray and Radium Protection," 1.

39. Taylor, "The Work of the National and International Committees on X-ray and Radium Protection," 1.

40. "Draft: Rules for Erection and Operation of Medical X-ray Plants." Taylor Papers, box 8, folder 22.

41. R. R. Newell to Members of Standardization Committee of the American Roentgen Ray Society—Comment on Report of Meeting of X-ray Units, New York, January 5, 1928. Taylor Papers, box 8, folder 21.

42. U.S. Department of Commerce: National Bureau of Standards, "X-ray Protection" (1931), 23.

43. U.S. Department of Commerce: National Bureau of Standards, "X-ray Protection" (1931), 26.

44. Taylor, "The Work of the National and International Committees on X-ray and Radium Protection," 1.

45. Minutes of Meeting of Advisory Committee on X-ray and Radium Protection at American Roentgen Ray Meeting in Atlantic City, September 1931, November 23, 1931. Taylor Papers, box 8, folder 23.

46. Taylor, Bureau of Standards Memo on Behalf of the X-ray and Radium Protection Committee, November 21, 1931. Taylor Papers, box 8, folder 26.

47. Taylor to Dr. Curtis Burnam, Howard A. Kelly Hospital, Baltimore, February 4, 1932. Taylor Papers, box 8, folder 27.

48. Taylor, Bureau of Standards Memo on Behalf of the X-ray and Radium Protection Committee, November 21, 1931. Taylor Papers, box 8, folder 26.

49. Taylor to Steinkamp, James Picker, February 11, 1932. Taylor Papers, box 8, folder 27.

50. Taylor to Steinkamp, James Picker, February 11, 1932. Taylor Papers, box 8, folder 27.

51. Werner, "X-ray Protection from the Manufacturer's Viewpoint," 6.

52. Letter from Steinkamp, Secretary and General Manager of James Picker, to Taylor, January 29, 1932. Taylor Papers, box 8, folder 27.

53. Don S. Brown, Westinghouse X-ray Corporation, to Taylor, January 27, 1932. Taylor Papers, box 8, folder 27.

54. Mr. Edwin R. Goldfield to Taylor, December 8, 1931. Taylor Papers, box 8, folder 26.

55. Letter from Steinkamp, Secretary and General Manager of James Picker, to Taylor, January 29, 1932. Taylor Papers, box 8, folder 27.

56. Vice President of General Electric X-ray Corporation to Taylor, February 22, 1932. Taylor Papers, box 8, folder 27.

57. W. S. Werner, Vice President of Kelly-Koett, to Taylor, February 17, 1932. Taylor Papers, box 8, folder 27.

58. R. S. Landauer, Standard X-ray Company, to Taylor, February 16, 1932. Taylor Papers, box 8, folder 27.

59. R. S. Landauer, Standard X-ray Company, to Taylor, February 16, 1932. Taylor Papers, box 8, folder 27.

60. Don S. Brown, Westinghouse Co., to Taylor, January 27, 1932, Taylor Papers, box 8, folder 27.

61. Steinkamp, Secretary and General Manager of James Picker, to Taylor, February 23, 1932. Taylor Papers, box 8, folder 27.

62. R. S. Landauer, Standard X-ray Company, to Taylor, February 16, 1932. Taylor Papers, box 8, folder 27.

63. Vice President of General Electric X-ray Corporation to Taylor, February 22, 1932. Taylor Papers, box 8, folder 27.

64. Taylor to Mr. W. S. Kendrick, General Electric X-ray Corporation, January 11, 1932. Taylor Papers, box 8, folder 27.

65. Report of Meeting of Advisory Committee with X-ray Manufacturers GE, Kelley-Koett, Standard X-ray, James Picker, and Westinghouse Held December 2, 1931, St. Louis, MO, January 16, 1932. Taylor Papers, box 8, folder 23.

66. "Discussion," 34.

67. "X-ray Protection" (1931), 23.

68. "X-ray Protection" (1931), 23.

69. "X-ray Protection" (1931), 23.

70. "Discussion," 35.

71. Wilhelm Stenstrom, "Protection in X-ray Therapy," 9.

72. Dr. Dejardins in "Discussion," 36.

73. "Discussion," 32.

74. Letter from Newell to Taylor, December 10, 1930. Taylor Papers, box 8, folder 24.

75. Chamberlain, "Protection in Diagnostic Radiology," 22.

76. Chamberlain, "Protection in Diagnostic Radiology," 23.

77. Newell to Taylor, January 22, 1931. Taylor Papers, box 8, folder 26.

78. "Draft: Rules for Erection and Operation of Medical X-ray Plants." Taylor Papers, box 8, folder 22.

79. "Discussion," 32.

80. "Discussion," 38.

81. Wanvig, "Legal and Insurance Aspects of Protection in Radiology."

82. "X-ray Protection" (1931), 4–5.

83. "Discussion," 33.

84. "Draft B: Rules for Erection and Operation of Medical X-ray Plants." Taylor Papers, box 8, folder 22.

85. Walker, *Permissible Dose*, 8.

86. Walker, *Permissible Dose*, 10–11.

87. Greene, *The Woman Who Knew Too Much*, 78–93.

BIBLIOGRAPHY

Archival Sources

Lauriston Sale Taylor Papers, 1904–1999 (inclusive), 1928–1989 (bulk). H MS c334. Harvard Medical Library, Francis A. Countway Library of Medicine, Boston, MA.

Published Sources

Brown, Percy. *American Martyrs to Science through the Roentgen Ray*. Baltimore: Charles C. Thomas, 1936.

Burrows, E. H. *Pioneers and Early Years: A History of British Radiology*. St. Anne, Alderney, Channel Islands: Colophon Limited, 1986.

Butcher, W. Deane. "The Future of Electricity in Medicine." *Proceedings of the Royal Society of Medicine: Section of Electrotherapeutics* 1 (1908): 1–14.

Butcher, W. Deane. "The Measurement of X-rays: The Standardisation of Röntgen Light." *Journal of the Röntgen Society* 4 (1908): 36–45.

Caldwell, Eugene. "Skiagraphy of the Accessory Sinuses of the Nose." *American Journal of Roentgenology* 5 (1918): 569–74.

Chamberlain, Edward. "Protection in Diagnostic Radiology: Avoiding the Dangers of X-ray Exposure and High Tension Shock." *Radiology* 19 (1932): 22–28.

"The Danger of X-ray Therapeutics." *Times (London)*, March 30, 1921, 9.

"Dangers of X-ray: New Investigation Following Recent Deaths to Insure Scientists' Protection." *New York Times*, May 15, 1921, 85.

"Discussion." *Radiology* 19 (1932): 32–40.

Dodd, W. J. "Treatment of Acute Roentgen Ray Dermititis." *American Journal of Roentgenology* 1 (1914): 430–31.

Goldberg, Daniel. "Suffering and Death among Early American Roentgenologists: The Power of Remotely Anatomizing the Living Body in Fin de Siècle America." *Bulletin of the History of Medicine* 85 (2011): 1–28.

Gorton, W. S. "Roentgen Ray Protective Materials." *American Journal of Roentgenology* 5 (1918): 472–78.

Greene, Gayle. *The Woman Who Knew Too Much: Alice Stewart and the Secrets of Radiation*. Ann Arbor: University of Michigan Press, 1999.

Herzig, Rebecca. *Suffering for Science: Reason and Sacrifice in Modern America*. New Brunswick, NJ: Rutgers University Press, 2005.

Kaye, G. W. C. "Protection and Working Conditions in X-ray Departments." *British Journal of Radiology* 1 (1928): 295–312.

Kaye, G. W. C. "The Story of Protection." *Radiography* 6 (1940): 41–60.

Kny-Scheerer Company. *Illustrated and Descriptive Catalogue: Roentgen X-ray Apparatus and Accessories*. 3rd ed. New York: n.p., 1905.

Kuchinskaya, Olga. "Twice Invisible: Formal Representations of Radiation Danger." *Social Studies of Science* 43 (2012): 78–96.

Lavine, Matthew. "The Early Clinical X-ray in the United States: Patient Experiences and Public Perceptions." *Journal of the History of Medicine and Allied Sciences* 67 (2012): 587–625.

Lund, Peer. "Procédés De Protection." *American Journal of Roentgenology* 5 (1918): 447–48.

Medical Correspondent. "An X-ray Crisis: The Danger to Reproduction, New Epoch in Radiography." *Times (London)*, March 31, 1921, 8.

Russ, Sidney. "A Personal Retrospect." *British Journal of Radiology* 26 (1953): 554–55.

R. V. Wagner Company. *Catalogue of Electrical Instruments for Physicians and Surgeons*. 7th ed. Chicago: R. V. Wagner, 1905.

Starr, Paul. *The Social Transformation of American Medicine*. New York: Basic Books, 1982.

Stenstrom, Wilhelm. "Protection in X-ray Therapy." *Radiology* 19 (1932): 7–11.

Taylor, Lauriston. "The Work of the National and International Committees on X-ray and Radium Protection." *Radiology* 19 (1932): 1–4.

Taylor, Lauriston. *X-ray Measurements and Protection, 1913–1964*. Edited by National Bureau of Standards. Washington, DC: United States Government Printing Office, 1981.

U.S. Department of Commerce: National Bureau of Standards. "Radium Protection." Washington, DC: United States Government Printing Office, 1938.

U.S. Department of Commerce: National Bureau of Standards. "Radium Protection for Amounts Up to 300 Milligrams." Washington, DC: United States Government Printing Office, 1934.

U.S. Department of Commerce: National Bureau of Standards. "Safe Handling of Radioactive Luminous Compound." Washington, DC: United States Government Printing Office, 1941.

U.S. Department of Commerce: National Bureau of Standards. "X-ray Protection." Washington, DC: United States Government Printing Office, 1931.

U.S. Department of Commerce: National Bureau of Standards. "X-ray Protection." Washington, DC: United States Government Printing Office, 1936.

Victor X-ray Corporation. *A Little Journey into the Realms of the X-ray*. Chicago: Victor X-ray Corporation, 1926.

Waite and Bartlett Manufacturing Company. *X-ray and High Frequency Apparatus*. 22nd ed. New York: Waite and Bartlett, 1910.

Walker, J. Samuel. *Permissible Dose: A History of Radiation Protection in the Twentieth Century*. Berkley: University of California Press, 2000.

Wanvig, H. V. "Legal and Insurance Aspects of Protection in Radiology." *Radiology* 19 (1932): 29.

Werner, Wilbur. "X-ray Protection from the Manufacturer's Viewpoint." *Radiology* 19 (1932): 5–6.

"X-ray and Radium Protection." *Journal of the Röntgen Society* 17 (1921): 100–103.

2

Contested Knowledge

THE TRINITY TEST RADIATION STUDIES

Janet Farrell Brodie

At 6 a.m. on July 16, 1945, Maria Clemens, blind from birth, saw, for the first time in her life, a bright flash of light. Others in that desert area of south-central New Mexico saw it too. A farmer called to his wife that the sun was rising in the wrong place. For miles throughout New Mexico, windows shook and doors rattled. Later, army officials explained that an ammunition dump had blown up, causing strange lights to appear and the earth to shake.[1] Nearby scientists from the secret facility at Los Alamos, where the atomic bomb was being created, along with a select group of high-ranking government officials observed in stunned silence as the first atomic device—not yet considered an actual bomb but nicknamed "the gadget"—exploded with the equivalent force of almost nineteen thousand tons of dynamite. With the detonation, a fireball rose from the desert floor and then a column of dust and debris swept up from below the glowing fireball, quickly topped by a mushroom cap of swirling detritus. What became the iconic image of the atomic bomb—that now familiar mushroom shape—rose for the first time before the awestruck observers.[2] Twenty-one days later, a simpler uranium bomb destroyed most of the Japanese city of Hiroshima. Three days after that, on August 9, Americans dropped a plutonium bomb similar to the one that exploded in New Mexico on the Japanese city of Nagasaki. Shortly after that, the Japanese surrendered, and the war in the Pacific ended. Until the Limited Nuclear Test Ban Treaty of 1963, the United States built nuclear weapons and tested them at sites in the Pacific Ocean and the Nevada desert.

This chapter explores conflicting reports about the amount and persistence of radiation from that first nuclear detonation in the New Mexico desert, code-named the "Trinity" test. Scientists from Los Alamos prepared for the test for months in advance, but immediate full-scale studies after the test were hampered by the shift in attention and personnel to the two atomic bombs dropped in Japan. Studies were also hindered by the secrecy surrounding everything connected to the atomic bomb, and especially to its radiation.[3] For decades after the war, experts from different fields in science and medicine, and from diverse institutions, studied the radiation from the Trinity test and wrote memos and reports. They drew divergent conclusions about the types and amounts of radiation, its mapping, and its toxicity. There are numerous explanations for these deviating analyses, including differing disciplinary approaches, problems with radiation-measuring instruments, the politics around atomic bomb tests, and authorities' fears of lawsuits. But the ways that knowledge was censored became a key factor, too, in the conflicting conclusions drawn by experts. Secrecy and restrictions about the circulation of information added layers of complexity to the already fraught knowledge production around atomic radiation.

Gamma, beta, and alpha radiation as well as neutrons from America's nuclear bomb tests were a source of intense interest and study by military and civilian groups from the Trinity test on. Significant discoveries about what came to be known as "fallout" illustrate how little factual data even the most knowledgeable experts had, well into the mid-to-late 1950s, about the patterns of dispersal, the distances radiation could travel if it reached the stratosphere, the phenomenon of "close-in" fallout (radiation coming back to earth within one hundred miles of the detonation), and the unexpected ways that radioactive debris could be deposited far from test sites.[4] The Trinity radiation studies were not formally conducted in conjunction with other studies such as, for example, those from the Nevada Test Site, but in the 1950s there was often considerable overlap among experts and institutions studying bomb effects at both sites.

In the first part of this chapter, I examine the main individuals and institutions studying the Trinity bomb's radiation as well as the written record—formal and informal—they produced. The second section is focused on their complicated and divergent findings. In the final section, I analyze some of the reasons for the discrepant arguments about radiation from Trinity, particularly the role of secrecy and censorship.

THE EXPERTS AND THEIR TRINITY TEST RADIATION FINDINGS

At least a dozen reports and memoranda written over six decades contain information about the radiation from the Trinity test, both from the initial blast and the residual aftermath. Some documents exist only from references in other documents; others are still formally classified and unavailable without the lengthy process of a Freedom of Information Act request. Some appear on the internet but in mangled format, with pages, authors, and other pertinent information missing. Others are available through traditional historical research methods in archives and special collections. This chapter is based on as many of those documents as I was able to track down but not, unfortunately, all of those known to have been written.

Experts from several institutions conducted the main Trinity radiation studies: researchers from Los Alamos (where the atomic bomb was invented, which after the war became one of the new National Nuclear Laboratories); the Crocker Radiation Laboratory at the University of California–Berkeley (later renamed the Lawrence Radiation Laboratory on the campus); the U.S. Atomic Energy Commission (AEC), especially those in its Division of Biology and Medicine; and, importantly, the Atomic Energy Project at the medical school at UCLA.[5] In the 1970s and 1980s, scientists from the Office of Radiation Programs at the Environmental Protection Agency (EPA), and later from the Defense Nuclear Agency, also produced reports about the persistence of radiation from the Trinity test. Fields of expertise varied. Traditional radiologists and those who during the war began to identify themselves as "health physicists" wrote some of the reports as they expanded their expertise from traditional medicine to the impacts of atomic radiation.[6] Quickly, however, specialists in new fields emerged: radiochemistry, radioecology, radiobiology, and, by the 1960s, environmental radiation. At the UCLA Atomic Energy Project in the early 1950s, scores of personnel worked in new fields revolving around ionizing radiation: eighteen in radiobiology, eighteen in pharmacology and toxicology, and fourteen in radioecology. In addition, eleven worked in "special problems," particularly the section blandly titled "Alamogordo," which was founded to study the Trinity radiation.[7]

Kermit Larson emblematizes some of these changes. Hired to help monitor radiation at Oak Ridge, Tennessee, during World War II, he then served with the scores of other radiation monitors at the Crossroads tests of atomic bombs in the Pacific in 1946 (see chapter 3 in this volume). He worked toward a PhD in soil chemistry at University of California–Berkeley, but it is unclear if he completed the degree. He then worked as an environmental analyst of radiation at the UCLA Atomic Energy Project, where for over a decade he analyzed the effects of nuclear tests' radiation on soil, plants, and animals. He coauthored important reports about the Trinity radiation, radiation from numerous 1950s tests in Nevada, and fallout in general. Larson helped develop methods for assessing small amounts of radiation from "close" and faraway fallout; he collected radioactive soils to assess impacts on plants (working out of the "hot lab" at the UCLA Atomic Energy Project); and he published widely in both the open and classified scientific literature on radiation and fallout.[8]

Larson's radiation reports straddled the line between alarm and reassurance. He could be prickly and hard-edged about minimizing the danger from radiation, especially from low levels of exposure. Asked to comment on draft chapters of historian Ferenc Szasz's account of the Trinity test, Larson challenged several of Szasz's critical assertions. Szasz wrote, "It was generally understood that any excessive amount of radiation damaged the body." Larson's marginal comments included: "Above what level?" and "Depends on individual."[9] Larson's reports about Trinity were among the first studies of bomb radiation to describe the unpredictability of where and how much radioactive material fell back to the ground. Radiation from atomic bombs surprised the early experts in not being dispersed in predictable patterns; it could not be traced in ever smaller amounts in concentric circles from ground zero. Instead, perplexingly, radioactive material fell back to the ground in amounts and places no one (at least in the late 1940s through the early 1960s) could predict ahead of time. In particular, large amounts sometimes fell hundreds of miles from the detonation site, creating what came to be known as "hot spots." Speaking later about an unpublished study he and Ralph Bellamy conducted in 1948, Larson noted that they had found no "hazard from external total body exposure to ionizing (gamma) radiation" in the fenced area. But they did not rule out an "internal emitter problem." What they discovered was an area in "flux with respect to distribution and biological availability of radioactive fission products and plutonium." And they had no idea how many years would pass before some kind of biological equilibrium would occur.[10]

TRINITY RADIATION REPORTS: CONTRADICTORY FINDINGS

The very first report about the Trinity radiation, an eight-page memo five days after the test, came from Manhattan Project radiologist Stafford Leake Warren, written for his immediate boss, the army commander heading the entire Manhattan Project, General Leslie Groves. Warren's memo presented a harrowing account of the radioactivity, but few people ever saw it. Groves stored it in the top secret files of the Manhattan Project, where it remained with other highly classified materials in Groves's possession for years and then in the national archives for decades.[11] Warren himself sought never to have it publicized because his career success rested significantly on his skill at minimizing lawsuits from radiation victims and because, after serious alarm about radiation exposure dangers early in his work with the Manhattan Project, he came to believe in the relative safety of radiation exposures and the ability of professionals to contain any danger.

Warren's 1945 memo provided details about the unanticipated fallout "intensities," tracking on pages of maps where and when radioactive material fell to earth. Warren emphasized that populated areas had not been endangered: "The highest intensities, fortunately, were only found in deserted regions . . . Intensities in the deserted canyon were high enough to cause serious physiological effects."[12] However, the map of radiation at what he labeled the "Hot Canyon" included his handwritten note that "House (with family)" located 0.9 miles beyond the canyon received "an accumulated total dose of 57–60r."[13] Warren observed that radioactive dust "could be measured at low intensities 200 miles north and northeast of the site on the 4th day [after the test]. There is still a tremendous quantity of radioactive dust floating in the air."[14] He also noted, "While no house area investigated received a dangerous amount, i.e., no more than an accumulated two weeks dosage of 60r, the dust outfall from the various portions of the cloud was potentially a very dangerous hazard over a band almost 30 miles wide extending almost 90 miles northeast of the site."[15]

Early predictions about the Trinity radiation were based on Los Alamos physicists' mathematical calculations and on simple prior tests including one conducted with smoke that scientists hoped would provide clues about the trajectories of radioactive debris from the actual test. According to the account of the radiologist in charge of health physics at the test, Louis Hempelmann, Manhattan Project scientists had largely overlooked radiation in their preparations for the Trinity test.[16] Rectifying that earlier oversight during the actual test, radiologists Stafford Warren, Louis Hempelmann, and Hymer Friedell fanned radiological monitors around the site and in towns ninety miles away; in addition, they placed measuring equipment at strategic areas around the site and ordered six dozen dosimeter badges for personnel to wear. The most serious immediate result of the safety measures came minutes after the test, when the monitors at a post near ground zero were ordered to evacuate, but the high radiation readings were later found to have been caused by faulty equipment.[17]

Over the years, the AEC shifted its stance about remaining Trinity radiation dangers. Although in the early 1950s some scientists criticized the AEC's Division of Biology and Medicine for favoring laboratory studies over field studies, it funded the earliest studies of the Trinity radiation.[18] At its twelfth meeting, in October 1948, the AEC's Advisory Committee on Biology and Medicine was concerned about the continuing radiation from the Trinity test. One member had recently visited the site and reported that "the grasses show a fairly high level of activity [radioactivity]." He added that the "droppings of animals also have an activity." The committee considered "the problem presented by the disintegration of the trinitite which covers the area in a layer one-half inch or so in thickness of one-half mile radius" and concluded "that active measures should be taken to remove this as a potential hazard."[19] In the early 1950s, the AEC authorized a crew to dig up the "trinitite" (the fused sand from the heat of the blast) and to rebury it elsewhere.

For the first three decades after the Trinity bomb test, some investigators drew alarming conclusions about the persistence of different types of radiation, while others downplayed any danger. The most numerous and detailed studies came from Los Alamos and from the UCLA Atomic Energy Project, with conclusions varying greatly. Los Alamos experts' findings might have sounded alarms had their contents been made available beyond the lab. Two and a half months after the test, Los Alamos radiologists traced "a swath of fairly high radioactivity on the ground covering an area of about 100 miles long by 30 miles wide."[20] Conversely, in an undated (probably 1947) study, Los Alamos scientist Joseph Hoffman stated that no alpha radiation had been detected from the Trinity test, adding that "while it is possible that a localized 'hot spot' might have occurred, none was found, and the actual contamination indicated a uniform distribution at a concentration which is considered not to constitute a health hazard."[21] There were internal disagreements within Los Alamos about that issue. In 1950 the general manager at Los Alamos told the secretary of the air force that at "ground zero" at Trinity and for six hundred feet there was "enough radioactivity to present a strong presumption of a hazard."[22] One year later, Thomas Shipman, head of the new postwar Health Division at Los Alamos, protested strongly to the director of biology and medicine at the AEC that there was no radiation danger at the Trinity site, arguing that AEC actions in burying the trinitite foolishly alarmed the public and were "inconsistent with previous publicity releases."[23] In June 1952, the director of security and a lawyer for Los Alamos wrote a memo expressing concern about public safety and legal liability if the site were to become a national historic site.[24] A little-circulated 1972 Los Alamos study found contamination of the soil at the Trinity site and claimed that the "increasing migration of Pu [plutonium] into the soils" constituted a major change.[25]

The UCLA Atomic Energy Project conducted half a dozen studies of the Trinity radiation after 1948, and their findings also varied greatly. The UCLA Atomic Energy Project became involved because Joseph Hamilton, one of the country's leading radiation experts, sent a four-page memo in the fall of 1947 urging serious study of the test site's radiation.[26] In response, Stafford Warren and Albert Bellamy, both newly hired faculty at the recently founded medical school at UCLA (with Warren concurrently negotiating with the AEC for what would become the generously funded, secret contract that established the Atomic Energy Project), contacted the AEC, emphasizing the danger at the Trinity site and the dire need for study of its radiological contamination. Bellamy emphasized that the wind and dust made the alpha radiation from plutonium especially dangerous.[27] In response, the AEC authorized what became the first UCLA Atomic Energy Project report (UCLA-32), issued in 1949, coauthored by Kermit Larson.[28] Some of the Atomic Energy Project investigators collected samples and traced the radioactivity caused by the fallout within thirty-two miles of the detonation site; others brought specimens of plants and animals back to the UCLA "hot labs." They found radiation in soil and water and reported that the "windblown dust . . . does have considerable activity associated with it."[29] The tone of the report aimed for scientific objectivity and expertise in its technical explanations of methods, and in its carefully modulated conclusions. The authors took care not to sound alarms: "Due to the low levels of activities found in the animals from the Alamogordo Area and the absence of controlled laboratory data for such activity levels, no conclusions can be reached at this time concerning the presence or absence of biological hazards in the contaminated areas."[30] From the animals studied, they found no "gross evidence" of radiation damage.

The report's conclusions surely heartened the AEC and other officials who wanted the United States to continue its growing reliance on atomic weapons and to increase the testing of those weapons. It also would have supported those who denied dangers from exposure to low levels of radiation. Of the 402 small animals studied, the report noted "strong evidence that radioactivity passed thru the digestive tracts of some of the specimens" but with "little or no evidence for absorption of active materials into the tissues and organs."[31] Plants from two areas (ground zero and the Chupadera Mesa some twenty miles northwest) had the highest readings of radioactivity, but in general the investigators found no significant differences in radioactivity in crop plants from contaminated areas compared to those from uncontaminated areas.

The UCLA group continued over the next decade to study the Trinity radiation, issuing reports that saw only limited circulation. Issued in 1951, UCLA-140 resurveyed three areas studied in 1948. The (unidentified) authors found continuing, even increasing radioactivity, which they minimized. Their language, in places, was convoluted and confusing: "It was demonstrated that vegetation is the most important influence in decreasing the removal of wind-borne material." In other words, vegetation appeared to aid in the persistence of radioactivity. Investigators found that radioactivity from dried plant material in soil samples had increased, but they expressed the findings in technical language: the ratio of soil to plant beta-gamma radioactivity of residual fission products in Area 21. In 1949 the activity of a gram of dried plant material was 3.85 percent of the radioactivity in a gram of soil, but in 1950 it was 5.59 percent.[32]

A 1957 UCLA report gave less reassurance.[33] The investigators, one of whom was Kermit Larson, resurveyed areas studied earlier and corrected earlier findings. They found that the area originally contaminated by fallout from the Trinity detonation "was greater than the 1,100 square miles estimated in the 1948 survey." They also wrote that "the relatively slight change in plutonium level from 1948 to 1956 in these locations indicates that no appreciable decrease in plutonium content of the soils is occurring due to erosional factors."[34] In a 1963 study, Larson and coauthors reported having found significant levels of alpha and gamma radiation in areas of New Mexico distant from the test site some four and five years after the Trinity test.[35] In his one-thousand-page history of post–World War II research into "internal emitters" (internally absorbed or ingested radiation), radiologist James Newell Stannard managed to obtain some of the UCLA studies (although he was annoyed at not finding all that he wanted). He noted that the UCLA scientists found that radioactivity outside the crater increased with distance as much as eighty-five miles, although the maximum radiation was found at about twenty-eight miles from ground zero. This was true of the surface soil and in animals and plants: "Even after many years and scores of rainstorms, activity deposited on the leaves of some plants was almost impossible to remove."[36] The data also suggested some sort of accumulation process because "concentrations in animal tissues that were very low in the first two years began to be detectable in 1949 and 1950. . . . Activity in plants rose similarly, but somewhat earlier."[37]

Thirty years after the Trinity test, federal government agencies still studied its residual radiation. In 1973, for reasons unexplained in the extant document, the EPA's Office of Radiation Programs in Las Vegas measured plutonium in soil samples in some seventy-five areas around the Trinity test site. What they found was negligible.[38] In 1982 the Defense Nuclear Agency issued an equally reassuring study that reported finding no residual radiation in surrounding areas.[39]

In the 1990s, new stakeholders entered into the long-lived contestations about the Trinity radiation, highlighting conflicts between groups credentialed as experts and members of the general public. In the mid-1990s, the Trinity test's effects began to receive sustained attention from New Mexican "downwinders" claiming damages from exposure to the test's radiation. The newly founded Tularosa Basin Downwinders Consortium (TBDC) began to advocate for members' inclusion in the Radiation Exposure Compensation Act (RECA) of 1990. A decade earlier, downwinders in Nevada and Utah sought legal redress for their radiation-induced losses of health and livestock after the many bomb tests at the Nevada Test Site. Congress passed RECA in response to those escalating demands but did not include the citizens affected by the Trinity test. The TBDC activists mount protests at the biannual Trinity test site open house and seek wider newspaper coverage, emphasizing their exposure to bomb fallout for decades.[40] The most recent and most detailed study of the Trinity radiation (cosponsored by the Centers for Disease Control and Prevention), the Los Alamos Historical Document Retrieval and Assessment report (LAHDRA), in 2010 asserted that radiation exposure rates in "public areas" near the Trinity blast "were measured at levels 10,000 times higher than currently allowed."[41] The study concluded, however, that "too much remains undetermined about exposures from the Trinity test to put the event in perspective as a source of public radiation exposure or to defensibly address the extent to which people were harmed."[42] It further noted that there is insufficient evidence to draw conclusions about radiation effects because no one at the time and in subsequent decades paid sufficient attention to the internal doses "received by residents [near the Trinity test] from intakes of airborne radioactivity and contaminated water and food."[43] So, after seventy years, experts continue to disagree about the Trinity radiation.

INTERPRETING THE DISAGREEMENTS ABOUT THE TRINITY RADIATION

The conflicting reports about the radioactivity from the Trinity bomb illustrate several important issues about post–World War II scientific work on nuclear energy. First, the differing reports underscore the novelty and complexity of trying to map and measure different kinds of radiation over decades in a remote high desert environment. In other fields, too, experts faced significant challenges when they moved from laboratories into field research. As historian Linda Nash demonstrates, California health officials in the late 1940s and 1950s faced complex problems when dealing with the toxicity of pesticides that could contaminate an entire landscape: they had to rethink procedures when there were no borders to the spatial aspects of some diseases.[44] So, too, with the study of environmental radiation, research personnel, equipment, methodologies, and terminologies changed over the years.

Radiation-detection instruments contributed to the problematic measurements. The LAHDRA study concluded that the radiation-detection instruments used at Trinity were "crude, ill-suited to field use, and incapable of effectively measuring alpha contamination from about 4.8 kg of unfissioned plutonium" that was dispersed after the Trinity blast.[45] Experts for decades had voiced similar complaints. From the teams trying to measure the radiation from the bombs in Japan, to the teams at the tests in the Pacific, to the studies at the Nevada Test Site, complaints about the equipment fueled attempts in the postwar years to invent more sensitive instruments to better measure exposure to different kinds of external and internal atomic radiation. That was one reason for the UCLA Atomic Energy Project's focus on developing more sensitive measuring equipment for field studies as well as for medical lab work.[46] Unfortunately, only a few of the Trinity radiation studies specify either the instruments or the methodologies used in obtaining the radiation counts. Kermit Larson told an interviewer that poor, insensitive, and inadequate instruments had created problems in measuring radiation in the early tests. This was especially a problem in trying to measure alpha radiation because the field instruments were so inadequate that investigators had to measure on their hands and knees, close to the source. No one, said Larson, wanted to do that.[47]

Contradictions in the Trinity radiation studies also illuminate broader political issues about American officials' pronuclear stances in the immediate years after Hiroshima and Nagasaki. Studies of fallout and residual radiation carried political consequences; civilians and military personnel in the nuclear establishment did not want public alarm and resultant threats to nuclear funding. The stark disagreements among experts about whether the radiation at Trinity posed safety hazards foreshadowed what in the next half century became bitterly divisive conflicts about whether exposure to low-level radiation causes physical harm or whether a threshold exists below which exposure is not harmful. No issue created more dissention among the nuclear energy establishment, worldwide medical personnel, and the public in the ensuing decades, and it continues to be divisive today.[48]

Ambitious individuals in the postwar radiation research world learned that their professional stance toward radiation safety could further or hinder a career. Stafford Warren, for example, did an about-face concerning the Trinity radiation. Completely contradicting his alarmed post-test report to Groves (as discussed earlier), Warren claimed in his 1960s oral history that early Trinity studies found fallout only on the test site. He claimed that not until later "was it realized that fallout from 60 to 100 miles from the test site had been significant."[49] Warren's repudiation of his own 1945 findings reflects his changing political awareness and ambition. During and immediately after the war, he expressed public and private trepidation about the health dangers of exposure to ionizing radiation. Those fears led him to becoming one of the high-ranking proponents of the earliest human radiation experimentation in the hospitals of the Manhattan Engineer District.[50] In 1947 he gave public speeches about the alarming radiation dangers of nuclear war. However, by the late 1950s and 1960s, Warren's public and private pronouncements reflected his recognition of the political benefits of minimizing the dangers from exposure to nuclear radiation. Placating the AEC and its pronuclear testing/pronuclear weapons stances brought continued funding to the UCLA Atomic Energy Project. Warren's colleague, Albert Bellamy, underwent a similar change from intense postwar fear about radiation health dangers to nonchalance a decade later.[51]

As historians have long argued, some of the secrecy surrounding fallout may have also stemmed from officials' fears of lawsuits. The Trinity studies, like much about the nuclear tests at the Nevada Test Site and in the Pacific, were kept undercover so as not to inflame public fears about radioactive fallout and in hopes of minimizing legal problems based on downwinders' health claims.[52] Manhattan Project officials selected the site of the Trinity test because they believed it to be an isolated desert area far from towns and ranches. Even so, they prepared for emergency evacuation of the closest towns but did not even alert residents when the radioactive cloud passed overhead.[53]

The controlled circulation of the Trinity studies provides insights into the evolving culture of Cold War secrecy in the United States. A full-blown system for classifying nuclear material did not emerge instantly, even though the Atomic Energy Act of 1946 declared that everything pertaining to nuclear matters was "born secret." The Trinity radiation studies illustrate the different paths that limiting circulation of nuclear information could take in the early decades of the Cold War. Some of the Trinity radiation studies were stamped "secret" when first issued, and a few were stamped "confidential." Others were treated as "grey literature," that is, reports, memos, and conference proceedings—whose printing came from government funds—that were not subjected to classification procedures but whose information was restricted simply because they were not being made readily available. Sometimes the reports were not even catalogued, or were shelved in obscure places, their circulation deliberately circumscribed.[54] In a 1978 interview, Joseph Ross, who took over the UCLA Atomic Energy Project from Stafford Warren, recalled that Larson and others brought "materials" from the test site and analyzed them for plutonium and uranium by "sophisticated chemical analysis." Ross noted that "a vast amount of that data was probably not published, which was most unfortunate." He did not know where the data was stored. He noted that he tried to get material published before he retired in 1965, but he could not overcome "many of the obstacles." Ross believed that much of the information ended up in AEC Gen-12 reports, and he said that during his ten years as head of the lab he scrupulously had them kept in the library at UCLA. This included "many things that were never published in the open press, although I don't believe that they should be considered classified."[55] None of those reports can be located today in the remote storage areas of the vast UCLA library system.

When Kermit Larson left the UCLA Atomic Energy Project, he took with him more than his samples of "hot soil." He also, in his words, "pirated" copies of reports he had coauthored. When Larson willingly loaned some of those reports to historian Ferenc Szasz decades later, he crossed out and made illegible some contents. If a report was labeled "secret," Larson crossed out that classification delineation at the bottom of every page.[56]

The circulation of reports was also controlled by the often used expedient of printing limited copies and safeguarding distribution lists. An eighty-two-page 1951 report with no authors listed, titled "Biological Field Surveys from Areas Contaminated by Fall-out from First Atomic Bomb Detonation in New Mexico," had a "special distribution" of twenty-three copies, of which #4, #5, and #6 went to "McCormack" and #14 went to "McLean at the Toxicity Lab." McLean's involvement leads to the concluding argument of this chapter: the role of the Trinity radiation studies in America's post–World War II radiological warfare work.[57]

A rarely discussed reason for the secrecy about fallout, residual radiation, and neutron-induced radioactivity came from interest in their potential as weapons in radiological warfare. American officials researching chemical and biological warfare early in World War II became deeply interested in the feasibility of radiation in the form of pellets, sprays, or aerosols rather than as bombs to be used as a weapon of war. In the decade from 1945 to 1955, interest and funding increased dramatically for radiological warfare. It came to be closely linked with chemical and biological warfare.[58] The UCLA-32 Trinity report of 1948 noted that one of the field assignments was to collect "data useful to Military and Defense organizations."[59] The UCLA-140 report in 1951 stated that it was essential to establish a basis for evaluating "chronic and acute radiological hazards to man and other biological systems." Tests of potential chemical and radiological weapons, especially "dusts" and aerosols, were carried out at the Dugway Proving Grounds in Utah in the late 1940s and early 1950s. In addition to the military involvement, radiological warfare research attracted think tank experts such as those at the RAND Corporation, scientists from Oak Ridge (and possibly Los Alamos), and university personnel. Some of the research at the UCLA Atomic Energy Project had dual uses for medicine and civil defense, and ambiguous but possible uses in radiological warfare.[60] U.S. interest in radiological warfare declined in the mid-1950s (in part because attention shifted to the hydrogen bomb), but until then, the potential of radiation as a war weapon held considerable allure and generated much of the secrecy around radioactive fallout.

The involvement of two experts on the Trinity radiation in particular provides a link between those studies and America's postwar radiological warfare research. Joseph Hamilton, as mentioned previously, became interested very early in studying the Trinity radiation. Hamilton was one of the foremost civilian experts in radiological weapons during and after World War II until his untimely death from leukemia at age fifty-seven in 1956. He conducted medical experiments with radiation at University of California–Berkeley as part of the Donner Lab; brought to the Met Lab in Chicago during World War II, he was one of the experts focused on radiological warfare. After the war, he worked with numerous military and civilian groups on radiological warfare, particularly at the Dugway test site. When Hamilton urged Stafford Warren in November 1947 to conduct the first scientific study of Trinity radiation, he wrote that as much as one kilogram of plutonium might have been spread over a one-hundred-square-mile area from the Trinity site. It was exceedingly important, he argued, to understand and to be able to predict the chain of events if large land areas became contaminated with fission products because of accidents "or military action."[61]

Franklin McLean provides another link between the Trinity radiation and radiological warfare. As noted earlier, McLean was on the distribution list on an early Trinity report, listed as affiliated with the "Toxicity Lab." That lab, located at the University of Chicago, became a significant center during World War II for research and development of products for chemical and biological warfare. After the war, with funding from the AEC, the lab worked closely with the Army's Chemical Warfare branch, which now also carried out investigations into radiological warfare.[62] McLean, the first director of the Toxicity Lab, resigned to become a lieutenant colonel in the Medical Corps of the U.S. Army, assigned to the Medical Division of the Chemical Warfare Service. He also served as director of toxicology at Edgewood Arsenal, Maryland, the army's principal site for chemical warfare work.[63] It is noteworthy that in October 1948, when Albert Bellamy wrote to persuade the AEC to study the Trinity radiation in part because the wind and dust made the alpha radiation especially dangerous, he sent a copy to Franklin McLean. Those in the know understood that the Toxicity Lab (also known under the pseudonym "Respiratory Project") focused on toxicity and respiration.[64] Radiological warfare experts sought knowledge about the effectiveness or ineffectiveness of windborne dispersal of fission products, but they did not want to make it public.

CONCLUSION

Like other chapters in this volume, this chapter examines competing claims about a particular toxicity and specific forms of expertise. Here, scientists and medical personnel in differing fields issued contradictory findings about the long-lived environmental radiation from the first atomic bomb test, though these divergent conclusions were seldom debated, either openly or behind closed doors. Experts rarely clashed openly because so much knowledge remained censored. Profoundly important issues about the effects of atomic fallout did not reach a public forum. Gradually, the secrecy surrounding the atomic bomb and its effects expanded in the postwar years to encompass broader categories of knowledge and what we recognize as the culture of Cold War secrecy crystallized. New ways of communicating knowledge in this deepening culture of pervasive secrecy emerged. There began to be approved ways for experts with equivalent levels of security classification to communicate with each other at classified conferences or by circulating classified reports through approved classified channels.[65] But this took time. Very little knowledge about fallout from atomic bomb tests circulated outside of closely guarded networks in the United States until the mid-to-late 1950s, and even then, knowledge about the Trinity radiation remained little known and little publicized.[66]

The different ways that the Trinity radiation reports were handled illustrate this early stage of the culture of Cold War secrecy and classification. By the mid-to-late 1950s, and in the 1960s, systems had been created for transmitting highly classified information, and systems had been put in place for classifying and controlling the spread of certain kinds of information. The silos around nuclear knowledge had become more impenetrable. But as this chapter has demonstrated, many of the Trinity radiation reports do not fall into these categories of Cold War secrecy. Instead, those studies remained little circulated because experts (even those within the same institution) did not yet have ways to talk to each other, nor well-oiled mechanisms for reading each other's reports. They conducted studies and then classified them and sat on them.

The silencing and censorship, however, did not close off debate about official responsibilities for the site or public awareness of residual radiation. One of the ironic consequences of the nuclear secrecy documented in the preceding pages is the growth of a mythology to fill the vacuum left by a lack of specific knowledge. Interviews conducted by anthropologist Jake Kosek reveal a widely held suspicion among New Mexican residents that the scientists who for decades collected data for their reports about the Trinity radiation were also responsible for the hundreds of cattle mutilations in the areas around Los Alamos in the 1970s and 1980s. Kosek found that New Mexicans believed that these scientists used the soft tissue of animals to test radiation levels. New Mexico citizens associated the removal of the cattle's sexual organs, udders, and tongues with the long history of environmental studies of radiation.[67] On the opposite end of the spectrum, lighthearted tourists flock by the thousands to the Trinity site on the two days it is open to the public annually. As with other popular "atomic tourism" sites, part of Trinity's allure is its radiation—historically and in the present. At an open house on May 1, 2016, tourists gleefully measured radiation with handheld dosimeters, while others posed smilingly by the sign warning "Radiation Danger."

ACKNOWLEDGMENTS

I thank the Southwest Research Center at the University of New Mexico Library for permission to quote from the Ferenc Szasz Papers and the archives at UCLA for permission to quote from the School of Medicine Administrative Files of Jeanne Williams (Series 255) and from the Stafford Leake Warren Administrative Files 1925–1968 (Series 300 Papers). Thanks to Special Collections at UCLA for permission to quote from the Stafford Leake Warren Collection #987 and to the Bancroft Library at the University of California–Berkeley for permission to quote from the Records of the President of the University of California (Special Problem Folders, 1899–1954, CU-5 Series 4 Inventory). I am also grateful to Special Collections in the Regenstein Library at the University of Chicago for permission to quote from the Biographical Files and the Office of the President, Hutchins Administration Records.

Special thanks to Martha E. DeMarre, manager at the Nuclear Testing Archive, Las Vegas, Nevada, for prompt help in locating materials online. I had excellent research assistance from graduate students in history at Claremont Graduate University: Michelle Hahn, Clark Noone, and Katrina Denham. Finally, I want to acknowledge what a pleasure it has been to work with Claremont consortium colleagues Vivien Hamilton and Brinda Sarathy on this project, from organizing the initial workshop until now.

NOTES

1. Else, *The Day after Trinity*; Szasz, *The Day the Sun Rose Twice*; and Szasz Papers, box 9, folder 11. (Szasz Papers are listed in the bibliography under Archival Sources, Southwest Research Center.)

2. Rhodes, *The Making of the Atomic Bomb*, 675–77.

3. Brodie, "Radiation Secrecy and Censorship after Hiroshima and Nagasaki."

4. Hacker, *Elements of Controversy*; Fradkin, *Fallout*.

5. The UCLA Atomic Energy Project remains little known even today because officials and project personnel so successfully kept its activities unpublicized and little scrutinized for decades. It brought the campus at least 1.5 million dollars annually (and even more in 1950s dollars) for over three decades for research into the impacts of radiation on humans, plants, and animals. I have pieced together information about the UCLA Atomic Energy Project from the Warren Administrative Files and Warren's Papers in Special Collections as well as the Jeanne Williams files. There is also material in box 22, CU-5 Series 4, Inventory of Records of the President of the University of California: Special Problem Folders, 1899–1954, at the Bancroft Library. The only other historical account of the project appears in Lenoir and Hays, "The Manhattan Project for Biomedicine," 39–42.

6. This group included Joseph Hamilton, MD, working at the UC Berkeley Crocker Lab, Stafford Warren (MD/radiologist), and MDs at Los Alamos such as Louis Hempelmann and Thomas Shipman. Westwick, "Abraded from Several Corners," 131–61, is especially helpful regarding the lack of clear genealogical lines in biophysics, molecular biology, molecular genetics, microbiology, nuclear medicine, and radiobiology after World War II.

7. Typewritten, nine-page alphabetical list of UCLA Atomic Energy Project personnel ca. 1951–1952. The actual typescript, undated, is in a folder dated 1951–1952, box 31, Stafford Leake Warren Administrative Papers (listed in the bibliography under Archival Sources, Center for Research Libraries). Research priorities shifted quickly. By 1956 two divisions not even listed in the earliest 1950s budget now received the largest percentages of the total: radioecology (17 percent) and biophysics (15 percent).

8. Larson coauthored reports after nuclear tests in Nevada: #182 after the Jangle test; #243 (1953), #247, and #438 (1960). For Larson reports, see Szasz Papers, box 7; see also J. Newell Stannard 1979 interview with Larson (on Opennet).

9. Larson, critique of chapters 4 and 5, 1982. Szasz Papers, box 7, folder 33.

10. Larson, "Continental Close-In Fallout," 19–25.

11. Stafford Warren to Major General Groves, "Report on Test II at Trinity, 16 July 1945." Microfilm at Center for Research Libraries. The document is confusing because the subject line makes it appear to have been written July 16, one day after the Trinity test of July 15. But in the body of the report Warren comments about measuring radioactive dust four days after the test.

12. Warren, "Report on Test II," 2.

13. Warren, "Report on Test II." Five (unnumbered) pages of maps depicted the trajectory of the radioactive cloud, the wind patterns, and the velocities. The final map, hand-drawn, depicts the location of the "hot canyon" (twenty miles northeast of ground zero). Warren wrote that "By 0800 hours the monitors reported an area of high intensity in a canyon 20 miles northeast of zero. . . . At no house in this whole north and northeast area between 20 miles and 40 miles from zero was a dangerous intensity found."

14. Warren, "Report on Test II," 2.

15. Warren, "Report on Test II," 3.

16. In later interviews about the Trinity test, nuclear scientists Magee and Hirshfelder stated that almost no one at the time gave serious thought about the radiation or the possibility of fallout. See Szasz's handwritten notes from interview with Magee, April 15, 1982, and April 30, 1982, box 7, folder 29, Szasz Papers. There is also information in Hempelmann Deposition, 1979, Opennet.

17. *Los Alamos Historical Document Retrieval and Assessment Report*, chapter 10, 30–33.

18. Lauren Donaldson, in a letter to Stafford L. Warren, August 1951, complained about the AEC's dislike of field studies. CIC 0050232. This document used to exist on the website of what was formerly called the Coordination and Information Center in Las Vegas (now the Nuclear Testing Archive). However, it is no longer available online and is not listed as a reference in this chapter's bibliography. Please contact Janet Brodie, author of this chapter, for a hard copy of the document.

19. AEC Advisory Committee for Biology and Medicine, Minutes October 8–9, 1948. Twelfth meeting, held at Hanford. Box S09F01B194; DOE 082294-B, folder 8 of 11. ACHRE Collection.

20. Interoffice memo from Weisskipf, Hoffman, Aebersold, and Hempelmann to Kistiakowsky, September 5, 1945. Szasz Papers, box 8, folder 48.

21. Los Alamos Report #626 n.d. [1947]. Joseph G. Hoffman, "Nuclear Explosion 16 July, 1945." In Szasz Papers, box 8, folder 39. Only some pages are available—not the whole report.

22. Boyer, General Manager [of Los Alamos], to Finlatter, Secretary of the Air Force. December 1, 1950. Szasz Papers, box 8, folder 48.

23. Shipman to Shields Warren, July 9, 1951; https://www.osti.gov/opennet/index.jsp?: NV0070246. Additionally, when Ernest J. Sternglass, professor of radiation physics at the University of Pittsburgh, charged in the early 1970s that infant mortality rates rose because of nuclear tests, Shipman and others at Los Alamos "stonewalled" giving him information about the Trinity fallout. See Hacker, *Elements of Controversy*, 252. It is also noteworthy that Shipman's Health Division at Los Alamos shared responsibility with the military for monitoring health safety in the early years at the Nevada Test Site. Hacker found evidence that the military authorities wanted more stringent radiological monitoring than the Los Alamos scientists, who valued scientific data over personnel safety (20).

24. James L. McCraw, Director of Security, and Chester Brinck, Counsel, Memo (June 25, 1952). Szasz Papers, box 8, folder 48.

25. Tho E. Hakonson and LaMar J. Johnson (Los Alamos), *Radioecology: Distribution of Environmental Plutonium in Trinity Site . . . after 27 Years*. Szasz Papers, box 8, folder 48.

26. Joseph Hamilton to Stafford Leake Warren, November 20, 1947, four-page memo.

CIC 95082.042. This document used to exist on the website of what was formerly called the Coordination and Information Center in Las Vegas (now the Nuclear Testing Archive). However, it is no longer available online and is not listed as a reference in this chapter's bibliography. Please contact Janet Brodie, author of this chapter, for a hard copy of the document. I have a CIC-reproduced copy and will forward it to anyone interested.

27. Albert Bellamy to James Jensen, Medicine and Biology, AEC, October 6, 1948. Szasz Papers, box 8, folder 48.

28. UCLA-32 (1949). *The 1948 Radiological and Biological Survey of Areas in New Mexico Affected by the First Atomic Bomb Detonation.* Szasz Papers, box 8, folder 40.

29. UCLA-32.

30. UCLA-32, possibly page 36; the page numbers are obliterated.

31. UCLA-32.

32. UCLA-140 (1951). Classified "confidential." Szasz Papers, box 14, folder 5.

33. UCLA-406 (1957). "Confidential." Thirteen copies. Szasz Papers, box 7, folder 34. Many pages are missing, although it is unclear why. Perhaps Szasz only photocopied parts of the report. It is listed as having twenty-eight pages in the bibliography of Olafson and Larson, "Plutonium Biology," in Schultz and Klement, *Radioecology.*

34. UCLA #406 (1957). "Confidential." Thirteen copies. Szasz papers, box 7, folder 34, 17.

35. Schultz and Klement, *Radioecology*, 45–49.

36. Stannard, *Radioactivity and Health*, 930.

37. Stannard, *Radioactivity and Health*, 930.

38. Measuring nanocuries per square meter (a measurement rarely used today; the nanocurie has been replaced by the becquerel), the highest record was at Trinity ground zero (1,100 nanocuries, which roughly converts to 0.0092 R); they measured 86 nanocuries (0.00072 R) at the Monte Prieto Ranch (never before identified in any Trinity-related literature) and several readings from 48 to 64 nanocuries around the small community of Bingham, twenty-three miles south/southwest of ground zero. U.S. Environmental Protection Agency, Office of Radiation Programs, Las Vegas Facility. Trinity Site Soil Sampling Locations and Results. Map figure 2. [1973]. Opennet. NV0051304.

39. Defense Nuclear Agency, Executive Agency for DOD Report Number DNA 6028F. Final Report. December 15, 1982.

40. Tularosa Basin Downwinders Consortium, "Unknowing/Unwilling and Uncompensated," http://www.trinitydownwinders.com (accessed June 2017).

41. *Los Alamos Historical Document Retrieval and Assessment Report*, chapter 10, 31.

42. *Los Alamos Historical Document Retrieval and Assessment Report*, chapter 10, 50.

43. *Los Alamos Historical Document Retrieval and Assessment Report*, chapter 10, 50.

44. Nash, *Inescapable Ecologies*, chapter 5.

45. *Los Alamos Historical Document Retrieval and Assessment Report*, executive summary, 31.

46. Blahd, "Benedict Cassen." Potchen, "Reflections on the Early Years of Nuclear Medicine," 628, credits "the UCLA group"—George Taplin, Ben Cassen, and Bill Blahd—with pioneering early developments in thyroid scanning, liver and kidney scanning, and brain scanning. The UCLA Atomic Energy Project's William Tappan (who also participated in investigations of radiation at the Nevada Test Site) invented a device for disseminating

aerosols. It aided patients with lung problems in inhaling medicines but also stimulated interest among radiological warfare experts who, in the early 1950s, saw potential in weaponized radioactive "dusts" and wanted ways to measure them.

47. J. Newell Stannard 1979 interview with Larson (on Opennet). The problems of measuring plutonium created particular demand for reliable alpha particle detectors, although a commentator in 1949 complained that "even today the ideal alpha survey instrument is far from a reality." Lapp, "Survey of Nucleonics Instrumentation Industry," 101.

48. Among the most controversial was British physician Alice Stewart, who argued in the 1970s that exposure to low levels of radiation was far more dangerous than officials believed. Stewart challenged the reliance on the linear dose-effect model: the assumption of a proportional relationship between dose and effect. See also the controversies around Thomas Mancuso, a professor of occupational health at the University of Pittsburgh, and John Gofman, who headed a division at the Livermore Nuclear Lab. Both became pariahs for arguing that no level of exposure was safe. See Greene, *The Woman Who Knew Too Much.* More sympathetic to the AEC, Hacker, in *Elements of Controversy*, neglects the AEC-sponsored biomedical research in the 1950s and dates the fears about low-level radiation and I-131 to the early 1960s, overlooking concerns in the 1950s.

49. "The skip of 20 to 30 miles that occurred in the fallout pattern, about 20 to 30 miles from point zero, has been found, in subsequent U.S. tests, to be rather common. . . . The failure to detect this skip in the first test gave the false impression [in early Trinity studies] that the fallout had been restricted to the test area and also produced the impression that it was less dangerous than subsequent investigations proved it to be" (Warren, "The Role of Radiology in Development of the Atomic Bomb," 885).

50. Welsome, *The Plutonium Files.*

51. Warren's fears about radiation can be seen in his Japan diaries: box 298, folder "Diary 1945." Bellamy, a zoologist with expertise in genetics, chaired the biophysics department at UCLA and later became associate dean of the medical school. He radically changed from being alarmed about human radiation exposure to giving soothing public pronouncements about radiation safety.

52. Ball, *Justice Downwind.*

53. Maps of the Trinity area were poor in 1945, and officials missed scores of farms and ranches in a thirty-mile radius of ground zero. Los Alamos health physics personnel made several unofficial visits over the years to check on the health of one family in the "hot canyon," whose ranch received considerable fallout. They did not disclose the reasons for the visits. See *Los Alamos Historical Document Retrieval and Assessment Report*, chapter 10, 28.

54. Eden, in *Whole World on Fire*, states that she could not have completed her study without the invaluable grey literature informants told her about. Stannard, too, discusses the value of the grey literature in radiation research, including materials stored in the "gold mine" basement of the first AEC building in Germantown, PA. J. Newell Stannard 1979 interview with Larson (on Opennet).

55. J. Newell Stannard interview with Joseph Ross, UCLA 1978. https://www.osti.gov/opennet/index.jsp?: NV0702790.

56. Many Larson reports can be found in Szasz Papers, box 7, folder 34.

57. For the report, see Szasz Papers 552, box 14, folder 4. James McCormack was the first

director of the AEC's Division of Military Application (DMA), created to be a link between the civilian AEC and the military in matters pertaining to nuclear energy. In late 1940s, and possibly later, it had a radiological branch that worked extensively in the field of radiological warfare. In 1949–1950 the DMA and the AEC's Division of Biology and Medicine jointly funded biomedical research at the University of Chicago's Toxicity Laboratory. In the 1950s, that lab helped fund biomedical research on fallout and possibly other biomedical topics. Advisory Committee on Human Radiation Experimentation, *Final Report*, Supp. 2:8.

58. Chemical and Biological Warfare Collection, National Security Archive, box 2, folder: Radiological Warfare Documents, 1940s.

59. UCLA-32.

60. Karl Z. Morgan believed that military concern with the feasibility of using radioisotopes "as an adjunct to chemical warfare" and the desire to "use fission products concurrently with chemicals" lay behind experiments conducted at Dugway. He also believed that the intentional release of radioactive materials into the air and water in the 1950s through the 1980s was related to chemical and radiological warfare experimentation. See his 1995 oral history on DOE website.

61. See endnote 26. Another radiological warfare memo is Hamilton to Stone, Opennet. Other sources are Hamilton Papers, carton 8, folder 25, and many citations in the Advisory Committee on Human Radiation Experimentation, *Final Report*.

62. Geiling, "History of the University of Chicago Toxicity Lab (Respiratory Project)," Office of the President, Hutchins Administration Records, folder 5, box 215.

63. McLean served on the DOD's Research and Development Board; he served as deputy chair of the Joint Panel on Medical Aspects of Atomic Warfare from 1949 to 1953 and on the Secretary of Defense's Technical Advisory Panel on Biological and Chemical Warfare from 1955 to 1959. See Archival Biographical File, University of Chicago Series VI M-O, Franklin McLean, folder 1.

64. Szasz Papers, box 8, folder 48.

65. Gusterson, "Secrecy, Authorship and Nuclear Weapons Scientists."

66. Not until the accidental exposure of Japanese fisherman to fallout from the testing at Bikini Atoll in 1954 did widespread public knowledge emerge. Congressional hearings regarding fallout in 1957 also provided fuller public understanding. Yet, even as knowledge about fallout from the Nevada and Pacific bomb tests reached the public, knowledge about the Trinity radiation remained little known and little publicized.

67. Kosek, *Understories*, 256.

BIBLIOGRAPHY

Archival Sources

Center for Research Libraries

Warren, Stafford Leake. Top Secret Correspondence of the MED, Subseries I, File 5, Subfile G, "Radiological Effects of the Special Bomb." Stafford Warren to Major General Groves, Report on Test II at Trinity, July 16, 1945. Microfilm.

Opennet Sources: https://www.osti.gov/opennet/index.jsp?
Hamilton to Stone, "A Brief Review of the Possible Applications of Fission Products in Offensive Warfare," May 26, 1943: NV0724782.
Louis Hempelmann Deposition, 1979: NV0708202.
J. Newell Stannard 1979 interview with Larson: NV0702908.
Shipman to Shields Warren, July 9, 1951: NV0070246.
U.S. Environmental Protection Agency, Office of Radiation Programs, Las Vegas Facility. Trinity Site Soil Sampling Locations and Results. Map figure 2. [1973]: NV0051304.
J. Newell Stannard interview with Joseph Ross, UCLA 1978: NV0702790.
ACHRE interview with Karl Z. Morgan, 1995: NV0751119.
National Archives and Record Administration, College Park, Maryland
Advisory Committee on Human Radiation Experimentation Collection
National Security Archive, Georgetown University, Washington, DC
Chemical and Biological Warfare Collection, box 2, folder: Radiological Warfare Documents, 1940s
Southwest Research Center, University of New Mexico Library, Albuquerque
Szasz, Ferenc Papers, Collection 552
University of California–Berkeley, the Bancroft Library
Records of the President of the University of California. Special Problem Folders, 1899–1954. CU-5 Series 4 Inventory
University of California–Los Angeles Archives
School of Medicine Administrative Files of Jeanne Williams, Series 255.
Warren, Stafford Leake Administrative Files 1925–1968, Series 300 Papers.
University of California–Los Angeles Special Collections
Warren, Stafford Leake Collection 987
University of California–San Francisco Medical Archives
Hamilton, Joseph Gilbert Papers, carton 8, folder 25
University of Chicago Special Collections, Regenstein Library
Archival Biographical File, University of Chicago Series VI M-O, Franklin McLean, folder 1.
Office of the President. Hutchins Administration Records, folder 5, box 215. E. M. K. Geiling, "History of the University of Chicago Toxicity Lab (Respiratory Project)," November 17, 1945.

Other Websites

https://ehss.energy.gov/ohre/roadmap/histories/0475/0475a.html
"Human Radiation Studies: Remembering the Early Years." Oral History of Karl Z. Morgan. Accessed April 2018.
http://www.trinitydownwinders.com
Tularosa Basin Downwinders Consortium. "Unknowing/Unwilling and Uncompensated: Effects of the Trinity Test on New Mexicans." Revised February 2017; accessed June 2017.

Secondary Sources

Advisory Committee on Human Radiation Experimentation. *Final Report.* Washington, DC: U.S. Government Printing Office, 1994.

Ball, Howard. *Justice Downwind: America's Atomic Testing Program in the 1950s.* Oxford: Oxford University Press, 1986.

Blahd, William H. "Benedict Cassen: The Father of Body Organ Imaging." *Cancer Biotherapy and Radiopharmaceuticals* 15, no. 5 (2000): 423–29.

Brodie, Janet Farrell. "Radiation Secrecy and Censorship after Hiroshima and Nagasaki." *Journal of Social History* 48, no. 4 (June 2015): 842–64.

Eden, Lynn. *Whole World on Fire: Organizations, Knowledge, and Nuclear Weapons' Devastation.* Ithaca, NY: Cornell University Press, 2006.

Else, Jon, dir. *The Day after Trinity.* Documentary film. 1981. San Jose, CA: KTEH.

Fradkin, Philip L. *Fallout: An American Nuclear Tragedy.* Tucson: University of Arizona Press, 1989.

Greene, Gayle. *The Woman Who Knew Too Much: Alice Stewart and the Secrets of Radiation.* Ann Arbor: University of Michigan Press, 1999.

Gusterson, Hugh. "Secrecy, Authorship and Nuclear Weapons Scientists." In *Secrecy and Knowledge Production*, edited by Judith Reppy, 57–75. Ithaca, NY: Cornell University Press, 1999.

Hacker, Barton C. *Elements of Controversy: The Atomic Energy Commission and Radiation Safety in Nuclear Weapons Testing, 1947–1974.* Berkeley: University of California Press, 1994.

Kosek, Jake. *Understories: The Political Life of Forests in Northern New Mexico.* Durham, NC: Duke University Press, 2006.

Lapp, Ralph E. "Survey of Nucleonics Instrumentation Industry." *Nucleonics* 4, no. 5 (May 1949): 100–104.

Larson, Kermit H. "Continental Close-In Fallout: Its History, Measurement and Characteristics." In *Radioecology: Proceedings of the First National Symposium on Radioecology Held at Colorado State University, Fort Collins, September 10–15, 1961*, edited by Vincent Schultz and Alfred W. Klement Jr. New York: Reinhold Publishing Corporation, 1963.

Lenoir, Timothy, and Marguerite Hays. "The Manhattan Project for Biomedicine." In *Controlling Our Destinies: Historical, Philosophical, Ethical, and Theological Perspectives on the Human Genome Project*, edited Phillip R. Sloan, 39–42. Notre Dame, IN: University of Notre Dame Press, 2000.

Los Alamos Historical Document Retrieval and Assessment Report. 2010. Accessed June 2017, http://www.lahdra.org/Nov2010.html.

Nash, Linda. *Inescapable Ecologies: A History of Environment, Disease, and Knowledge.* Berkeley: University of California Press, 2006.

Potchen, E. J. "Reflections on the Early Years of Nuclear Medicine." *Radiology* 214, no. 3 (March 2000): 623–29.

Rhodes, Richard. *The Making of the Atomic Bomb.* New York: Simon and Schuster, 1986.

Schultz, Vincent, and Alfred W. Klement. Radioecology: Proceedings of the First National Symposium on Radioecology Held at Colorado State University, Fort Collins, Colorado, September 10-15, 1961. New York: Reinhold, 1963.

Stannard, James Newell. *Radioactivity and Health: A History.* Edited by Raymond W. Baalman Jr. Washington, DC: U.S. Department of Energy, 1988.

Szasz, Ferenc Morton. *The Day the Sun Rose Twice: The Story of the Trinity Site Nuclear Explosion.* Albuquerque: University of New Mexico Press, 1984.

Warren, Stafford L. "The Role of Radiology in Development of the Atomic Bomb." In *Radiology in World War II,* edited by U.S. Army Medical Department, 831–921. Washington, DC: Department of the Army, Office of the Surgeon General, 1966.

Welsome, Eileen. *The Plutonium Files: America's Secret Medical Experiments in the Cold War.* New York: Dial Press, Random House, 1999.

Westwick, Peter J. "Abraded from Several Corners: Medical Physics and Biophysics at Berkeley." *Historical Studies in the Physical Sciences* 27, no. 1 (1996): 131–61.

3

Crossroads in San Francisco

THE NAVAL RADIOLOGICAL DEFENSE LABORATORY AND ITS AFTERLIVES

Lindsey Dillon

In April 2015, a marine expedition conducted by the National Oceanic and Atmospheric Administration (NOAA) found the USS *Independence,* a radioactive World War II–era light aircraft carrier, roughly forty miles off the coast of San Francisco. The discovery of the *Independence* was not a surprise. Researchers have long known that in 1951, the U.S. military sank the warship in an area that is today known as the Monterey Bay National Marine Sanctuary, near California's Farallon Islands.[1] The *Independence* had been contaminated with radiation in 1946 as part of Operation Crossroads, the United States' first postwar nuclear weapons test in the Bikini Atoll in the Marshall Islands. Brought back to the Hunters Point Naval Shipyard in San Francisco in 1947, the *Independence* became a floating lab for the Naval Radiological Defense Laboratory (NRDL), which was established in 1947, and operated on the San Francisco naval base for nearly twenty years. The *Independence* also became a storage facility for the radioactive waste produced by radiation laboratories in the San Francisco Bay Area. In late January 1951, the navy towed the *Independence* out to sea and sank it by underwater explosion.

The images of the nuclear-bombed *Independence* taken by NOAA haunt the imagination. They raise the specter of ongoing contamination—of the toxic afterlives of the United States' decades-long program of nuclear weapons development. The re-surfacing of the *Independence* into the public spotlight in 2017 also points to the history of militarization and nuclear weapons development in the San Francisco Bay Area, a history that seems incongruous with the region's environmentally progressive and "green" reputation. Why did the military bring radioactive ships back to San Francisco? Why did it establish a radiation lab in the city, and what other experiments took place there? And what are the social and ecological legacies of the NRDL today?

This chapter follows the journey of the *Independence* to the bottom of the ocean as a way of examining a particular moment in the history of nuclear weapons and the landscape of environmental toxicity in the San Francisco Bay Area. I focus on the early work of the NRDL, which operated at the Hunters Point shipyard in southeast San Francisco between 1947 and 1969. From its initial, ad hoc experiments on ships used in Operation Crossroads, the NRDL grew into the U.S. Navy's largest applied radiation lab.[2] The lab developed expertise in radiological safety and defense, including more precise instrumentation and techniques for radiation detection, protective gear for soldiers in the field, and military manuals for decontaminating ships and for recovering from a nuclear attack. As part of producing radiological safety, the NRDL also exposed shipyard workers to radiation and released radioactive waste into the Pacific Ocean and the San Francisco Bay. When the lab closed in 1969, it left radioactive waste in Hunters Point.[3] This history of the NRDL reveals how the project of radiological safety had its own toxic consequences and reminds us that the fallout of the bomb remains an ongoing socioecological condition.[4]

AN ATOMIC CROSSROADS

The NRDL's primary objective, as explained to the Atomic Energy Commission in 1947, was to "study the consequences and defensive aspects of atomic warfare."[5] Although the NRDL ostensibly defended the U.S. public from nuclear war, I argue that the lab is better understood as having a more specific function: that of defending the U.S. military's nuclear weapons program through the production of knowledge and technologies supporting the idea that nuclear weapons could be safely tested. In the summer of 1946, the United States exploded two atomic bombs in the Bikini Atoll of the Marshall Islands—a small ring of islands 2,500 miles southwest of Hawai'i—as part of a military project called Operation Crossroads. They were the fourth and fifth such bombs in history, preceded only by Trinity (the bomb dropped in New Mexico in 1945 as part of the wartime Manhattan Project, and detailed in chapter 2) and the two bombs dropped on Hiroshima and Nagasaki a month later (see chapter 9).[6] The architects of Operation Crossroads intended to test three atomic bombs (named Shot Able, Shot Baker, and Shot Charlie), but the experiment was cut short by the disastrous outcome of Shot Baker, which I detail later in this chapter. The NRDL emerged from the fallout of Shot Baker. According to James Newell Stannard, author of *Radioactivity and Health: A History,* "the raison d'etre for this laboratory [the NRDL] lay in the tons of highly contaminated sea water that showered down upon the assembled flotilla of warships in Test Baker of Operation Crossroads."[7]

Operation Crossroads was both a massive experiment and a display of postwar U.S. power.[8] The event was conducted by an army-navy task force, which in May 1946 assembled 242 ships, 156 airplanes, and 45,000 people in Bikini to test the material, biomedical, and ecological effects of the bomb and its radioactive aftermath. The army also brought trucks, radar equipment, seeds, bacteria samples, and live animals to Bikini, which it subjected to the two bombs. During the explosions, the animals were placed on twenty-two different ships, simulating a human crew during battle.[9]

One of the biggest research questions motivating Crossroads was the effect of the bomb on naval ships, and specifically the dilemma—created by the U.S. military itself, through the Manhattan Project—of whether the navy could survive a nuclear attack. Some headlines from *The Masthead,* the official newspaper for Naval Station Treasure Island, located in the middle of the San Francisco Bay, include: "All Because of Atom Bomb: Congressional Committee Cuts Navy Budget by One Billion" and "Atom Bomb Blast May Determine Future Naval Plans and Tactics."[10] Could a warship survive a nuclear attack? To address the viability of the navy in the atomic age, the Crossroads task force placed nearly one hundred ships in the middle of the Bikini Atoll (including the *Independence*) as the primary target of its atomic weapons test.[11]

While the United States moved its research project to Bikini, it simultaneously removed 170 Bikinians to the island of Rongerik, 130 miles away.[12] Through this relocation of people, the military produced a seemingly blank space for its nuclear weapons testing program.[13] A year after Operation Crossroads, in 1947, the United States annexed—with United Nations approval—the Marshall Islands as a trust territory and in 1948 designated Enewetak Atoll, east of Bikini, as the Pacific Proving Ground, its outdoor laboratory where military scientists could "prove" theoretical advances in nuclear weapons technology. The Atomic Energy Commission also provided funds for basic scientific research in the Marshall Islands. The well-known ecologists Eugene and Howard Odum conducted research in the Marshall Islands that relied on radiotracers, which had been introduced into the environment through weapons testing, to develop the field of ecology.[14] While ecological thought would influence the U.S. environmental movement, as well as opposition to nuclear weapons, the conditions of possibility for this body of knowledge included the displacement of Marshall Islanders, their exposure to the bomb's fallout, and the ecological degradation of the region. The Bikini Atoll remains uninhabited.

The first bomb of Operation Crossroads, Shot Able, was dropped on July 1, 1946. It missed its target by half a mile, sinking only a few ships. Still, Shot Able's fallout, drawn into the stratosphere, marked the beginning of "the subtle but total transformation" of the planet.[15] The United States would test over two hundred atmospheric (or aboveground) nuclear weapons until the Partial Test Ban Treaty in 1963. The test ban pushed nuclear weapons testing underground, where the United States continued its weapons program until 1992, ultimately testing and releasing the toxic by-products of over one thousand nuclear weapons.[16]

Three weeks after Shot Able, the Crossroads task force lowered a second bomb, Shot Baker, to ninety feet below the lagoon's surface. The bomb erupted from underneath the target ships as a tall, wide column of water followed by a circular wall of ocean mist that pushed outward, blanketing the lagoon with radiation and compelling the military to terminate Operation Crossroads.[17] According to Neal Hines, a researcher from the University of Washington who studied the effects of atomic bombs in the Marshall Islands, "the target area became a maelstrom of radioactive debris, and at the bottom of the lagoon was a shallow basin half a mile wide from which the force of the explosion had scooped hundreds of thousands of tons of sludge and coral algal sediment. . . . Within the waters, and particularly in the tons of sludge again settling to the lagoon floor, were radioactive contaminants whose disposition would present problems of greater complexity than anyone at that point in time might have guessed."[18]

At first, navy crewmembers were ordered to decontaminate the target ships at Bikini. In a written testimony to the U.S. Congress, Jonathan Weisgall, a researcher and author of *Operation Crossroads: The Atomic Tests at Bikini Atoll*, notes that the first patrol boats entered the Bikini lagoon forty-one minutes after Shot Baker, followed by other boats filled with people who boarded the radioactive ships (a salvage group, the radiological safety monitors, and technicians), all despite drone readings that showed radiation levels more than twice the legal dose.[19] Radioactive fallout in the Bikini lagoon was concentrated within the marine life that stuck to the ship's hulls, to the wood of their decks and the paint on their sides. Concentrations of fallout also varied widely across a ship—decks might be considered "safe" on one end and contaminated on the other.[20] According to Weisgall, "for weeks men routinely boarded these ships, ate on these ships, slept on these ships and did what they could to get rid of this radioactivity, which of course was something brand new."[21]

Radiation had also begun to threaten the tens of thousands of people on the operation's support fleet, outside the lagoon. As David Bradley, a doctor with the Radiological Safety Section, who wrote a best-selling memoir of Operation Crossroads, *No Place to Hide,* recalls, "We had found by repeated measurements that the water was steadily increasing in radioactivity, owing presumably to an up-welling of material from the bottom [of the lagoon]. By noon [on July 29, 1946,] the intensity was such to endanger our water intakes and evaporators, and so at the request of the Radiological Safety Section the entire live fleet up-anchored and sailed to a point nearer Enyu channel."[22]

By mid-August 1946, Operation Crossroads was relocated from Bikini to the nearby atoll of Kwajalein. By the end of the month, the navy sank the most radioactive ships in Kwajalein's waters and began to tow the rest back to West Coast naval bases for further decontamination and study. Some of the ships were sent to Hawai'i's Pearl Harbor and to Puget Sound in Washington. San Francisco received the majority of Crossroads ships, including eighteen target ships (those that had been in the center of Bikini's lagoon) and fifty ships from the support fleet. In the ensuing months, Hunters Point emerged as "the center of research and expertise on the problem of decontamination" and set standards for these procedures at other naval bases.[23] According to an internal navy memo from November 18, 1946: "With the dissolution of Joint Task Force One, a modified organization for continuing the monitoring and clearance [of Crossroads ships] has been established . . . at Naval Shipyard San Francisco."[24] The NRDL was thus formed to continue the work of Operation Crossroads in San Francisco.

Before detailing the early years of NRDL in San Francisco, it is important to address a question of geography and location. Why send most of the ships from the Marshall Islands back to San Francisco? Why did it make sense to establish a scientific laboratory at the Hunters Point shipyard—an active, industrial shipyard? In the following section, I explore the long-standing relationship between the U.S. military and the San Francisco Bay Area as a way of explaining how the city became entangled in the geopolitics of nuclear weapons development.

"A NAVAL LAKE"

Although today the San Francisco Bay Area is synonymous with tech capital, green urbanism, and progressive politics, urban development in the region has been deeply entwined with U.S. military power. Historian Roger Lotchin calls this relationship a "metropolitan-military complex," to describe the impact of defense spending and related industries on the regional economy.[25] Waves of militarization have etched themselves into the Bay Area's social and physical landscape, especially after the Spanish-American War in 1898 and the twentieth century's two World Wars.

The San Francisco Bay was a strategic site for the navy by the late nineteenth century. The United States acquired a number of Pacific islands in the second half of the nineteenth century, including Howland, Jarvis, Midway, and Wake Islands; Samoa; Guam; and the Philippines. These island territories formed a network of coaling stations for U.S. military and commercial ships, and a basis for the projection of U.S. imperial power. Social scientist Ruth Oldenziel refers to U.S. island territories as a network of "naval nodes for control of ocean space," writing that "expansion through control of the ocean—politicizing and militarizing ocean space—thus created a global system of international relations in which islands, peninsulas, and littoral spaces played a key geopolitical role."[26] The connections between San Francisco and U.S. territories in the Pacific Ocean was solidified by the end of the nineteenth century, as the navy installed a large number of naval bases throughout the San Francisco Bay Area. A large, private shipbuilding industry also developed in southeast San Francisco, near the Hunters Point shipyard.[27] The density of naval operations in the Bay Area led to several of its nicknames at the time, including "the Navy's lake" and "American Singapore."[28] Later, through its use as a weapons test site, the Marshall Islands would join the United States' Pacific island geography.

By 1940 the San Francisco Bay Area was a key site of national defense activity as well as the most productive shipbuilding region in the country. Six months before the United States entered World War II, the navy had invested $1 billion in the San Francisco Bay Area, including $650 million in shipbuilding industries.[29] San Francisco Bay Area naval bases included the Benicia Arsenal and Mare Island naval station, in Contra Costa County; Treasure Island naval station, in the middle of the San Francisco Bay; the naval air station in Alameda; Moffitt Field, in today's Silicon Valley; and the Hunters Point naval shipyard. The army also maintained bases at Fort Baker in the Bay Area's Marin County and at Fort Mason, Fort Funston, Fort Point, and the Presidio in San Francisco—forming, along with Bay Area naval bases, what Richard Walker calls a "military archipelago."[30] Large, private shipbuilders during World War II included Moore Dry Dock in Oakland, Kaiser Shipyards in Richmond, and Marinship, in Sausalito. During World War II and for several decades afterward, the Hunters Point shipyard was the largest naval ship repair yard in the country.[31]

Several of the key support ships involved in Operation Crossroads embarked for their fateful journey from the Bay Area. "For us Operation Crossroads began officially at San Francisco on a glorious morning," writes David Bradley, from the deck of the USS *Haven*, with other members of the radiological safety group. En route to the Marshall Islands, Bradley watched as "the big bridge [and] Legion of Honor, finally slipped by. The coast line to the south, that marvelous stretch of beach, was lost in the white turmoil of the breakers, and the white city of San Francisco, sprawling over the sand to the foot of the hills, was obscured by the spume."[32] One of the ships that departed from Hunters Point was the USS *Burleson*, which carried the animals used as test subjects at Bikini. In an interview with the *San Francisco Chronicle* on June 2, 1946, an army general referred to the *Burleson* as a "great, dirtless farm, a palatial hotel for animals." The general assured the *Chronicle*'s journalist that the goats, pigs, mice, and rats would be "pampered and cared for until A-day," perhaps seeking to dispel concerns by San Franciscan residents about the sacrifice of animals as nuclear tests subjects.[33]

For these reasons, it would have made sense, from the military's perspective, to bring radioactive ships from Operation Crossroads back to Hunters Point for "repair." But why did Hunters Point become the "center of research and expertise on decontamination," and why did the navy establish an important radiation laboratory there? The decision to site the NRDL at the Hunters Point shipyard has much to do with its proximity to University of California–Berkeley, located just across the San Francisco Bay. Berkeley scientists were deeply involved in the making of the atomic bomb during World War II. Ernest Lawrence developed the cyclotron at Berkeley, which aided in the production of plutonium in 1941. J. Robert Oppenheimer, the head of the wartime Manhattan Project, also taught at Berkeley.[34]

Perhaps the most important Berkeley scientists for the navy after Operation Crossroads were Joseph Hamilton and his colleague Kenneth Scott. Hamilton's research focused on the biomedical effects of radiation, using both rats and humans as test subjects. In 1994 the United States admitted to having secretly used thousands of people as radiation test subjects, many of them without consent or adequate knowledge of the nature of these experiments.[35] Human radiation subjects included hospital patients at places such as the University of California–San Francisco, where Hamilton conducted some of his experiments. Eileen Welsome explores the scale of human radiation experiments in *The Plutonium Files,* noting that while the U.S. public had no knowledge of human radiation testing, many of the scientists conducting those experiments "sat on the boards that set radiation standards, consulted at meetings where further human experimentation was discussed, investigated nuclear accidents, and served as expert witnesses in radiation injury cases."[36] In other words, the scientific community responsible for delineating radiological "safety" from "harm" was also active in placing some people in harm's way.

Both Hamilton and Scott attended a meeting at the Hunters Point naval shipyard in October 1946, a few months after Shot Baker. At the time, Hamilton was a leading expert on the biological metabolism of plutonium, and Scott worked in Hamilton's lab. They were invited to Hunters Point to advise the navy on what to do with Crossroads ships, and what kinds of safety standards to set for navy crewmembers who were actively decontaminating those ships.[37] By then, Hamilton had already conducted radiation experiments on twenty-nine patients at the University of California–San Francisco hospital.[38] The meeting in Hunters Point was contentious, with few conclusions except that more research was needed. This is precisely what the navy did on ships such as the *Independence* at Hunters Point—first in collaboration with Berkeley scientists and later through the formally established NRDL. Through the NRDL and its subsequent research activities, the San Francisco Bay Area became a new kind of "naval lake" and a key site within an emerging geography of military nuclearism.

A "SIXTH SENSE" FOR THE ATOMIC AGE

Different forms of radiation exhibit distinct characteristics, or ways of moving through the world. This point is important in understanding one of the "crossroads" confronted by the U.S. military in Bikini in 1946, and later at the Hunters Point shipyard in San Francisco. Gamma and x-rays are a form of electromagnetic radiation, similar to high-energy light. As a health threat, they are classified as external radiation, because they move across long distances and pass through matter easily, readily penetrating bodies. Shielding and other barriers are never completely effective against gamma and x-rays. Alpha and beta radiation, as particles, behave much differently. Both exhibit a distinct range of movement, traveling only short distances, and can be stopped by particular barriers. Beta particles are high-energy electrons. They can pass through human skin, although they settle quickly in body tissues and their effects remain local to that site. Beta particles are most harmful when inhaled or ingested, radiating from deep within. Alpha particles are heavy—they cannot penetrate the outer layer of human skin and are not considered much of a threat as an external source of radiation. Yet their two protons give alpha particles a large positive charge that pulls strongly at nearby electrons, ionizing atoms easily. Because of this large mass and charge, alpha particles exert a strong effect on biological tissues as internal radiation (or as an "internal emitter"), causing significant damage.[39]

Although scientists who had worked on the Manhattan Project were aware and deeply concerned about the fallout from an underwater blast, radiological safety practices at Bikini in 1946 were largely based on prewar occupational experiences with x-rays and radium.[40] In part, this is because the bomb's plutonium was a new element in the 1940s, and, as scientists were only beginning to understand, it presented a distinct health threat. Like radium, it imitates calcium, seeking the body's bones, but persists longer in the body and fixates in the lungs, so that the dangers of inhalation are greater.[41] Until the first atomic bomb in 1945, scientists had only encountered comparatively small amounts of it, in relatively controlled laboratory settings.[42] In their history of radiological safety standards for plutonium, W. H. Langham and J. W. Healy describe how scientists had only begun to worry about the biomedical effects of plutonium several years after its production. First produced in 1941, it was not until 1944 that G. T. Seaborg, head of the Manhattan District's chemistry division, wrote to R. S. Stone, the project's medical director: "It has occurred to me that the physiological hazards of working with plutonium and its compounds may be very great . . . In addition to helping to set up safety measures in handling so as to prevent the occurrence of accidents, I would like to suggest that a program to trace the course of plutonium in the body be initiated as soon as possible."[43] According to Gioacchino Failla, a U.S. physicist and radiobiologist who submitted a statement to the U.S. congressional hearings in 1960 titled "Radiation Protection Criteria and Their Standards," "after the war it became evident that the radiation protection problem had reached proportions undreamed of before."[44]

At the time of Operation Crossroads in 1946, no one had ever worried about plutonium in the waters of a lagoon or spread across the decks of a ship. Nor had anyone attempted to measure the exposure of thousands of people to large amounts of plutonium, let alone monitor plutonium outside of a laboratory setting. No one had ever tried to decontaminate a ship (or an aircraft carrier), and safety standards for this kind of work were nonexistent.[45] According to a 1947 report on the ship decontamination immediately after Shot Baker: "Men walked through it [radioactive material], tracked it around, and got it on their clothing and hands and faces. There was some tendency on the part of the men to disregard a danger which they could not see, nor touch, nor smell."[46]

Moreover, enforcing existing safety standards required equipment that the military either did not have at the time or had only in short supply. Since radiation eludes the human senses, instrumentation works as a kind of prosthesis, or what David Bradley, in *No Place to Hide,* calls a "sixth sense" and a "prerequisite for survival in the atomic age." One of the biomedical crises at Operation Crossroads was the staggering and tragic inadequacy of field monitoring instruments used by the radiological safety group. The instruments—even those designed to measure beta and gamma radiation—were not well equipped to work outdoors and were certainly not designed to measure the intense amounts of radiation produced by an atomic bomb. Nor could any of the field instruments in Bikini in 1946 detect plutonium's alpha particles.[47]

The two primary field instruments used at Bikini in 1946 were Geiger-Mueller X-263 counters and Victoreen Model 247 ionization meters. The Geiger-Mueller X-263 measured beta and gamma radiation, yet it could not measure the extreme radiation fields produced by Shot Baker.[48] As described in the NRDL's own *Radiological Safety Manual,* published in 1947, "At high radiation intensities, counts may be lost, or the meter may become 'paralyzed.' For example, if the upper limit of the X-263's range of 0.48 r/day is exceeded, the meter will swing off scale and audible sound will change from clicks either to a steady buzz or the meter will become 'paralyzed' and no sounds will be heard."[49] The Victoreen Model 247 ionization meter could measure higher exposure rates than the X-263 and was also designed to be light and "spray-resistant." Like the X-263, it was one of the few instruments designed—albeit untested—for the "field," or the outdoor, aquatic environment of Bikini. Still, the Victoreen Model 247 only measured gamma radiation.[50] There were a handful of instruments for measuring alpha radiation brought to Bikini, but none of these worked outside of the closed, air-conditioned space of the medical support ship, the USS *Haven*.[51] As safety monitor David Bradley recollects, "since [plutonium] is an alpha-particle emitter, our Geiger-Mueller counters and ionization chambers will not detect its presence. A special and extremely delicate 'alpha counter' is necessary for such work, and here at Bikini, under field conditions, no such alpha counter would survive a single day's work."[52]

Along with ship decontamination, one of the NRDL's first research projects was developing more precise instrumentation. In the late 1940s, the lab developed the military's portable field monitoring instruments, including instruments for detecting and measuring alpha emitters, such as plutonium. At first, NRDL researchers sought to modify the instruments used at Bikini, such as "tropicalizing" the X-263 to operate in the humid climate of the Marshall Islands. By 1948 the Instruments Division of the NRDL had begun to develop new portable alpha monitoring instruments, which included the AN/PDR-10 survey meter and an alpha probe counter "to count in inaccessible locations such as nostril, corners, and crevices."[53] The NRDL continued to develop radiation detection and monitoring equipment throughout the 1950s and 1960s, in part by testing its new instruments during subsequent nuclear weapons explosions in the Marshall Islands and at the Nevada Test Site.

Although the NRDL's research was theoretically aimed at protecting military personnel, in practice, its research supported the idea that atmospheric atomic bombs could be safely tested, as long as the right technology was in place.

LOCATING THE NRDL

Hunters Point—as an active naval shipyard—was an unusual site for a radiation laboratory that conducted sensitive experiments using nuclear fallout. Until a large, modern laboratory building for the NRDL was constructed at the shipyard in 1955, the NRDL's activities took place in surplus military buildings, fashioned as temporary lab spaces, which were scattered across the five-hundred-acre shipyard. During this period, the "lab" included a former barracks, where the animal colony was housed, and an old mess hall, where the physics department, instrumentation, and supply materials crowded together.[54] In 1948 NRDL biologists worried that they could not perform their research tasks, since the biomedical lab at the time was temporarily located in a building that was outfitted for the chemistry division. The biologists did not have adequate room for their experiments. More importantly, as they detail in a written request for a new biology building, radioactivity from neighboring chemistry labs "renders the delicate detection incident to biological investigations impossible."[55] Finding a secure location for radiation experiments remained an issue throughout the late 1940s. For example, in agreeing to the temporary use of one building in 1948, the shipyard's commander wrote to the NRDL's scientific director: "This building is not under any circumstances to be considered for possible permanent use by the laboratory."[56]

NRDL scientists anticipated these problems and lobbied for a change in location. In October 1947, the NRDL's Facilities and Equipment Committee met to discuss the lab's long-range plans and concluded that the lab should move elsewhere. Vibration from heavy shipyard equipment interfered with scientific instruments, and scientists were unable to find spaces remote enough at the working shipyard for hazardous experiments. Moreover, all the buildings leased by the shipyard for laboratory work were too small.[57] Adequate space was also an issue for the samples of radioactive fallout arriving from the Marshall Islands and of manufactured radioisotopes, which the NRDL used for its experiments. According to a monthly progress report from 1947, "the problem of storage of samples, proper security of samples, and sufficient separation of active samples from sensitive research instruments is a very serious problem."[58] During the late 1940s, these radioactive samples were kept in lead boxes in a small storage shack outside one of the laboratory's temporary buildings, which, as NRDL scientists themselves note in memos requesting better facilities, put lab and shipyard workers at risk. In a request to construct a "proper isotope storage building" in 1950, NRDL scientists explain: "from 1947 to date, radioisotopes have been received in various quantities from the AEC [Atomic Energy Commission], during which time storage was improvised in lead caves in many miscellaneous locations such as abandoned head facilities, locker rooms outside work shacks, and laboratories." The scientists also note that radiation from the storage shacks had risen to an unsafe level, exceeding permissible radiation exposure thresholds of the time.[59]

Why wasn't the NRDL relocated? The historical record offers no specifics. However, an internal report from 1955 explains that the Hunters Point shipyard was chosen as the primary base for continuing decontamination work after Operation Crossroads because of its proximity to other radiation laboratories—referring to University of California–Berkeley and Stanford—and because San Francisco was expected to be "the natural staging point for future Pacific Weapons tests."[60] The NRDL's research and activities thus bound San Francisco to the Marshall Islands, through the circulation of people, ships, technologies, and fallout.

CROSSROADS IN SAN FRANCISCO

From its initial studies of Crossroads ships, the NRDL soon developed a broad, multidisciplinary research program. In the immediate aftermath of Crossroads, however, the lab focused its efforts on the historically new problem of how to decontaminate a radioactive warship. Did a handblown torch remove radiation from the ship's external paint layer, Outside War Grey? Could an acidic solution, flushed through a ship's saltwater system for several hours, clean out its interior pipes and hoses? At a meeting in San Francisco in October 1947, while military personnel in Kwajalein were still working on Crossroads ships, the navy decided to begin experimental decontamination studies on the USS *Rockbridge*. The *Rockbridge* had already been brought back to San Francisco from the Marshall Islands and was considered to be the most radioactive ship on the Hunters Point shipyard at the time.[61]

On October 4, 1946, scientists selected an area on the port side of the *Rockbridge*'s hull, "previously determined as representative of the worst condition available," to test whether hand-scraping marine life and layers of paint could decrease radioactivity. Square sections along the sides of the ship were marked and monitored for beta and gamma radiation, using radiation detection equipment that had returned from Bikini and was, according to a report issued a few months later, "for the most part in poor condition."[62] The square sections were scraped by hand and monitored for radiation, then scraped and monitored again.[63] Experimental decontamination studies on other Crossroads ships continued through the late 1940s. On May 10, 1947, NRDL scientists targeted another ship, the USS *Crittenden*. Many different radioactive surfaces on the ship were scrubbed with an acidic solution for eighteen minutes. The NRDL concluded that the solution had decreased radiation but did not remove it entirely, and that this particular method worked best on rusted surfaces but was less effective with paint. The conclusion also notes that for "best results, the surface should be scrubbed vigorously as much as possible with a stiff brush."[64] Based on these studies, the NRDL produced a series of manuals on radiological safety, which circulated across military divisions.[65]

In the sterile, scientific language in the safety manuals and in its dutifully recorded experiments, it is difficult to conjure up the bodies of the workers in San Francisco, laboring in intimate relationship with Crossroads ships. Who vigorously scrubbed the surface of the *Crittenden* with a stiff brush? Who hand-scraped and later sandblasted the *Rockbridge*? The bodies of these workers disappear in the NRDL's technical reports and in the historical record more broadly. These same bodies were, like the ships and the animals brought to Bikini, also test subjects and a terrain on which knowledge of the bomb and its effects was produced.

WASTING THE BAY

Producing knowledge about radiological safety had other toxic effects. As part of its research program, the NRDL generated a tremendous amount of radioactive waste. During the 1950s, the NRDL also emerged as the center of radioactive waste disposal in the Bay Area, until this work was contracted out to a private company in 1959.[66] Most of the radioactive waste handled by the NRDL was released into the ocean, a standard practice for radiation facilities at the time.[67] Between 1954 and 1958, the NRDL sank 3,762 fifty-five-gallon drums and sixty-seven concrete blocks of radioactive waste—totaling 1,750 tons—in the ocean, near the Bay Area coastline.[68] The NRDL's own records are detailed—like its experiments on Crossroads ships—providing the precise latitude, longitude, and depth of these disposal operations. Still, the waste disposal procedure did not always go as planned. In a 1954 letter, the NRDL admonishes University of California–Berkeley's Radiation Lab for overpacking radioactive waste in its fifty-five-gallon disposal drums. Twenty-nine waste drums from the Berkeley lab had accidentally fallen into the San Francisco Bay during the NRDL's routine process of loading the waste disposal barges at the Hunters Point shipyard. The drums were salvaged but, as the letter notes, "not without a certain degree of difficulty, and unacceptable hazard to the safety of personnel, due to the weights involved."[69] Years later, the U.S. Environmental Protection Agency estimated that 25 percent of the NRDL's waste drums likely broke under water pressure before they reached the bottom of the ocean. In 1980 marine biologists from University of California–Santa Cruz also reported high levels of radioactive strontium and cesium, and some plutonium, in fish from the Farallon Islands, where much of the lab's waste was deposited.[70]

Most of the target ships brought back to Hunters Point were sunk in Southern California in 1948, but the NRDL retained the *Independence* until 1951 for use in experimental decontamination studies and to store radioactive waste.[71] The NRDL also stored waste from other radiation labs in the Bay Area, ultimately filling the entire aircraft carrier. In November 1950, as the NRDL drew up plans to sink the *Independence*, University of California–Berkeley's Radiation Lab requested to stow more radioactive waste on board the ship. In its response to this request, the NRDL's scientific director wrote: "Since presently accessible areas are nearly filled with waste, the acceptance of waste drums from [Berkeley] would require cutting through the deck of the INDEPENDENCE in order to gain access to the port side of the after [*sic*] engine room."[72] That is, the *Independence* had become so filled with waste drums that it was almost impossible to accept any more.

In 1949 the NRDL debated whether to sell the *Independence* as scrap but found it more cost-effective to sink the ship instead. In line with its prior use, the sinking of the *Independence* was also an experiment. According to a detailed plan for the event: "This Task Element will tow the ex-CVL 22 [USS *Independence*] to sea and sink it by underwater explosion, making photographic record of its movement under explosion stresses, to dispose of the hulk and to obtain technical information useful in the design of ships and weapons."[73] Although the navy gained operational expertise about ships from the *Independence*'s final test, countless stories and experiences disappeared along with the radioactive hulk underneath the ocean waters.

THE BURDEN OF THE BOMB

Anthropologist Hugh Gusterson reflects on the ways immense human pain and suffering in Hiroshima and Nagasaki became obscured through scientific studies and technologies—what he calls the "disappearing body" in war. Referring to the subjects of nuclear tests more broadly, Gusterson writes: "Scientists have methodically metamorphosed the mutilated and suffering bodies of these people and animals into tiny bodies of data used in myriad strategic calculations, for example, to help determine the efficiency of radiation and other nuclear weapons."[74] Japanese survivors in Hiroshima and Nagasaki were studied by U.S. scientists as part of the Atomic Bomb Survivors Life-Span Study (see chapter 9 in this volume). This study is ongoing and played a central role in the history of radiation protection standards in the United States.[75] U.S. scientists also studied acute and long-term exposure in Marshallese people, beginning in 1954 after the Bravo test bomb in Bikini, and the NRDL was deeply involved in these studies. According to Stannard, the historian and author of *Radioactivity and Health*, studies of exposure in the Marshallese "gave them [the NRDL] first-hand experience that could influence laboratory research and vice versa."[76]

In San Francisco, as the bodies of naval crewmembers and lab technicians at the NRDL worked to clean Crossroads ships and develop knowledge about instrumentation and radiobiology, their bodies also "worked" for the military in other ways. As the NRDL's archived files show, during the late 1940s, the lab kept detailed records on the beta and gamma exposures of its personnel. In the 1950s, more systematic monitoring of lab personnel included regular urine tests and physical examinations. While these practices were routine aspects of health safety, they also furthered one of the NRDL's objectives, as stated in its founding document: "*On the basis of clinical records and observation of personnel in regular contact with ex-target vessels* and elaboration thereof, [to] develop and improve diagnostic procedures and field analysis methods of exposure" and collect data from "studies of overexposure."[77] In other words, the NRDL would develop expertise in radiological safety based on exposure of its own workers. This practice was not confined to the NRDL. Robert Stone, the director of the Health Division of the Manhattan Project's lab at the University of Chicago, wrote in 1943: "It must be remembered that the whole clinical study of the personnel is one vast experiment. Never before has so large a collection of individuals been exposed to so much irradiation."[78]

When the NRDL closed in 1969, it left the by-products of radiological defense in Hunters Point, buried in landfills on-site and through the shipyard's sewer system.[79] Today, Hunters Point is part of a broader geography of nuclear industrialism, which includes plutonium factories such as Hanford in eastern Washington and in the Marshall Islands. This geography also includes the more intimate scale of the bodies and livelihoods of military personnel working on Crossroads ships, of the Marshallese, and of shipyard and NRDL lab workers exposed to radiation. They also include residents of the Hunters Point neighborhood, which surrounds the Hunters Point shipyard, who had little knowledge, if any, of the NRDL.

In 1989 the shipyard was classified as an EPA Superfund site, and today it is undergoing remediation as part of a large urban redevelopment project. The redevelopment project will transform the old naval base into a landscape of condominiums, offices, and waterfront parks.[80] In this process of excavating the shipyard's toxic past, local residents have sought a fuller accounting of the NRDL's environmental history and its potential effects on their lives. In this moment of social and ecological transformation, the discovery of the *Independence* in 2015 offers more than a lesson in marine biology. It offers a window into San Francisco's transoceanic connections, as a node within Pacific Ocean imperialism. It is also a reminder of the environmental legacy of the Cold War–era project of atmospheric nuclear testing, including the ecological afterlives of radiological safety and defense.

ACKNOWLEDGMENTS

I would like to acknowledge James Lyons, as well as Elena Kim and other participants of the Military Ecologies workshop at University of California–Irvine in 2017, for their insightful feedback.

NOTES

1. The Monterey Bay National Marine Sanctuary was designed by the federal government (through the National Oceanic and Atmospheric Administration, or NOAA) in 1992, under the National Marine Sanctuaries Act. The National Marine Sanctuaries Act was passed in 1972, the same year as the establishment of the NOAA and the same year as the United Nations Convention on the Prevention of Marine Pollution by Dumping of Wastes and Other Matter, commonly called the London Convention. Ironically, the London Convention prohibited the ocean disposal of radioactive waste.

2. Davis, "Fallout," *SF Weekly*, July 31, 2002.

3. U.S. Environmental Protection Agency, "Superfund: Hunters Point Naval Shipyard," U.S. Environmental Protection Agency, Pacific Southwest, Region 9: Superfund: Hunters Point Naval Shipyard, accessed June 20, 2017, https://yosemite.epa.gov/r9/sfund/r9sfdocw.nsf/vwsoalphabetic/Hunters+Point+Naval+Shipyard?OpenDocument.

4. Masco, "The Age of Fallout."

5. U.S. Naval Radiological Defense Laboratory, "Establishment of NRDL as an Atomic Energy Commission Project."

6. The Manhattan Project was a research project (1942–1945) led by the U.S. government, which produced the first atomic bomb.

7. Stannard and Baalman, *Radioactivity and Health*, 1065.

8. Masco, *Nuclear Borderlands*.

9. Weisgall, *Operation Crossroads*; "Veterans and Radiation," Department of Veterans Affairs, August 2004. http://www.publichealth.va.gov/docs/vhi/radiation.pdf.

10. *The Masthead*, May 25, 1946; *The Masthead*, June 15, 1946.

11. Delgado, *Ghost Fleet*.

12. Miller, *Under the Cloud*.

13. DeLoughrey, "The Myth of Isolates.

14. Hagen, *An Entangled Bank*.

15. Masco, *Nuclear Borderlands*. On the effects of nuclear testing in the United States, see Solnit, *Savage Dreams*.

16. Solnit, *Savage Dreams*.

17. Defense Nuclear Agency, *Operation Crossroads 1946*; Hines, *Proving Ground*; Weisgall, *Operation Crossroads*; Hacker, *The Dragon's Tail*.

18. Hines, *Proving Ground*, 40–41.

19. U.S. Congress, *Radiation Exposure from Pacific Nuclear Tests*, 16.

20. Welsome, *The Plutonium Files*; Defense Nuclear Agency, *Operation Crossroads 1946*; Weisgall, *Operation Crossroads*.

21. U.S. Congress, *Radiation Exposure from Pacific Nuclear Tests*, 5.

22. Bradley, *No Place to Hide*, 83. Bradley is talking specifically about measurements of beta and gamma radiation.

23. Stannard and Baalman, *Radioactivity and Health*.

24. U.S. Department of the Navy, "Radiological Decontamination of Target and Non-Target Vessels."

25. Lotchin, "The City and the Sword"; Lotchin, *Fortress California, 1910–1941*.

26. Oldenziel, "Islands," 16.

27. Brechin, *Imperial San Francisco*.

28. Cherny and Issel, *San Francisco, Presidio, Port and Pacific Metropolis*; Arbona, "Trial by the Bay."

29. Lotchin, "The City and the Sword."

30. Walker, "The Island at the Center of the Bay."

31. Lotchin, "The City and the Sword." On private shipbuilders in the Bay Area, see Wollenberg, *Marinship at War*; Johnson, *The Second Gold Rush*.

32. Bradley, *No Place to Hide*, 7.

33. "Animal Ship Bikini Bound," *San Francisco Chronicle*, June 2, 1946.

34. Welsome, *The Plutonium Files*; Creager, *Life Atomic*.

35. See Welsome, *The Plutonium Files*, and Stannard and Baalman, *Radioactivity and Health*.

36. Welsome, *Plutonium Files*, 486.

37. U.S. Department of the Navy, "Radiological Decontamination of Target and Non-Target Vessels."

38. Welsome, *Plutonium Files*.

39. Stannard and Baalman, *Radioactivity and Health*; Parr, *Sensing Changes*.

40. Hacker, *The Dragon's Tail*. On the concern voiced by Manhattan Project scientists regarding an underwater blast, see Jonathan Weisgall's testimony in U.S. Congress, *Radiation Exposure from Pacific Nuclear Tests*.

41. Creager, *Life Atomic*; Langham and Healy, "Maximum Permissible Body Burdens and Concentrations of Plutonium."

42. Langham and Healy, "Maximum Permissible Body Burdens and Concentrations of Plutonium."

43. Langham and Healy, "Maximum Permissible Body Burdens and Concentrations of Plutonium," 572–73.

44. U.S. Congress, *Radiation Protection Criteria and Standards*, 56.

45. Weisgall, *Operation Crossroads*; Hacker, *The Dragon's Tail*; Jessee, "Radiation Ecologies."

46. U.S. Congress, *Radiation Exposure from Pacific Nuclear Tests*, 18.

47. Defense Nuclear Agency, *Operation Crossroads 1946*; Bradley, *No Place to Hide*; Stannard and Baalman, *Radioactivity and Health*.

48. Defense Nuclear Agency, *Operation Crossroads 1946*.

49. U.S. Department of the Navy, "Radiological Safety Manual."

50. Moreover, the radiological safety group only had twenty working ionization meters in the aftermath of Shot Baker. Defense Nuclear Agency, *Operation Crossroads 1946*.

51. According to a report on Operation Crossroads published by the Defense Nuclear Agency (later the Department of Energy) in 1984, "CROSSROADS requirements for radsafe instruments turned out to be far greater than had been expected when planning for the operation began. No comprehensive program existed for the development and manufacture of rugged instruments for use under field conditions; thus the head of the Radsafe Section had to make do with what the Manhattan Engineer District could provide from its inventory and what the Victoreen Instrument Company could manufacture quickly" (*Operation Crossroads 1946*, 49).

52. Bradley, *No Place to Hide*, 74.

53. U.S. Naval Radiological Defense Laboratory, "Reports in Preparation, 19 November 1948."

54. U.S. Department of the Navy, "U.S. NRDL History 1946–1955.

55. U.S. Naval Radiological Defense Laboratory, "Request for New Facilities, Building 507.

56. U.S. Naval Radiological Defense Laboratory, "Building 508 Temporary Assignment."

57. U.S. Naval Radiological Defense Laboratory, "Minutes of Meeting of Facilities and Equipment Committee."

58. U.S. Department of the Navy, "Monthly Progress Report, March 1947."

59. U.S. Department of the Navy, "Request for Approval to Construct Isotope Storage Building."

60. U.S. Department of the Navy, "U.S. NRDL History 1946–1955."

61. Defense Nuclear Agency, *Operation Crossroads 1946*.

62. "Monthly Progress Report, March 1947," 1.

63. U.S. Naval Radiological Defense Laboratory, "Monitors Report of Experimental Sand Blasting of Hull—USS *Rockbridge*."

64. U.S. Naval Radiological Defense Laboratory, "USS *Crittenden* Experimental Decontamination."

65. U.S. Department of the Navy, "Radiological Safety Manual."

66. U.S. Navy, *Final Hunters Point Naval Station Historical Radiological Assessment.*

67. Hamblin, *Poison in the Well.*

68. U.S. Department of the Navy, "Letter to Mr. Arnold B. Joseph, Sanitary Engineer."

69. U.S. Naval Radiological Defense Laboratory, "Letter to Mr. Ray O'Day."

70. "Washington Notes," *Los Angeles Times*, September 24, 1980, B26. In 1980, the EPA estimated that 47,500 containers, containing 13,500 curies of radioactive waste, were sunk near the Farallon Islands between 1946 and 1970.

71. U.S. Navy, *Final Hunters Point Naval Station Historical Radiological Assessment.*

72. U.S. Department of the Navy, "Stowage of Radioactive Wastes Aboard the U.S.S. *Independence.*"

73. U.S. Department of the Navy, "Disposition of USS *Independence* (CVL-22)."

74. Gusterson, *People of the Bomb*, 69.

75. See Cram, "Becoming Jane."

76. Stannard and Baalman, *Radioactivity and Health*, 1065.

77. U.S. Naval Radiological Defense Laboratory, "Topics for Investigation by Radiation Laboratory"; emphasis added.

78. Quoted in Welsome, *Plutonium Files*, 45.

79. U.S. Environmental Protection Agency, "Superfund: Hunters Point Naval Shipyard."

80. Dillon, "Race, Waste, and Space."

BIBLIOGRAPHY

Arbona, Javier. "Trial by the Bay: Treasure Island and Segregation in the Navy's Lake." In *Urban Reinventions: San Francisco's Treasure Island*, edited by Lynne Horiuchi and Tanu Sankalia, 125–39. Honolulu: University of Hawai'i Press, 2017.

Bradley, David. *No Place to Hide*. Boston: Little, Brown and Company, 1948.

Brechin, Gray. *Imperial San Francisco: Urban Power, Earthly Ruin*. Berkeley: University of California Press, 2006.

Cherny, Robert W., and William Issel. *San Francisco, Presidio, Port and Pacific Metropolis*. Heinle and Heinle Publishers, 1981.

Cram, Shannon. "Becoming Jane: The Making and Unmaking of Hanford's Nuclear Body." *Environment and Planning D: Society and Space* 33, no. 5 (2015): 796–812.

Creager, Angela. *Life Atomic: A History of Radioisotopes in Science and Medicine*. Chicago: University of Chicago Press, 2013.

Davis, Lisa. "Fallout: The Past Is Present." *SF Weekly*, July 31, 2002. https://archives.sfweekly.com/sanfrancisco/fallout-the-past-is-present/Content?oid=2145516.

Defense Nuclear Agency. *Operation Crossroads 1946: United States Atmospheric Nuclear Weapons Tests, Nuclear Test Personnel Review*. Washington, DC: U.S. Department of Defense, 1984.

Delgado, James P. *Ghost Fleet: The Sunken Ships of Bikini Atoll*. Honolulu: University of Hawai'i Press, 1996.

DeLoughrey, Elizabeth M. "The Myth of Isolates: Ecosystem Ecologies in the Nuclear Pacific." *Cultural Geographies* 20, no. 2 (2013): 167–84.

Dillon, Lindsey. "Race, Waste, and Space: Brownfield Redevelopment and Environmental Justice at the Hunters Point Shipyard." *Antipode* 46, no. 5 (2014): 1205–221.

Gusterson, Hugh. *People of the Bomb: Portraits of America's Nuclear Complex*. Minneapolis: University of Minnesota Press, 2004.

Hacker, Barton C. *The Dragon's Tail: Radiation Safety in the Manhattan Project, 1942–1946*. Berkeley: University of California Press, 1994.

Hagen, Joel B. *An Entangled Bank: The Origins of Ecosystem Ecology*. New Brunswick, NJ: Rutgers University Press, 1992.

Hamblin, Jacob. *Poison in the Well: Radioactive Waste in the Ocean at the Dawn of the Nuclear Age*. New Brunswick, NJ: Rutgers University Press, 2008.

Hines, Neal O. *Proving Ground: An Account of the Radiobiological Studies in the Pacific, 1946–1961*. Seattle: University of Washington Press, 1963.

Jessee, Emory Jerry. "Radiation Ecologies: Bombs, Bodies, and Environment During the Atmospheric Nuclear Weapons Testing Period, 1942–1965." PhD diss., Montana State University, 2013.

Johnson, Marilyn S. *The Second Gold Rush: Oakland and the East Bay in World War II*. Berkeley: University of California Press, 1996.

Langham, W. H., and J. W. Healy. "Maximum Permissible Body Burdens and Concentrations of Plutonium: Biological Basis and History of Development." In *Uranium, Plutonium, Transplutonic Elements*, edited by Harold C. Hodge, J. N. Stannard, and J. B. Hursh, 569–92. Berlin: Springer, 1973.

Lotchin, Roger W. "The City and the Sword: San Francisco and the Rise of the Metropolitan-Military Complex 1919–1941." *Journal of American History* 65, no. 4 (1979): 996–1020.

Lotchin, Roger W. *Fortress California, 1910–1961: From Warfare to Welfare*. Urbana: University of Illinois Press, 1992.

Masco, Joseph. "The Age of Fallout." *History of the Present* 5, no. 2 (2015): 137–68.

Masco, Joseph. *Nuclear Borderlands: The Manhattan Project in Post–Cold War New Mexico*. Princeton, NJ: Princeton University Press, 2006.

Miller, Richard Lee. *Under the Cloud: The Decades of Nuclear Testing*. The Woodlands, TX: Two-Sixty Press, 1986.

Oldenziel, Ruth. "Islands: The United States as Networked Empire." In *Entangled Geographies: Empire and Technopolitics in the Global Cold War*, edited by Gabrielle Hech, 13–42. Cambridge, MA: MIT Press, 2011.

Parr, Joy. *Sensing Changes: Technologies, Environments, and the Everyday, 1953–2003*. Vancouver: University of British Columbia Press, 2010.

Solnit, Rebecca. *Savage Dreams: A Journey into the Hidden Wars of the American West*. Berkeley: University of California Press, 1994.

Stannard, James Newell, and R. W. Baalman Jr. *Radioactivity and Health: A History*. Richland, WA: Pacific Northwest National Laboratory, 1988.

U.S. Congress, House of Representatives, Subcommittee on Oversight and Investigations of the Committee on Natural Resources. *Radiation Exposure from Pacific Nuclear Tests.* 103th Cong., 2nd sess., 1994.

U.S. Congress, Special Subcommittee on Radiation of the Joint Committee on Atomic Energy. *Radiation Protection Criteria and Standards: Their Basis and Use.* 86th Cong., 2nd sess., 1960.

U.S. Navy. *Final Hunters Point Naval Station Historical Radiological Assessment: History of the Use of General Radioactive Materials, 1939–2003.* U.S. Naval Facilities Engineering Command, 2006.

Walker, J. Samuel. *Permissible Dose: A History of Radiation Protection in the Twentieth Century.* Berkeley: University of California Press, 2000.

Walker, Richard. "The Island at the Center of the Bay." In *Urban Reinventions: San Francisco's Treasure Island,* edited by Lynne Horiuchi and Tanu Sankalia. Honolulu: University of Hawai'i Press, 2017.

Weisgall, Jonathan, M. *Operation Crossroads: The Atomic Tests at Bikini Atoll.* Annapolis, MD: Naval Institute Press, 1994.

Welsome, Eileen. *The Plutonium Files: America's Secret Medical Experiments in the Cold War.* New York: Random House, 1999.

Wollenberg, Charles. *Marinship at War: Shipbuilding and Social Change in Wartime Sausalito.* Berkeley, CA: Western Heritage Press, 1990.

Primary Sources

"Building 508 Temporary Assignment." U.S. Naval Radiological Defense Laboratory. Box 1, Appropriations 1948; General Correspondence, 1946–1948; Records of Naval Districts and Shore Establishments, Record Group 181; National Archives Branch Depository, San Francisco.

"Disposition of USS *Independence* (CVL-22)" (HRA-1708). U.S. Department of the Navy, Naval Sea Systems Command. *Historical Radiological Assessment, Volume II, Use of General Radioactive Materials, 1939–2003, Hunters Point Shipyard* (2004), Appendix D.

"Establishment of NRDL as an Atomic Energy Commission Project." U.S. Naval Radiological Defense Laboratory. Box 1, Plans, Projects, and Politics, 1948; General Correspondence, 1946–1948; Records of Naval Districts and Shore Establishments, Record Group 181: National Archives Branch Depository, San Francisco.

"Letter to Mr. Arnold B. Joseph, Sanitary Engineer" (HRA-1350). U.S. Department of the Navy, Naval Sea Systems Command. *Historical Radiological Assessment, Volume II, Use of General Radioactive Materials, 1939–2003, Hunters Point Shipyard* (2004), Appendix D.

"Letter to Mr. Ray O'Day." U.S. Naval Radiological Defense Laboratory. General Correspondence, 1946–1948; Records of Naval Districts and Shore Establishments, Record Group 181: National Archives Branch Depository, San Francisco.

"Minutes of Meeting of Facilities and Equipment Committee." U.S. Naval Radiological Defense Laboratory. General Correspondence, 1946–1948; Records of Naval Districts and Shore Establishments, Record Group 181; National Archives Branch Depository, San Francisco.

"Monitors Report of Experimental Sand Blasting of Hull—USS *Rockbridge*," October 14, 1946. U.S. Naval Radiological Defense Laboratory. General Correspondence, 1946–1949: Box 1, Plans, Project, and Policies, 1948; Records of Naval Districts and Shore Establishments, Record Group 181: National Archives Branch Depository, San Francisco.

"Monthly Progress Report, March 1947." U.S. Department of the Navy, Naval Sea Systems Command. *Historical Radiological Assessment, Volume II, Use of General Radioactive Materials, 1939–2003, Hunters Point Shipyard* (2004), Appendix D.

"Radiological Decontamination of Target and Non-Target Vessels" (HRA-596). U.S. Department of the Navy, Naval Sea Systems Command. *Historical Radiological Assessment, Volume II, Use of General Radioactive Materials, 1939–2003, Hunters Point Shipyard* (2004), Appendix D.

"Radiological Safety Manual," 1947 (HRA 2715). U.S. Department of the Navy, Naval Sea Systems Command. *Historical Radiological Assessment, Volume II, Use of General Radioactive Materials, 1939–2003, Hunters Point Shipyard* (2004), Appendix D.

"Request for Approval to Construct Isotope Storage Building" (HRA 418). U.S. Department of the Navy, Naval Sea Systems Command. *Historical Radiological Assessment, Volume II, Use of General Radioactive Materials, 1939–2003, Hunters Point Shipyard* (2004), Appendix D.

"Request for New Facilities, Building 507." U.S. Naval Radiological Defense Laboratory. Box 1, Appropriations, 1948; General Correspondence, 1946–1948: Records of Naval Districts and Shore Establishments, Record Group 181: National Archives Branch Depository, San Francisco.

"Reports in Preparation, 19 November 1948." U.S. Naval Radiological Defense Laboratory. General Correspondence, 1946–1948: Records of Naval Districts and Shore Establishments, Record Group 181: National Archives Branch Depository, San Francisco.

"Stowage of Radioactive Wastes Aboard the U.S.S. *Independence*" (HRA 1910). U.S. Department of the Navy, Naval Sea Systems Command. *Historical Radiological Assessment, Volume II, Use of General Radioactive Materials, 1939–2003, Hunters Point Shipyard* (2004), Appendix D.

"Topics for Investigation by Radiation Laboratory." U.S. Naval Radiological Defense Laboratory. General Correspondence, 1946–1948: Records of Naval Districts and Shore Establishments, Record Group 181: National Archives Branch Depository, San Francisco.

"U.S. NRDL History 1946–1955" (HRA-587). U.S. Department of the Navy, Naval Sea Systems Command. *Historical Radiological Assessment, Volume II, Use of General Radioactive Materials, 1939–2003, Hunters Point Shipyard* (2004), Appendix D.

"USS *Crittenden* Experimental Decontamination," May 10, 1947. U.S. Naval Radiological Defense Laboratory. General Correspondence 1946–1949: Records of Naval Districts and Shore Establishments, Record Group 181: National Archives Branch Depository, San Francisco.

4

Born Opaque

INVESTIGATING THE NUCLEAR ACCIDENT AT THE SANTA SUSANA FIELD LABORATORY

William Palmer

On top of Old Susie, all covered in rust,
There was an old reactor that must have gone bust.

Nestled atop the Simi Hills, thirty miles from downtown Los Angeles, an experimental nuclear reactor experienced a major accident in July 1959. Thirty percent of the reactor core was damaged. Between five and ten thousand curies of radioactive fission products escaped the fuel elements in the core.[1] But the trajectory of those radioactive products remains a matter of controversy. Did these products decay safely inside the reactor, or did a significant amount of that radiation reach the surrounding environment? Aside from the small number of people involved with the Sodium Reactor Experiment (SRE) at the Santa Susana Field Laboratory (SSFL), almost no one knew anything about the accident until 1979. Even half a century after the accident, experts can only speculate about the consequences of the partial meltdown of the reactor, although it may have been one of the worst nuclear reactor accidents in American history.[2]

The SSFL was operated by North American Aviation (NAA), a major manufacturer of aviation technology. Immediately following WWII, NAA began nuclear reactor research in their Aerophysics Lab in Los Angeles. As their work expanded, NAA created the Atomic Energy Research Department (AERD), based in Downey, California, which investigated a number of "dual-use" nuclear reactors—those that produced energy for electricity as well as plutonium for the U.S. weapons program.[3] In 1955 AERD became

Atomics International (AI), with headquarters in Canoga Park, California, and most of their nuclear reactor research was conducted at the SSFL, atop the Simi Hills in what was called the Nuclear Development Field Lab. Nuclear research occupied only a small part of the SSFL; NAA utilized the majority of the site, which grew to almost three thousand acres, for research and testing of massive rocket engines. Issues of nuclear waste and contamination nonetheless impacted the entire facility. There were many research reactors in operation at SSFL, but the SRE became one of the more notorious as people learned of the accident that occurred in 1959.

The sodium reactor used in the SRE was the product of a new partnership between the Atomic Energy Commission (AEC) and AI. By the mid-1950s, the American strategy for reactor development had shifted from exclusively military aims to applications for the peaceful use of nuclear energy. As a result, the AEC created a series of power reactor demonstration programs that sought to enlist the private sector in the nascent nuclear energy industry by proving the feasibility of producing electricity with nuclear reactors. AI participated in the first two rounds of these demonstration projects. The goal of the SRE in particular was to explore and improve the design of a liquid-sodium-cooled, graphite-moderated reactor. During the SRE project, which ran from the late 1950s to the early 1960s, it was the only such reactor in operation in the United States. Results from the SRE did successfully inform the design of a similar sodium graphite reactor built at the Hallam Nuclear Power Facility, near Lincoln, Nebraska, in the early 1960s.

As it was used entirely for experimental research, AI operated the SRE reactor in a series of "power runs"—intermittent periods of operation for varying lengths of time. The accident occurred during power run 14, in July 1959. Even though much about the accident has become contested in recent years, AI investigators argued soon thereafter that within the first few days of run 14, the sodium coolant had become blocked, which caused a power excursion—a situation in which the reactor power began to increase exponentially, outside of the control of the operators. It seems likely that the severely high temperatures, resulting from a loss of coolant, and the subsequent power excursion caused damage to one-third of the reactor's core fuel elements. Despite this significant event, AI continued to run the reactor for almost two weeks before finally terminating the run. It then began the lengthy process of removing the damaged core, decontaminating the facility, and investigating the causes and consequences of the accident. It remains unclear, and especially controversial, however, what the consequences of

Table 4.1. Timeline of major events and publications associated with the SRE incident

July 1959	*Los Angeles Times* journalist visits AI
August 1959	AI issues press release
November 1959	AI, *SRE Fuel Element Damage, Interim Report*
January 1960	AEC, *Status Report on Sodium Graphite Reactors as of 1959*
December 1961	AI, *SRE Fuel Element Damage, Final Report*
March 1962	AI, *Distribution of Fission Product Contamination in the SRE*

the accident were for personnel, the surrounding communities, and the environment.

We may never know for certain what happened, and this is perhaps the inevitable result of the culture of secrecy surrounding nuclear technology, a situation one AEC member referred to as the "twilight zone."[4] However, just prior to this accident, the revised Atomic Energy Act of 1954 had lifted this secrecy somewhat, and while the circle of insiders was still small, public and private entities involved in nuclear research were forced to navigate a new level of transparency. I suggest that AI's reports reflect this effort to navigate potential scrutiny while maintaining control over the interpretation and meaning of the accident. In this way, the reports offer an opportunity to engage in "agnotology," a term proposed by Robert Proctor and Londa Schiebinger to describe the analysis of the production of ignorance.[5] The significant gaps, omissions, uncertainties, and tentative conclusions in AI's reports have produced an unnerving lack of knowledge about the accident and its consequences. Unlike the many classified documents of the time that were "born secret," these unclassified documents were born opaque. Even though the reports about the accident seem to offer extremely detailed information, much remains unknown about the consequences of the accident because the reports leave a great deal of information unclear.

BORN OPAQUE

Eisenhower's Atoms for Peace initiative motivated new transparency concerning the details of nuclear energy but also created new challenges for those working in the nuclear industry. In a powerful 1953 speech at the United Nations, Eisenhower sought to win hearts and minds by promoting the peaceful, rather than destructive, uses of the atom. Hoping to capture the loyalties of the many newly independent postcolonial countries around the world, he argued that American expertise in nuclear technology could improve peoples' lives by providing electricity and new technologies for agriculture and medicine. The U.S. Congress followed the president's lead the next year by passing the revised Atomic Energy Act, which lifted many constraints on nuclear secrecy. The new legislation changed the rules surrounding classification and patenting of nuclear innovation, allowing for greater access to the new technologies. The AEC hoped to enlist as many private firms as possible into the new nuclear energy industry, and that was only possible by increasing transparency and access to information. In practice, however, this was neither obvious nor straightforward.[6]

The movement toward transparency created new challenges as nuclear projects entered the public stage. For example, the SRE became a high-profile reactor project. Not only had the AEC promoted this reactor concept to Congress as one of five pilot demonstration programs, but two senior scientists from AI promoted its design, efficacy, and safety at the Atoms for Peace conference in Geneva in 1955.[7] John Krige has likened this 1955 conference to a "panopticon," noting that "international exchange is at once a window and a probe, an ideology of transparency and, by virtue of that, an instrument of control, a viewpoint from which to look in and watch over."[8] To the extent that the conference, and the Atoms for Peace program more generally, functioned as a panopticon—with American scientists watching the Soviets and American allies—it seems likely that the AEC and AI managers and scientists were cognizant that the window and probe could work in both directions. With these multiple possible audiences in mind, the authors of the reports about the accident created a narrative that emphasized the important technological lessons that could be gleaned from the event while mitigating as much as possible any wrongdoing on the part of the operators and minimizing potential threats to people and places neighboring the facility.

The possibility of public backlash against reactor development was very real, especially since severe accidents in the UK, Canada, and elsewhere had been the topic of numerous newspaper articles.[9] Hoping perhaps to forestall this kind of criticism, AI issued a press release a month after the termination of run 14 claiming that "a parted fuel element was observed." The press release went on to state that the incident posed no threats to personnel or neighbors, noting simply that "the occurrence is of importance from a technical standpoint."[10] AI did not complete and publish its initial investigation

Figure 4.1. Secrecy at Atomics International. Al Markado, copyright ©1959. *Los Angeles Times*. Fred Baumberger, "100,000 in Area Work on Secret Projects," *Los Angeles Times*, July 19, 1959, 1.

into the accident until November 1959, several months after the termination of run 14. The press release, therefore, seems particularly premature. Even though the statement understated the severity of the accident, it nonetheless demonstrated the perceived need to state something, however minor.

Even more remarkable, though, was the visit of a *Los Angeles Times* reporter to AI during run 14, within days of the power excursion that likely caused the damage to the reactor's core. Though the reporter noted that many companies in the area were working on top-secret projects for the government, the only company mentioned was AI. One large photograph is of two women posing in front of a poster on the wall warning them to keep quiet (see figure 4.1).

"She's got a secret," the caption reads. "Mrs. Irma Klein, left, secretary at Atomics International Valley plant, is eager to broadcast it despite warning posters in back of her, but Suzie Beldon reminds her of responsibility to keep it to herself. These women posed to dramatize the 'wartime security' picture in the area." Moreover, the text on the poster reads, "What you see here . . ." (the remainder of the text near the bottom of the poster is mostly obstructed but may read "let it stay here"). But next to the text is one large eye, reminiscent of George Orwell's *1984*. While Suzie points to the poster as a reminder of secrecy practices, her thumb actually points along a direct line toward the large eye, which might give the impression that they are being watched. Another photograph shows Harry Lange, presumably a security official at AI, supervising "the destruction of some classified papers" in the "Pulp Machine." The reporter advised his readers: "Let's just not talk about it."[11] Like the press release, this article seems, however strangely, to call attention to the facility through a desire to hide it. If AI wanted to stay off the public radar, it might have avoided any public statements and dealings with the press. Both the press release and the invitation for the journalist to visit AI indicate the company's efforts to map a new relationship with the public.

While the press release and the *Los Angeles Times* article demonstrate AI's commitment to increased communication with the public, I do not want to overemphasize the new degree of transparency during the 1950s. Even before the revised Atomic Energy Act, there was significant transfer of information between agencies, labs, and corporations conducting nuclear research. A number of senior scientists at AERD and then AI had worked in the military's nuclear weapons design and production facilities. For example, Chauncey Starr, the president of AI, had worked at both the Oak Ridge and Berkeley labs during World War II. Another senior scientist, who worked on the SRE program at the time of the accident, had come to AI from Hanford. Others had as well. As one former employee who had a master's degree in metallurgy from the University of Wisconsin explained: "We were a collection of people out of universities, government labs, and the military."[12] Researchers from AERD had been working on improvements to Hanford's reactors and likely incorporated some of the designs of the Hanford reactors into the SRE. The use of graphite moderators in the SRE, for instance, is a strong indication that the sodium reactor borrowed elements from the Hanford designs. Finally, an AEC report on the status of sodium-cooled, graphite-moderated reactors noted that AI derived much information about this concept from the Submarine Intermediate Reactor and the Experimental Breeder Reactor I—two sodium-cooled reactor projects begun earlier. Engineers at AERD thus combined information gleaned from Hanford regarding graphite and from Argonne and Knolls, two AEC labs, regarding liquid sodium.[13] Yet, continued information exchange between the National Reactor Testing Station in Idaho Falls and AERD would have been important since the Experimental Breeder Reactor I suffered a partial meltdown in 1955, the year construction of the SRE began. The flow of information also extended beyond the national labs. A manager of the SRE at the time of the accident recently conveyed this story: "I don't think there was a cover-up at SRE because after the accident I was asked to give a talk at the last minute. I asked IJL [initials maintain anonymity] if I could talk about the accident and use the slides I had been given and he said I could . . . It was an annual meeting [of] the East Coast Utilities. There were 1,500–2,000 people in the audience! I talked for over four hours! . . . I showed all the slides. That's not what I would call a cover-up."[14] While this presentation occurred after the passage of the revised Atomic Energy Act, it is consistent with a longer trend of an expanding nuclear community, which included private companies interested in the progress of the new technologies.

The AERD of the NAA grew out of the nuclear research program for the military, and the company had contacts among several of the national labs and production facilities. While the revised Atomic Energy Act of 1954 eased some of the constraints of nuclear secrecy, it did not represent a major shift in existing communications practices. It did, though, create new challenges for AERD (later AI), as the company needed to traverse the seemingly thin line between secrecy and transparency. This is the context in which I situate my reading of AI's reports about the accident. These reports were not classified, and seven hundred copies were made available for distribution.[15] The reports seem to have been written less for internal review than for external evaluation and consumption. In the introduction to the interim report, the authors state: "For the reader who is not familiar with the SRE, the report begins with a brief description and operating history of the reactor."[16] And in the preface to the section describing reactor design, they note: "The SRE was designed and constructed by Atomics International, a division of North American Aviation, Inc., as part of a joint program with the Atomic Energy Commission to develop a sodium-cooled, graphite moderated, thermal power reactor for civilian application."[17] As a participant in a broader nuclear reactor community—the AEC, labs, corporations, and utilities—AI must have been cognizant that their high-profile reactor would garner much attention and potential scrutiny. I suggest that the authors of these reports had this in mind, and I argue that their awareness of this scrutiny gave birth to a new kind of secrecy where information was on display yet still obscured.

ENVIRONMENTAL CONTAMINATION

One of the primary issues of contention regarding the accident pertains to the quantity of radioactive fission products that escaped the reactor system and entered the open environment. Much of this has focused on I-131, a particularly dangerous radioactive isotope of iodine, which may have vented as a gas following the accident, entering into the local atmosphere. The interim report offered "tentative conclusions" about this issue, and it emphasized a great deal of uncertainty about the methods and results of the investigation.[18] This report was written by members of the SRE safety committee who investigated the accident, and it constitutes the longest and most detailed account of the accident and the investigation by AI safety experts into the cause of the event and its effects. The committee analyzed the radioactivity of the liquid-sodium coolant, the cover gas (a layer of helium above the coolant), gas storage tanks, the stack monitors, various components of the reactor system, the building that housed the reactor, and readings from a number of environmental monitors that examined radioactivity in air, soil, vegetation, and water. Even though the committee investigated numerous possible routes of contamination, their methods and conclusions were very often circumscribed by significant deficiencies and limitations. Later reports supplemented this initial investigation but failed to add much clarity to central questions concerning the release of radioactive toxins.

The interim report acknowledged that a power excursion had caused fuel damage but provided an unclear and uncertain account of the implications of this incident for the area surrounding the reactor. The report concluded that tetralin (a secondary coolant for pipes, pumps, and seals) had entered the primary liquid-sodium coolant and decomposed into carbon, which blocked the coolant from reaching the fuel elements. The authors of the report suggested that the fuel damage occurred when the reactor experienced a power excursion because of the loss of coolant on the second day of power run 14 in July 1959. The *SRE Fuel Element Damage, Final Report* (1961) confirmed that the reactor experienced "severe thermal cycling" as a result of the excursion and that the cladding separating the fuel from the coolant burst. In some places, such high temperatures were reached that the iron in the cladding and the uranium from the fuel rod formed new metal alloys, which then melted.[19] While the interim report did not hesitate to acknowledge nuclear fuel damage, conclusions surrounding the results of this damage are much less clear. On the one hand, the report states that the liquid sodium absorbed all of the volatile fission products released from the damaged fuel elements and thus did not create any hazards for people or the environment. The authors write, "In spite of the cladding failure to at least 11 of the fuel elements, no radiological hazard was present to the reactor environs." But the report authors also admit that every area of investigation into possible contamination carried inherent deficiencies and uncertainties, warning that "any information relative to the extent of the original release of activity which can be deduced from this data is limited by a number of factors."[20] These factors include a number of concerns about the measurements that were taken following the incident. A report issued almost three years later titled *Distribution of Fission Product Contamination in the SRE* studied this issue specifically and claimed to have accounted for many volatile fission products released from the core fuel elements because of the accident. It could not, however, make any definitive conclusions about the extent and location of the radioactive iodine.[21]

One of the most important places to look for evidence of elevated levels of radioactivity was the sodium coolant; however, the committee noted significant sources of uncertainty introduced by the process they followed to make these measurements. In order to study the radioactivity of the sodium, they first performed chemical separation on their samples and then took gamma-ray scans. They noted, however, that measured activity after chemical separation could be significantly different from the activity prior to this chemical process. Furthermore, they acknowledged drawbacks in their statistical analysis of their data. The authors explained that "the very limited size of the sodium sample from which the chemical separations start, the fact that the final counting statistics are low due to the relatively low sample activity present at the start of the chemical separation, and the errors inherent in using subtraction techniques in analyzing gross gamma scans, would possibly produce variations of the order noted."[22] The authors concluded that radioactivity in the sodium was difficult to determine because: "1) Very meager experimental data is available concerning the extent of radioactivity in the primary sodium coolant. 2) The significance of radiometric analyses performed on the sodium samples is only good to within a factor of 10."[23]

The committee qualified their findings even more explicitly and beyond the problems associated with small sample sizes and chemical separation. They noted that the fission products with a short half-life would have decayed by the time they took the sample (I-131 has a half-life of eight days);

particles may have settled in various parts of the reactor and cooling system; some fission products may have been caught in the cold and hot traps; and, finally, and most worrisome, "the more volatile products may have escaped prior to the time the sample was taken."[24] The authors presented a list of sodium samples, and a significant gap exists between the last two: April 14, 1959, and August 2, 1959. There were no samples taken between April and August, even though the reactor showed major problems in June and July.[25] Despite these significant limitations, the interim report concludes confidently that because I-131 was not detected in the cover gas (helium), "it undoubtedly combined with the sodium as rapidly as it was evolved."[26] But it seems that the basis for this statement had more to do with the fact that I-131 was not found in other outlets to the degree expected. If it did not combine with the sodium, then there are other places it might have gone.

One of the possible avenues of escape for fission products like I-131 was through cold traps, devices used to decontaminate the liquid sodium by removing oxides and other impurities. When the committee analyzed the measured activity of these cold traps, they found that the radioactivity was one hundred times greater in August than in April, noting that this value might have been even higher in July, right after the incident. They explained that "measurements could not start until August 8 (due to the radiation hazard . . .), at which time the dose rate, extrapolated to near the surface, was about seventy r/hr. It is possible that initial cold-trap dose rates, had they been measured, would have yielded higher values."[27] Hiding behind their conclusion—that most of the fission products combined with the sodium—was the very real and openly acknowledged possibility that some of these products had reached the cold traps.

When the committee studied the cover gas, they noted very high levels of radioactive xenon and krypton, two noble gases with short half-lives, not considered as dangerous as other elements such as radioactive iodine. They remarked that the fact that "such an extremely high concentration [of radioactive Xe and Kr isotopes] did in fact exist is borne out by the *off-scale* readings of decay tank activity." On July 15, for example, the activity in the gas storage tanks was "*so high in fact that the counter had not been calibrated in that range*."[28] Regardless, the authors determined that since the allowable levels of effluents were averaged for a single year, given AEC guidelines, they saw no restrictions on venting large amounts of radioactivity into the atmosphere. They noted that "allowable stack effluent concentrations are averaged over a one-year period. Therefore, the consequences of this latter release were not serious in any way and the release did not disturb the ability of the SRE to continue to vent radioactivity to the atmosphere."[29] Here, where they do acknowledge the release of significant levels of radioactive gas into the atmosphere, the authors minimize the resulting environmental contamination by appealing to existing government regulations.

R. S. Hart of AI completed a study in 1962 that specifically analyzed fission product contamination from the SRE, and he made some of the limitations of investigating the contamination even more explicit. The first problem he raised concerned the timing of the samples taken. Hart noted that "it is apparent the vast majority of the fission product release to the primary coolant took place during run fourteen, July 12 to 26, 1959 . . . it would appear that at least some and perhaps most of the release occurred during the first few days of this period." The author noted that because the first sodium sample was taken three weeks after the excursion, when the fuel elements released the majority of the fission products, "this sample may not present a true picture of the initial fission product distribution" and that "it is not feasible to adjust this first sample quantitatively for deposition prior to its obtainment."[30]

In the discussion of the helium cover gas samples, Hart alluded to the narrow study of the two noble gases detected, xenon and krypton. Estimates of radioactivity in the cover gas at shutdown were calculated assuming that these were the only products released from the core. He acknowledged that additional shorter-lived isotopes "probably created a higher gross level at shutdown." As he explained, "it is difficult to interpret cover gas samples subsequent to the July 26 shutdown since bleeding and flushing operations to the gas decay tanks and out the stack were almost immediately commenced."[31] If the gases were purged and vented immediately following the power excursion, the samples taken later could not reasonably represent the actual fission products released when the fuel began to melt.

Besides noting the uncertainties introduced because of what was measured and when, Hart admits to having no clear explanation for the lower than expected values of I-131 found in the sodium coolant and the cover gas. He ranked the percentage of release of different isotopes according to their volatility, claiming that cesium, strontium, and iodine had the "highest release fractions." But he also noted that "some anomalies do exist. One would expect the release fraction of iodine to be considerably higher than that of cesium rather than one-third of it as actually found." He determined that "a possible answer to the low iodine value would be the escape of this element from the primary sodium to the cover gas system." Iodine has the unusual characteristic of transforming immediately from a solid to a gas, bypassing the liquid phase. But he stated that "iodine has never been found in the cover gas of the SRE either after this incident or in any of the many other gas samples taken during the operation of the reactor." He conceded that it could have gone undetected amid the high concentrations of xenon and krypton. And, as noted earlier, the purging of the reactor gases had begun immediately during run 14. Thus, the issue of determining the release of I-131 persisted, and in his conclusion he could not explain the low values for it. "The iodine release fraction seems unusually small on a volatility basis relative to the other elements. Since neither deposition in the primary system nor escape to the cover gas occurred in detectable quantities, the low sodium value for iodine remains to be explained."[32] This last comment remains true to this day.

Researchers at AI continued to study the issue, and a new research group was created for this purpose. They could not make definitive conclusions either.[33] In 1969 G. B. Zwetzig of AI surveyed the existing reports on fission product releases in reactors that used liquid sodium as a coolant. He concluded that all of the fission products released would be retained in either the cover gas, the coolant, or the primary system walls. While he admitted his conclusions were based on the available documents, not new studies, he ignored the many admissions of uncertainty peppered throughout these reports.[34]

Finally, the authors of the interim report claimed that AI had established a "continuous program of routine monitoring of soil, vegetation, water and airborne activity" since 1954. They noted high levels of radioactivity in 1958 but attributed these to nuclear weapons tests in Nevada and Russia, and seem to have dismissed the higher than normal soil activity readings recorded in June and July 1959, which correspond to the accident. They did not make any definitive conclusions about water contamination "due mainly to the relatively small number of samples taken."[35] This represents one of many unknowns and thus makes the interim report a document full of uncertainties, speculations, and assumptions.

The interim report made cautious and "tentative" conclusions about the nature and consequences of the accident for a number of reasons.[36] First, the investigators made only superficial and delayed studies. This is especially significant because the damaged reactor and related components were often too radioactive to observe immediately, and because radioactive iodine, which has a half-life of eight days, would be difficult to detect when AI analysts looked for it several weeks, months, and years later. Moreover, the environmental monitoring program that AI had set in place may not have been sufficient, nor did their analysis fully scrutinize the readings that were available. Finally, their tentative conclusions provided opportunities to conduct further studies (e.g., on the ability of liquid sodium to retain fission products) while simultaneously claiming that the reactor was inherently safe and stable. Overall, these reports lay out a detailed consideration of the measurements taken and the limitations of those measurements, but the repeated admissions of uncertainty work to obscure the effects of the accident.

WORKER EXPOSURE

Just as troubling was AI's discussion of the exposure of people at the facility. In recent years, P2 Solutions, a private company that contracted with the Department of Energy to facilitate reconciliation efforts over the environmental remediation at SSFL, conducted a series of interviews with former employees of AI. The comments from former employees in this chapter are drawn from those interviews, unless otherwise indicated. The employees discussed the SRE accident in often radically different terms: varied memories thus served to form additional layers of opacity and further obscured any comprehensive understanding of the accident. In the discussion of this topic in the final report, AI conceded that some workers had been exposed during cleanup after the incident, but that the amounts were of no significance. "Only very light contamination of personnel occurred. The few cases which did occur were quickly detected and remedied."[37] In 2006 Boeing, the current owner of SSFL, issued a brief summary and statement about the SRE accident. Concerning employees, the company argued that "personnel employed in operating the reactor and those employed during the post-accident recovery, cleanup and refurbishment were continually monitored for external and internal radiation exposure. No personnel exceeded annual exposure limits for radiation workers. The AEC reviewed these records closely."[38]

Yet, the contamination may have been more extensive than the company reports indicate. A senior mechanic recently gave a hint of this. "When I was hired, I went immediately to cleaning up the *outside* of the SRE, which was typical for new hires as those employees doing that job got burned out pretty quickly." If the coolant had retained the fission products, why would he have cleaned the outside of the SRE? It is possible that this person was referring to the inside of the building that housed the reactor, which had become contaminated during the removal of the damaged core. One former health physicist described the SRE complex as a "mess in terms of contamination leftover from the accident."[39] According to another health physicist working at the SRE complex, "there was a lot of contamination of Sr-90 [strontium]." One of the workers who had been involved in the cleanup had inhaled some amount of strontium. "His urinary excretion of Sr-90 was followed for several years . . . I ended up with a fairly lengthy body of research. He got an intake of Sr-90 such that we didn't have to worry about the effects on his body but it was large enough that for several years we could measure what was excreted in his urine. I presented several papers based on this research."[40]

One former worker, who was part of a group known as the Mavericks, noted: "None of us had college degrees and we didn't have our noses up in the air. We would do whatever it took to get a job done—short of killing ourselves!" This worker kept in touch with another one of the Mavericks who had worked on the SRE at the time of the accident. "I asked him about it and he said that nothing really happened, that it was no big deal."[41] Given that this worker had gained the reputation as a Maverick, it may not be surprising that a nuclear reactor accident would be insufficient to impress him. Many former employees have discussed the accident in a similarly casual manner. One employee commented that "it just did not seem like anything out of the ordinary." Another former worker confirmed that "any experimental reactor is always off-normal."[42]

In fact, among the many employees interviewed, no clear picture or consistent story of the accident emerges. Many deny that the SRE experienced a meltdown and that it was merely a few damaged fuel elements. Others believe that some fuel elements melted but that there was no cover-up. Others referred to the SRE accident as "dangerous" and very secret. Every possible variation of the SRE narrative was offered by the many different employees interviewed, who were present at the time of the incident.[43]

John Pace, a twenty-year-old technician-in-training at the time of the accident, told reporters in 2009 a radically different version of the events surrounding the accident and cleanup than the narrative contained in the 1959 interim report. Pace's shift began a few hours after the power excursion on July 13, 1959. He quickly learned that the gas storage tanks were full, so his supervisors decided to purge the radioactive gases straight to the atmosphere because they worried the reactor might explode. "I found out that the reactor had run away from them and they had to release the gases. . . . Those of us who were working there were, of course, very upset about what happened because on that particular shift, all of our families lived in the San Fernando Valley and all that radiation went over their homes in the west valley over there."[44] According to Pace, they were told not to tell anyone about anything, including their families, who lived downwind of the purged gases.

Pace said that radioactive gases were released each time the reactor was restarted and scrammed, roughly every twenty-four hours, for almost two weeks following the excursion, frequently straight into the atmosphere, "preferably at night."[45] Moreover, R. K. Owen, a health physicist at the SRE, had recommended shutting down the reactor in a letter dated July 17, 1959. He reported that "the condition occurring July 12, 1959, during power run fourteen" had caused the High Bay atmosphere (where technicians such as Pace worked) to become contaminated at "three hundred times the maximum permissible concentration in air for unidentified beta gamma emitters."[46] But the SRE ran for almost another week. Pace recalled that during the cleanup, he was instructed not to wear a film badge. He did not have a face mask while examining the damaged core, removing the fuel elements, and cleaning up the severe radioactive contamination. He also stated that AI had barricaded an area of about a mile around the SRE complex because of the extensive environmental contamination.[47] None of this information appeared in the interim or subsequent reports.

In August 2009, the Department of Energy organized the SRE Workshop to "educate" local residents. Three experts presented their opinions on the accident: Dr. Paul Pickard of the Sandia National Laboratory, Dr. Thomas Cochran of the Natural Resources Defense Council, and Dr. Richard Denning of Ohio State University. All agreed that the accident more than likely did not cause any significant health threats to people living below the research facility and that very little, if any, radioactive fission products escaped the SRE complex. They were all working from the same set of documents. Perhaps most problematic with regard to their conclusions was the fact that they did not have access to all of the documents, reports, and competing interpretations. It seems that, in addition to the original company reports, these experts also read reports by experts hired by Boeing, but not reports by experts enlisted by community activists and former workers. Only Dr. Cochran acknowledged and lamented the fact that none of those who presented their opinions in 2009 had access to all of the various interpretations.[48]

The SRE accident also factored into the determinations of SSFL employees' eligibility for compensation from a new law—the Energy Employees Occupational Illness Compensation Program Act—enacted in 2000. The National Institute of Occupational Safety and Health (NIOSH), which evaluates claimants' possible exposures to radiation and resulting illnesses, accepted Boeing's accounting at face value. Melton Chew, the NIOSH site profile team leader for SSFL, explained to a group of retired United Auto Workers from SSFL: "We have seen the newspaper articles that call this your Chernobyl or Three Mile Island. I'd like to give you my perspective." Chew downplayed how serious it might have been by saying: "They [Boeing] gave us a lot of information on the loss of the cladding and how much material actually escaped. We have that documented fairly well to our satisfaction. . . . Boeing did a thorough reporting of incidents that may potentially impact radiation exposure. We actually worked from that particular book. Everything was all included in one book. The cladding failure was well addressed in the documents."[49]

Lavonne "Bonnie" Klea, a former secretary at AI who survived bladder cancer and whose husband—also a former AI employee—died of cancer, spearheaded the effort to establish that AI workers qualified for compensation under the provisions of the act, by drawing attention to the SRE accident and criticizing the reliance on Boeing's reports.[50] As she argued at one public NIOSH hearing: "The site profile basically was written by the Boeing Company. One large notebook was used and the Boeing consultant was the company's own expert witness who has been fighting workers' compensation claims for years, and now he is involved intimately with the NIOSH program. This is an extreme conflict of interest." She concluded her testimony by affirming: "I will not give up."[51] A year later, S. Cohen and Associates, on behalf of NIOSH, reviewed many historical and contemporary documents and acknowledged a number of unresolved issues, including "concern whether the internal monitoring program was sufficiently robust to estimate exposures before and after January 1, 1959," "lack of information related to the potential exposures associated with facility 'incidents,'" and "lack of information on the environmental exposures."[52]

CONCLUSION

The opacity of AI's reports has had lasting effects. The SRE accident became the symbolic lightning rod for conflicts between activists, local residents, former workers, Boeing officials, and state and federal agencies, all of whom have participated in the negotiations, debates, and lawsuits concerning the twenty-five-year process of environmental remediation at SSFL. The accident became public in 1979, six months following the reactor accident at Three Mile Island, when local activists discovered documents about the SRE accident and brought them to the press. For the past several decades, the SRE accident specifically, and SSFL more generally, have become fiercely contested issues.

Recent attempts by researchers to understand the nature of the SRE accident have been especially difficult because of the limited information available. If secrecy in the early atomic energy program operated within a nebulous "twilight" zone, the boundaries of that zone became more entrenched and guarded the more people peered beyond the metaphorical chain-link fences. As activists and local residents learned about the hazardous practices and the resulting contamination at SSFL, they vigorously sought more details about the 1959 accident and enlisted the help of outside experts. These latter experts only had available to them the original documents created by the company at the time, and, still, they have questioned the conclusions of Boeings' experts about what might have really happened. In this way, the official narrative of the SRE accident continues to be contested.[53]

The 1959 interim report produced ignorance with its many omissions, gaps, uncertainties, and assumptions. It also remains the only documented account of the accident and its immediate aftermath. All subsequent interpretations and narratives of the accident have thus had to rely on this report. While AI successfully navigated the ambiguous parameters of secrecy in the early years of atomic energy, its successive parent corporations would later aggressively fortify the once nebulous zones of secrecy and uncertainty using the very same documents AI originally created in a remarkably different political milieu.

NOTES

Epigraph: *Santa Susana Field Laboratory Former Worker Interviews*, 331; henceforth *SSFLFWI*.

1. See, for example, Dickinson, "Coolant Block Damages SRE Fuel."

2. For example, experts who interpreted the SRE accident in later decades would argue that the 1959 accident released far more fission products to the environment than Three Mile Island. Three Mile Island released seventeen curies of radioactive iodine, and some analysts have recently estimated that the SRE may have released thousands of curies of radioactive iodine into the open environment. Makhijani, "Declaration"; Lochbaum, *An Assessment of Potential Pathways*.

3. On the "dual-purpose" reactors, see, for example, Allen, *Nuclear Reactors for Generating Electricity*. This is also discussed in Hewlett and Duncan, *Atomic Shield*.

4. Atomic Energy Commission, *Control of Radiation Hazards in the Atomic Energy Program*, Appendix 10, 228.

5. Proctor and Schiebinger, eds., *Agnotology.*

6. For a discussion of some of these issues then and more recently, see Green, "Information Control and Atomic Power Development"; Galison, "Secrecy in Three Acts."

7. Atomic Energy Commission, "Current Statement of the Atomic Energy Commission on the Five-Year Reactor Development Program"; Parkins, "Sodium Reactor Experiment"; and Starr, "The Sodium Graphite Power Reactor."

8. Krige, "Atoms for Peace, Scientific Internationalism and Scientific Intelligence," 167.

9. Robinson, "Can Your City Control an 'Atomic Accident'?" *Los Angeles Times,* May 11, 1958, K10; Peter Kihss, "Reactor Safety is Pressured by U.S.," *New York Times,* November 1, 1957, 8; Baer, "New Problems for the Profession."

10. Atomics International, "News Release," 1.

11. Fred Baumberger, "100,000 in Area Work on Secret Projects," *Los Angeles Times,* July 19, 1959, SF1.

12. *SSFLFWI,* 330.

13. Atomic Energy Commission, *Status Report on Sodium Graphite Reactors as of 1959.*

14. *SSFLFWI,* 109–10.

15. Jarrett, *SRE Fuel Element Damage, Interim Report,* ii.

16. Jarrett, *SRE Fuel Element Damage, Interim Report,* I-1.

17. Jarrett, *SRE Fuel Element Damage, Interim Report,* II-A-1.

18. Jarrett, *SRE Fuel Element Damage, Interim Report,* xi.

19. Fillmore et al., *SRE Fuel Element Damage, Final Report,* III-12.

20. Jarrett, *SRE Fuel Element Damage, Interim Report,* I-2, IV-C-2.

21. Hart, *Distribution of Fission Product Contamination in the SRE.*

22. Jarrett, *SRE Fuel Element Damage, Interim Report,* IV-C-1.

23. Jarrett, *SRE Fuel Element Damage, Interim Report,* IV-C-8.

24. Jarrett, *SRE Fuel Element Damage, Interim Report,* IV-C-2.

25. Jarrett, *SRE Fuel Element Damage, Interim Report,* IV-C-3.

26. Jarrett, *SRE Fuel Element Damage, Interim Report,* IV-C-4.

27. Jarrett, *SRE Fuel Element Damage, Interim Report,* IV-C-12.

28. Jarrett, *SRE Fuel Element Damage, Interim Report,* IV-C-15, 16; emphasis added.

29. Jarrett, *SRE Fuel Element Damage, Interim Report,* IV-C-25.

30. Hart, *Distribution of Fission Product Contamination in the SRE,* 10, 13.

31. Hart, *Distribution of Fission Product Contamination in the SRE,* 13.

32. Hart, *Distribution of Fission Product Contamination in the SRE,* 20–23.

33. Kunkel et al., *Fission Product Retention in Sodium.*

34. Zwetzig, *Survey of Fission and Corrosion-Product Activity.*

35. Jarrett, *SRE Fuel Element Damage, Interim Report,* IV-C-25, 27, 29.

36. Jarrett, *SRE Fuel Element Damage, Interim Report,* V-1.

37. Fillmore et al., *SRE Fuel Element Damage, Final Report,* II-1.

38. Boeing Company, "Sodium Reactor Experiment (SRE) Accident," 2.

39. *SSFLFWI,* 342, 223; emphasis added.

40. *SSFLFWI,* 174.

41. *SSFLFWI*, 351–52.

42. *SSFLFWI*, 331, 361.

43. *SSFLFWI*, 121, 136, 141, 147, 111, 151, 171, 224, 282, 294.

44. "Meltdown Man," *EnviroReporter*, June 25, 2009. http://www.enviroreporter.com/investigations/rocketdyne/meltdown-man/.

45. Teresa Rochester, "'59 Nuclear Reactor Accident Remains Vivid for Former Santa Susana Field Laboratory Worker." *Ventura County Star*, July 12, 2009. http://archive.vcstar.com/news/59-nuclear-reactor-accident-remains-vivid-for-former-santa-susana-field-laboratory-worker-ep-3715518-350756651.html/.

46. R. K. Owen to R. E. Durand, "Airborne Radioactive Contamination in SRE High Bay," July 17, 1959.

47. Rochester, "'59 Nuclear Reactor Accident Remains Vivid."

48. SRE Workshop, August 29, 2009. Video documenting the workshop can be found at https://www.etec.energy.gov/Library/Video/Current_Videos.php. There are some indications in the presentations that these experts had read the following: Daniel, "Investigation of Releases from Santa Susana Sodium Reactor Experiment"; Christian, *Chemical Behavior of Iodine-131*; Frazier, *Report of John R. Frazier*; and Krsul, *Review and Evaluation of Report of David A Lochbaum*.

49. National Institute for Occupational Safety and Health, "Worker Outreach Activities," 8.

50. The following is from the Executive Summary: "This Historical Site Assessment (HSA) is developed to summarize the operational history of Area IV for both the Department of Energy (DOE) and The Boeing Company (Boeing) from a radiological perspective. This activity is undertaken to identify areas of radiological operations, compile prior radiological cleanups and releases and to identify further actions needed to ensure that the radiological cleanup of Area IV is completed." "SSFL Area IV HSA (May 2005 Final)," ES-1. The full document can be found at http://www.etec.energy.gov/Library/Historical-Site-Assessment.php.

51. Klea, "Testimony before the NIOSH Advisory Board," 38–39.

52. Cohen and Associates, *Review of the Santa Susana Field Laboratory (SSFL) Area IV Special Exposure Cohort (SEC) Petition-00093 and the NIOSH SEC Petition Evaluation Report*, 10.

53. Makhijani, "Declaration"; Lochbaum, *An Assessment of Potential Pathways*"; Beyea, *Feasibility of Developing Exposure Estimates*, Beyea, *Section-by-Section Response*; *Report of the Santa Susana Field Laboratory Advisory Panel*, 14; Daniel, "Investigation of Releases"; Christian, *Chemical Behavior of Iodine-131*; Frazier, *Report of John R. Frazier*; Krsul, *Review and Evaluation of Report of David A Lochbaum*.

BIBLIOGRAPHY

Allen, Wendy. *Nuclear Reactors for Generating Electricity: U.S. Development from 1946 to 1963*. RAND, R-2116-NSF. June 1977. http://www.rand.org/content/dam/rand/pubs/reports/2007/R2116.pdf.

Atomic Energy Commission. *Control of Radiation Hazards in the Atomic Energy Program.* Washington, DC: Government Printing Office, 1950.

Atomic Energy Commission. "Current Statement of the Atomic Energy Commission on the Five-Year Reactor Development Program to the Subcommittee on Research and Development." Joint Committee on Atomic Energy, 1955.

Atomic Energy Commission. *Status Report on Sodium Graphite Reactors as of 1959.* Civilian Power Reactor Program Part III, TID-8518(6) (1960): 1–5. http://www.osti.gov/scitech/servlets/purl/4155357.

Atomics International. "News Release." August 29, 1959. http://www.etec.energy.gov/Library/Main/Doc._No._38_SRE_Press_Release_August_25_1959.pdf.

Baer, Eli. "New Problems for the Profession: What Will the Law Be Like in the Atomic Age?" *American Bar Association Journal* 45 (May 1959): 455–57.

Baumberger, Fred. "100,000 in Area Work on Secret Projects." *Los Angeles Times,* July 19, 1959, SF1.

Beyea, Jan. *Feasibility of Developing Exposure Estimates for Use in Epidemiological Studies of Radioactive Emissions from the Santa Susana Field Laboratory.* October 5, 2006. http://www.etec.energy.gov/Library/Main/Doc._No._41_Beyea_SSFL_October_5,_2006.pdf.

Beyea, Jan. *Section-by-Section Response to Critiques of Studies of the 1959 Accident at the Santa Susana Field Laboratory Made by John R. Frazier on Behalf of the Boeing Company.* http://www.cipi.com/PDF/Beyea_Response_to_Frazier_Report_Sect_by_Sect.pdf.

Boeing Company. "Sodium Reactor Experiment (SRE) Accident." March 23, 2006. http://www.etec.energy.gov/Library/Main/Doc._No._28_Short_Description_of_SRE_Accident_prepared_by_Boeing.pdf.

Christian, Jerry. *Chemical Behavior of Iodine-131 during SRE Fuel Element Damage in July 1959.* May 26, 2005. http://www.etec.energy.gov/Library/Main/Christian_Report_on_SRE.pdf.

Cohen, S., and Associates, for the Advisory Board on Radiation and Worker Health, National Institute for Occupational Safety and Health. *Review of the Santa Susana Field Laboratory (SSFL) Area IV Special Exposure Cohort (SEC) Petition-00093 and the NIOSH SEC Petition Evaluation Report.* Contract No. 200-2009-28555 SCA-SEC-TASK5-0066, Revision 1. October 2009. https://www.cdc.gov/niosh/ocas/pdfs/abrwh/scarpts/sca-ssflsec93-r1.pdf.

Daniel, John A. "Investigation of Releases from Santa Susana Sodium Reactor Experiment in July 1959." May 27, 2005. *Lawrence O'Connor et al. v. Boeing North America, Inc.* Case Number CV 97-1554 DT (RCx), U.S. District Court, Central District of California, Western Division.

Dickinson, Robert. "Coolant Block Damages SRE Fuel." *Nuclear Engineering* 18, no. 1 (January 1960): 107–11. https://www.etec.energy.gov/Library/Main/Doc._No._53_SRE_Accident_Nucleonics_1960-01.pdf.

Fillmore, F. L., et al. *SRE Fuel Element Damage, Final Report.* NAA-SR-4488 (suppl.), 1961. http://www.etec.energy.gov/Library/Main/Doc._No._2_SRE_Fuel_Element_Damage_Final_Report_1961_NAA-SR-4488_(suppl).pdf.

Frazier, John R. *Report of John R. Frazier.* November 4, 2006. https://www.etec.energy.gov/Library/Main/Doc._No._43_Frazier_Report_Final.pdf.

Galison, Peter. "Secrecy in Three Acts." *Social Research* 17, no. 3 (Fall 2010): 941–74.

Green, Harold. "Information Control and Atomic Power Development." *Law and Contemporary Problems* 21, no. 1 (Winter 1956): 91–112.

Hart, R. S. *Distribution of Fission Product Contamination in the SRE*. NAA-SR-6890, March 1, 1962, 10, 13. http://www.etec.energy.gov/Library/Main/NAA-SR-6890.pdf.

Hewlett, Richard G., and Francis Duncan. *Atomic Shield: A History of the United States Atomic Energy Commission*. Vol. 2. University Park, PA: Penn State University Press, 1969.

Historical Site Assessment of Area IV Santa Susana Field Laboratory Ventura County, CA. Prepared by Sapere Consulting Inc. and The Boeing Company for the DOE. Executive Summary. May 2005. https://www.etec.energy.gov/Library/Main/HSA_Executive_Summary.pdf.

Jarrett, A. A., ed. *SRE Fuel Element Damage: An Interim Report*. November 1959. Atomics International Division of North American Aviation. NAS-SR-4488.11/15/1959. http://www.etec.energy.gov/Library/Main/NAA-SR-4488-Interim.pdf.

Kihss, Peter. "Reactor Safety is Pressured by U.S." *New York Times*, November 1, 1957, 8.

Klea, Bonnie. "Testimony before the NIOSH Advisory Board on Radiation and Worker Health." April 8, 2008, 38–39. http://www.cdc.gov/niosh/ocas/pdfs/abrwh/2008/tr040808.pdf.

Krige, John. "Atoms for Peace, Scientific Internationalism, and Scientific Intelligence." *Osiris* 21, no. 1 (2006): 161–81.

Krsul, John R. *Review and Evaluation of Report of David A Lochbaum*. November 4, 2006. http://www.etec.energy.gov/Library/Main/Doc._No._44_Krsul_Review_and_Evaluation_of_Lochbaum_Report.pdf.

Kunkel, W. P., et al. *Fission Product Retention in Sodium—A Summary of Analytical and Experimental Studies at Atomics International*. NAA-SR-11766. 8/8/1965. http://www.osti.gov/scitech/servlets/purl/4510250.

Lochbaum, David. *An Assessment of Potential Pathways for Release of Gaseous Radioactivity following Fuel Damage during Run 14 at the Sodium Reactor Experiment*. October 5, 2006. http://www.etec.energy.gov/Library/Main/Doc._No._40_Lochbaum_SRE_Report_October_5,_2006.pdf.

Makhijani, Arjun. "Declaration." *Lawrence O'Connor et al. v. Boeing North America, Inc.* Case Number CV 97-1554 DT (RCx). U.S. District Court, Central District of California, Western Division.

"Meltdown Man." *EnviroReporter*. June 25, 2009. http://www.enviroreporter.com/investigations/rocketdyne/meltdown-man/.

National Institute for Occupational Safety and Health. "Worker Outreach Activities." Transcript. January 2006. http://www.cdc.gov/niosh/ocas/pdfs/tbd/outreach/w011906a.pdf.

Owen, R. K., to R. E. Durand. *Airborne Radioactive Contamination in SRE High Bay during Reactor Operation*. July 17, 1959. https://www.etec.energy.gov/Library/Main/Doc._No._54_Internal_Letter_Owen_to_Durand_1959-07-17.pdf.

Parkins, W. E. "Sodium Reactor Experiment." C. Starr, "The Sodium Graphite Power Reactor." Combined as chapter 7 in *Nuclear Power Reactors*, edited by James Pickard. Princeton, NJ: D. Van Nostrand, 1957.

Proctor, Robert, and Londa Schiebinger, eds. *Agnotology: The Making and Unmaking of Ignorance*. Stanford, CA: Stanford University Press, 2008.

Report of the Santa Susana Field Laboratory Advisory Panel. October 2006. http://www.etec.energy.gov/Library/Main/Doc._No._39_SSFL_Panel_Report_October_2006.pdf.

Robinson, Donald. "Can Your City Control an 'Atomic Accident'?" *Los Angeles Times*, May 11, 1958.

Rochester, Teresa. "'59 Nuclear Reactor Accident Remains Vivid for Former Santa Susana Field Laboratory Worker." *Ventura County Star*, July 12, 2009. http://archive.vcstar.com/news/59-nuclear-reactor-accident-remains-vivid-for-former-santa-susana-field-laboratory-worker-ep-3715518-350756651.html/.

Santa Susana Field Laboratory Former Worker Interviews. P2 Solutions, Idaho Falls, ID. November 2011. http://www.etec.energy.gov/Environmental_and_Health/Documents/WorkerHealthFiles/Former_Worker_Interview_Final_Report.pdf.

"SSFL Area IV HSA (May 2005 Final)." Executive summary. http://www.etec.energy.gov/Library/Main/HSA_Executive_Summary.pdf.

Zwetzig, G. B. *Survey of Fission and Corrosion-Product Activity in Sodium or NaK-Cooled Reactors*. AI-AEC-MEMO-12790. February 28, 1969. http://www.osti.gov/scitech/servlets/purl/4823059.

PART TWO

Industrial Toxins

5

Making Way for Industrial Waste

WATER POLLUTION CONTROL IN SOUTHERN CALIFORNIA, 1947–1955

Brinda Sarathy

> The limiting factor of water supply in California has been apparent to everyone. The far more limiting factor of waste disposal has been appreciated by few.
>
> —**Randal Dickey,** *Report of the Interim Fact-Finding Committee on Water Pollution*

In March 1978, the Santa Ana Regional Water Control Board in Southern California (henceforth the Regional Board) faced an emergency situation. Heavy rains that winter wreaked havoc in the form of flash floods, mudslides, and overtopping of local rivers and reservoirs. Particularly alarming was the state of the Stringfellow Acid Pits in an unincorporated area of Riverside County. In 1955 the Regional Board, in conjunction with other county and state agencies, had authorized the creation of Stringfellow in order to meet the growing waste disposal needs of the region's aerospace-industrial complex.[1] Between 1956 and 1972, the seventeen-acre site received over thirty-four million gallons of industrial effluent and was among the largest hazardous waste dumps in the nation.[2]

These conditions, combined with the deluge of 1978, set the stage for what would eventually become California's first Superfund site.[3] Toxic waters had overtopped their holding ponds and threatened to wash out the single earthen barrier that separated the dump from hundreds of families in the community of Glen Avon to the south. Confronted with the possibility of a full-scale blowout, the Regional Board *intentionally* released industrial effluent into local flood control channels, and these coursed through the

Figures 5.1 and 5.2. Spring floods in Glen Avon, California, 1978. Stringfellow Archive, University of California–Riverside, box 16, folder 6. Photographer unknown.

town and inundated pastures, residential yards, and playgrounds (see figures 5.1 and 5.2).[4] Government officials informed administrators at Glen Avon Elementary School of their decision ahead of time but asked that parents not be notified so as to avoid panic. Fortunately, several teachers shared their concerns about toxic waters with parents in the community and helped spread word about potential contamination.[5]

Flooding from Stringfellow eventually mobilized locals to successfully pressure the U.S. Environmental Protection Agency to designate the pits as California's first Superfund site in 1983.[6] Soon afterward, geologists would find that the dump was built on highly fractured bedrock, which had allowed for the long-term and invisible seepage of toxic substances into the groundwater below.[7] Officials have estimated that it will take at least four hundred years to clean up contamination at the site. Cases like the Stringfellow Acid Pits force us to grapple with how experts historically understood, rationalized, and managed hazardous wastes to justify toxic dumping prior to the 1960s, when U.S. state and federal agencies began to regulate such substances. Why did policy makers during the 1950s overlook threats to groundwater quality, and what does this reveal about the management of toxic wastes? Asking such questions not only allows us to understand the processes and politics leading to the siting of Stringfellow in particular but also provides a lens through which to understand the limits of state institutions and policies to meaningfully protect public health and natural environments.

I begin this chapter with a critical examination of the creation of regional water boards in California (of which the Santa Ana Regional Board was one), an action that came out of legislators' concerns about "the danger to humans from poisonous wastes of industry"[8] and postwar efforts to address ineffective systems of managing water pollution. I argue that although the regionalization of water pollution control initially held promise for greater oversight of industry, it eventually saw policy makers more narrowly construe threats to public health and water quality, and minimize the dangers posed by industrial waste. More specifically, I show how regional water boards were (1) institutionally structured to manage industrial wastes as a nuisance rather than as a hazard to public health, (2) inclined to cooperate with industry and support economic growth, and (3) oriented toward minding surface water quality rather than groundwater pollution. I then explore how various experts, including geologists and engineers, conducted and interpreted field studies to support the siting of a toxic dump at the Stringfellow rock quarry, as well as the ways in which officials did or did not monitor and enforce dump operations once the site had been established. All these factors combined set the stage for eventual disaster at the Stringfellow Acid Pits.

DEFINING AND REGULATING WATER POLLUTION IN CALIFORNIA

In California, as elsewhere in the United States, control of water pollution prior to the 1950s was primarily a local and community affair.[9] Federal legislation such as the Rivers and Harbors Act of 1899 (which prohibited the dumping of refuse and impeding of navigable waterways), the Public Service Health Act of 1912 (which authorized studies of waterborne diseases, sanitation, sewage, and pollution of navigable streams and lakes), and the 1924 Federal Oil Pollution Control Act (which forbade seafaring vessels from dumping oil into the ocean) all signaled early but largely ineffective attempts to manage water contamination. By the 1940s, deteriorating water quality, as a result of growing population densities, agricultural expansion, and industrialization, saw most U.S. states establish a regulatory patchwork to address pollution.[10] Yet, overlapping agency jurisdictions and conflicting mandates at the state level continued to stymie any comprehensive understanding or regulation of water pollution.[11] In California, too, there was no centralized regulatory agency during this period to oversee water pollution in the state as a whole (see figure 5.3).[12]

Threats to water quality from the post–World War II population and industrial booms were perhaps nowhere more direly felt than in the Golden State.[13] In 1946 waste generated by a growing population led to emergency appropriations of $45,000,000 by the state legislature to counties for sewerage works, and for the first eight months of 1948, construction of sewerage and waterworks in California was more than 30 percent of all construction—in comparison to a figure of about 13 percent for the nation as a whole.[14] In the counties of Orange, Riverside, and San Bernardino, in the Santa Ana River Basin of Southern California, the number of manufacturing establishments grew from a total of 305 to 695 between 1939 and 1947 (see table 5.1).[15]

Close to 45 percent of these establishments were involved in the processing of raw materials (fruit and vegetable processing, meatpacking, and milk processing), which led to significant amounts of putrescible waste. In addition, a growing number of steel and metal fabrication industries as well as numerous different types of chemical industries generated inorganic wastes.

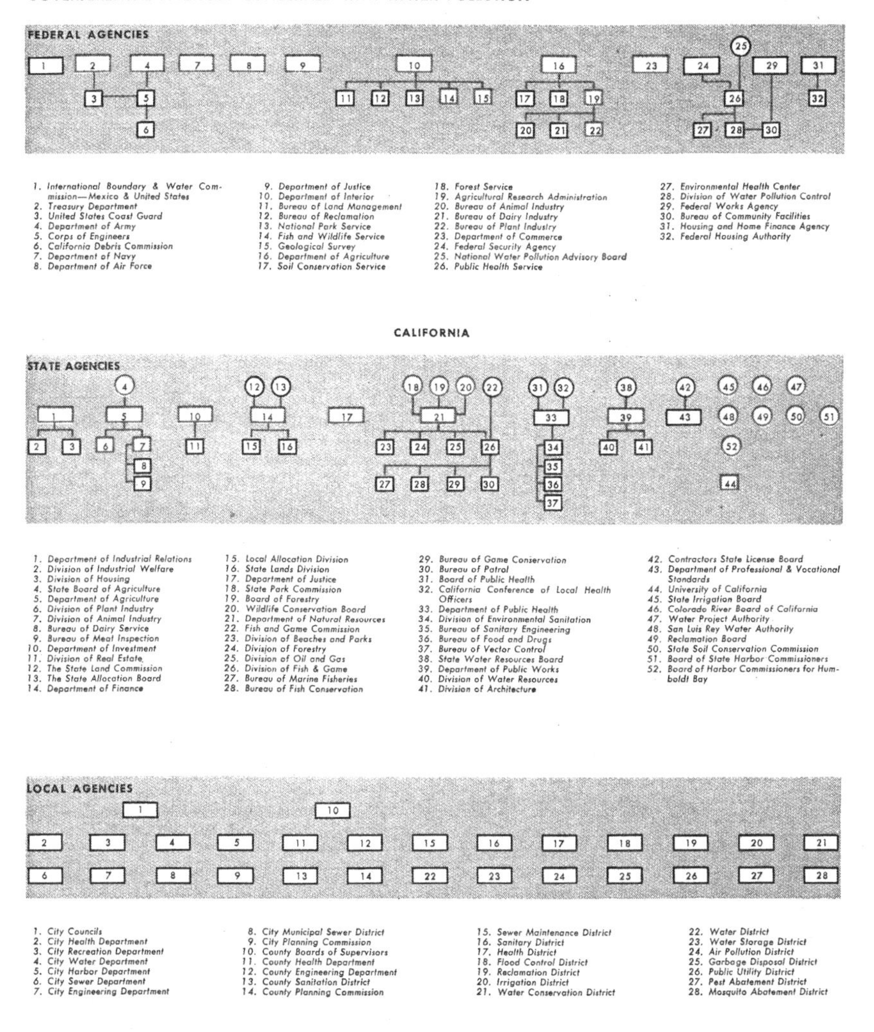

Many governmental agencies have direct responsibility or important indirect concern with problems of water pollution or waste disposal. This has been a logical development based on specialization and on geographic jurisdictions, but it has resulted in serious overlaps and gaps in authority and interest. From the standpoint of a city or industry, or a citizen with a complaint, the absence of any coordinating body or "clearing house" for the many conflicting interests is bewildering.

Figure 5.3. Governmental agencies concerned with water pollution. Randal Dickey, "Interim Fact-Finding Committee on Water Pollution," 33. Assembly of the State of California, 1949.

Table 5.1. Data on manufacturing in Riverside, San Bernardino, and Orange Counties

Counties	Number of establishments		Production workers		All employed	Value added by manufacturing
Year	1939	1947	1939	1947	1947	1947
Riverside	75	140	1,480	3,435	4,414	$27.6 million
San Bernardino	113	256	2,487	8,703	11,186	$75.8 million
Orange	117	299	2,083	4,723	21,700	$31.2 million
Totals	**305**	**695**	**6,050**	**16,861**	**37,300**	**$134.6 million**

Thus legislators not only had to contend with the contamination of surface waters by domestic sewage but also increasingly fretted about the "great wartime expansion of processing industries . . . and possible pollution of underground waters by mineral waste."[16] While cities worried about degraded water quality, they often had no funds for waste disposal infrastructure. Along with local chambers of commerce, city leadership also typically supported the continued expansion of industrial and population growth in their region. Policy makers, too, did not generally regard industrial waste as a hazard to public health and saw contamination by chemical waste as a "remote possibility."[17] Meanwhile, the combination of overlapping agency jurisdictions, historically lax enforcement of permit conditions by the state Department of Public Health, and, later, competing requirements by public health department officials to achieve generally stated results hampered new industries from addressing the problem of waste disposal altogether.[18]

In 1947 the California State Assembly acted on these concerns with the passage of House Resolution 27 and establishment of the interim fact-finding committee on water pollution. Chaired by pro-business Republican assemblyman Randal F. Dickey, this committee produced a report calling for the overhaul of the regulations that had allowed for "occasional outbreaks of intestinal disease from sewage-polluted public water supplies, unsanitary conditions" in recreational waters, and "the local degradation of the quality" of "underground water supplies."[19] The report also led to the passage of the Dickey Water Pollution Act in 1949. The act addressed one obstacle for water quality management in California: unclear authority structures. It established nine regional quality control boards to develop and enforce policies surrounding water quality—including those of surface, groundwater, and saline sources. These boards represented local economic interests and local governments and were collectively overseen by the centralized State Water Pollution Control Board. This structure was an extension of the seven watersheds outlined by the fact-finding committee's report and reinforced the need for context-specific water quality determinations.[20] The division of California into these jurisdictional basins acknowledged the fact that the use of water was (and still is) concentrated in the south, while most water resources are found in the north. More importantly, in creating the regional water boards, the Dickey Act separated jurisdiction over the economic benefits of waste discharge from control over effects on public health, which remained with the state Department of Public Health and local health officers. The act also eliminated the permit system and, instead, limited control to end results, with waste dischargers left to attain the prescribed results in a lawful manner.[21] As a result, permits issued by the Department of Public Health under the old law lost their effectiveness as privileges or grants of authority.[22]

The fact-finding committee's parsing of jurisdiction over water pollution—between economic versus public health impacts—held significant import for how the subsequently formed regional boards would go on to view and manage industrial wastes as an economic nuisance rather than a significant threat to public health. Three factors contributed to the minimizing of concerns about industrial waste and, consequently, the overall laxity in its management. First, both nationally and locally, the dominant focus and understanding of threats to public health prior to and continuing through this period were those posed by sewage and bacteriological contamination.[23] As historian Martin Melosi has documented, concern about pollution from the Progressive Era onward focused primarily on sewage and the degradation of water quality by bacteria. Only after World War I were industrial wastes thought to negatively impact water quality, and, even then, they tended to be regarded as an annoying nuisance at most. In some cases, such wastes were even "considered a germicide that, when added to water, would inhibit the putrescibility of organic material."[24]

This predilection, to frame industrial pollution as a nuisance with economic implications rather than as a public health hazard, was evident in the hearings held by the fact-finding committee. Attempts to recognize the potential health threats of industrial waste, moreover, were dismissed. During a December 1948 hearing in Los Angeles, for example, sanitary engineer E. A. Reinke advocated for a more nuanced approach, which sought to account for the context, content, and effects of domestic *and* industrial waste.

He argued for a less clear-cut separation between the categories of domestic sewage as public health menace and industrial waste as economic nuisance. Reinke noted,

> I don't see how you can separate it. . . . I think you would have a great deal of difficulty within an industry separating those things that you would define as related to health from those things you would define as related to economics. I may be dense, but I simply can't see the distinction. . . . Well, if I might comment on just one thing that is rather recent in public health reports. For many years we have considered nitrate in water as being just an evidence of old pollution or natural nitrate formation in the ground of little sanitary or public health significance. In the last two years, it has been definitely established that nitrates in drinking water, if used in making up the formula for infants under the age of six months, does cause blue babies, and is directly a health problem and that is why I say I think it would be very difficult to distinguish what is a nuisance from what is a health menace.[25]

Reinke's testimony was a clear attempt to highlight the uncertainty behind scientific data on water quality, arguing against legislators drawing hard lines between what should be considered a nuisance versus a health hazard. Despite his testimony, the fact-finding committee noted that Mr. Reinke's perspective ran contrary to that of the majority of witnesses (most of whom represented industrial interests) and recommended that "this distinction [between economic and health impacts] *can* be made on the basis of *reasonable* appraisal of the technical facts involved, and that it *must* be made in order to assure real progress in pollution abatement."[26]

The committee's exchange with Mr. Reinke pointed to a moment in which meanings and definitions of water quality were being contested and renegotiated by legislators. Rather than viewing all waste as potentially hazardous to public health, committee members attempted to distinguish between more and less harmful types of effluent. The fact-finding committee thus began with a broad question: what is pollution? According to their investigations, pollution emanated from a range of sources, including silt erosion, minerals, organic wastes, petroleum, sewage and drainage, industrial waste, agricultural waste, navigational barriers, recreational waste, and sea water intrusions. Whereas committee members continued to maintain historical regard for domestic sewage as "contamination" and a threat to public health, they categorized industrial waste as either "pollution" or "nuisance."

For the committee, pollution did not directly affect public health, but it did make water use for domestic, industrial, agricultural, navigational, and recreational purposes more difficult and potentially more costly. Pollution could thus inflict economic damage. Committee members also emphasized their difference from engineers in their desire to distinguish domestic from industrial waste. Engineers, their report explains, framed industrial waste in comparison to domestic sewage, converting "the organic portion of the industrial wastes into a number which is considered the equivalent strength of domestic sewage."[27] Doing so, the legislators believed, overdramatized the threat industrial wastes posed. In this way, the committee agreed with industry stakeholders, who mobilized to argue that industrial waste did not constitute a public health threat. In their summary of findings and conclusions from studies on industrial waste disposal practices, the California Association of Production Industries stated a position echoed in the final recommendations of the fact-finding committee itself: "Since industrial wastes, by and large, do not in themselves constitute a health hazard, they should be considered separately and realistically from the standpoint of the economic factors involved, rather than rigidly controlled by laws designed to prevent the spread of disease."[28] Defining waste categories—as contamination, nuisance, or pollution—and institutionally separating the management of sewage from industrial wastes thus became central to shaping the greater or lesser regulation and scrutiny over various streams of effluent. The water boards that followed on the heels of the fact-finding committee report were thus structured from the start to manage industrial wastes as a form of economic nuisance rather than a public health threat.

Second, although legislators were concerned about degraded water quality and frustrated with the exceeding complexity of jurisdiction over water's management, they largely blamed government for this rather than industry. And the inaction of state agencies in the realm of water pollution control certainly provided ample fodder for legislators' chagrin. The Department of Health, for instance, was both culpable and roundly critiqued for being tacit in its consent of no or minimal regulation of (industrial) waste disposal. Between 1907 and 1947, it had issued only thirty-four permits to industries, three-quarters of which were for sanitary wastes. Committee members further chastised the Department of Health for the significant time lag between when permits were issued and when any follow-up was conducted. For one permit issued by the Department of Public Health in El Segundo in 1921, the follow-up inspection only occurred in 1947, "the day on which

[the] committee was hearing testimony in Los Angeles with regard to that particular disposal."[29] Here, committee members perhaps rightly blamed various agencies at federal, state, and local levels for failing to "secure any final statement of what conditions" needed to "be met or of what waste disposal facilities" needed to be installed.[30] They clearly suggested that state—rather than private and industrial—interests were to be faulted for the consequences of poor wastewater management, and the less than stellar record of agency performance helped fortify their position. Solutions to managing waste and water pollution thus became aimed at reforming a convoluted state bureaucracy and ensuring greater predictability and ease of doing business for private industries.

Legislators' imperatives also reflected political desires for less burdensome regulation. As it was, governmental interests at the state level did not pressure dischargers to follow through with their obligations around registered permits. As a result, disposal practices went largely unchecked and the default was local control over sewage waste in the state's waters. Given strong political sentiments against overregulation, the fact-finding committee's approach to water pollution was oriented toward cooperating with industries and supporting economic growth.[31] Indeed, committee members explicitly supported economic growth and regarded confusing authority structures and inefficient regulations as barriers that "hampered or prevented" industrial development.[32] They noted that waste disposal was a necessity. In the hearings that informed the report, fourteen industrial groups (representing companies responsible for "fruit and vegetable processing, meat packing and fish canning, milk processing, sugar refining, paper manufacture, oil producing and petroleum refining, mining, and lumbering") met with the fact-finding committee to discuss their concerns about waste disposal.[33] The committee assured them that their intention was not to "harass or intimidate industries" and that their waste could largely be described as only a "nuisance." "Few of these wastes are hazardous to the public health," explains the report, adding that "there is no official record of any instance where proposed industries have been formally denied permits."[34]

Although legislators aspired for more "effective or equitable" control over water pollution, there were clear tensions between these stated aims and the ultimate conclusions of their investigations into water quality control. Committee members determined that water quality was a problem requiring "exhaustive analysis, intelligent treatment and widespread education of all concerned," and outlined the ambitious goals of their inquiry: to address water pollution not just at surface water levels but also belowground water sources and meeting points between surface water flows and the ocean.[35] Despite these aspirations, their report clearly favored economic growth and dismissed centralized regulatory authority as a means to control pollution. "Effective and equitable" control, in this light, was weighted to protect the interests of industrial stakeholders. As such, committee members noted:

> It would be easy to say that a solution to any and all of these problems of pollution from industrial wastes could be found in the simple procedure of establishing minimum standards of waste treatment which would be applicable throughout the State, by issuing orders on all industries, and by peremptorily closing those industries which did not comply. In the opinion of some persons, this is the answer to the problem, both with respect to correction of existing conditions and preventing of further water pollution. It is obvious, however, that such an attempt to correct overnight the conditions which have been developed for one hundred years, while it might greatly simplify the administrative duties of bureau officials, would result in inequities and restrictions which would severely affect the economy of the State.[36]

Because committee members viewed overregulation as a threat to the equitable treatment of industry, they effectively soft-pedaled regulation. Instead, their solution to industrial pollution control consisted of active cooperation with industry and streamlining onerous regulation: "Successful abatement of existing pollution from industrial waste, and successful control of preventive measures, will be realized only though a whole-hearted cooperative effort between the industries and all of the governmental agencies concerned. Real and continuing progress will be made under laws which make such cooperation not only possible, but which make it easy."[37] Contrary to the comprehensive investigative approach claimed by the fact-finding committee, legislators displayed a willingness to trust industrial miscreants, or forgive them for pollution practices. In the following passage, for example, egregious offenders such as the City of Vernon, whose very government was controlled by industry, and which had a record of openly flouting regulation, was described in a benign light:

> During the course of your committee's investigation this State has experienced its first real progress toward solution of its industrial waste disposal problems. This has been the result of bringing together all of the affected in-

> dustries, the governmental agencies, and all of the other interests concerned, for full discussion of the problems presented. Out of the interest stimulated by the committee's work, substantial progress has been made by the various industrial groups in study, analysis and planning of their common problems of waste disposal... This has been accomplished with a minimum of compulsion under any state or local law, and in fact the record was marred only by the *unwarranted resort to threats of 'crackdown'* by the State Department of Health in the instances of Vernon, San Jose, and Long Beach.[38]

For the report's authors, cooperation with industry and industry self-regulation became the solution to managing industrial waste; any government intervention was portrayed as "unwarranted crackdown," an overstepping of state authority in its efforts to protect public health. In the end, the fact-finding committee would sum up their studies and analyses of water pollution by recommending that laws and procedures for the protection of public health be *strengthened.* The report went on to argue for the "repeal of the faulty and antiquated state permit law, and enact[ment] of practical provisions to place responsibility for protection of the public health squarely upon the health authorities."[39] Yet, these recommendations were only aimed at waste defined as "contamination," and thus limited to regulating domestic sewage. By contrast, for the prevention and control of pollution deemed as "nuisance"—the category into which industrial waste fell—committee members recommended "establishing a mechanism to bring about *coordinated* and *cooperative* action and planning at local or regional levels, where the greatest number of affected interests are, and where the moral and legal responsibilities lie."[40] For California's regional water boards, this meant a pro-industry orientation from their inception. In general, then, legislative committee members favored easing water pollution enforcement to facilitate industrial waste disposal for economic benefit, leading one commentator at the time to note that the Dickey Act "would hardly appear to be the proper solution to the pollution problem."[41]

Third, and finally, water pollution control in California, as elsewhere in the United States, primarily focused on surface water quality and the monitoring of stream discharges.[42] Yet, committee members had key opportunities to recalibrate attention to groundwater, relative to that of surface waters. In one premonitory instance, Mr. Paul Bailey, consulting engineer for the Orange County Water District, warned committee members "that the most serious threat of future pollution lies in a slow increase in dissolved substances in underground supplies over a considerable period of time . . . and that the change might take place so slowly that it would not be noticeable until serious damage were done and remedy made difficult, unless special observation is adequately organized in advance."[43] Bailey was painfully aware of the implications of poor water quality for downstream users in Orange County and had the foresight to recognize the value of clean groundwater basins for a growing population.[44] However, Bailey's testimony came during a period in which the region was not yet built out and still had relatively abundant water supplies. In this context, his plea for vigilance around groundwater quality and its protection from industrial wastes fell on deaf ears.

Apart from Bailey's appeal, there were few references to groundwater quality in the interim fact-finding committee's report, and legislators noted that "very few instances of any widespread underground pollution . . . have been found."[45] While this may have accurately reflected the existing conditions at the time, it did not account for the increase in industrial waste during this period, a situation that the committee itself noted during its investigation. Moreover, when committee members did acknowledge industrial pollution of groundwater, they continued to maintain that "few of these wastes [were] hazardous to public health."[46] They were also quick to minimize any public health impacts of documented instances of industrial contamination:

> The danger to humans from poisonous wastes of industry is always a possibility. Specific chemicals may be toxic to humans in relatively low concentrations. Radioactive materials may some day be such a problem. However, from all the testimony received this danger appears to be remote. . . . The only instance cited to your committee of contamination by toxic wastes concerned a discharge of metal-plating waste, which was said to have increased the chromium content of water in a nearby well of the City of Los Angeles beyond allowable standards. Yet even in this case, although all agreed that the wastes should not enter the ground, the local health officer did not consider that an actual health hazard has resulted.[47]

In this way, the contamination of groundwater in California, like groundwater more generally, stayed out of sight and out of mind for regulators, including those who would soon work at the regional water boards.[48] In the following section, I chronicle how these factors shaped the Regional Board's management of the Stringfellow Acid Pits, the inland region's first industrial waste dump and later, tragically, the state's first Superfund site.

FINDING A DUMP

In creating regional water boards, the Dickey Act supplanted an approach to water pollution control by way of a permit system run by the state Department of Public Health. The newly minted boards also debuted during a period of rapid postwar industrial growth, related rise in federal government expenditures on aircraft construction, and largely unregulated waste disposal in the Golden State. By the early 1950s, the Cold War had further entrenched aircraft manufacture in the Southland and also added the fabrication of electronics and missiles to its repertoire. Los Angeles County had by far the largest share of prime military contracts in the seven counties of Southern California, accounting for 2,633 establishments, 54 percent of total state awards, and 12.5 percent of total national awards in 1960.[49] Termed the "aerospace-electronics-industrial complex," this sector was structured around hubs of major aircraft assemblers like Lockheed, McDonnell-Douglas, Northrop, Rockwell North American, and General Dynamics Convair, specialized missile manufacturers, electronics firms, and a vast agglomerated network of local contractors and subcontractors who served as suppliers of parts to these larger corporations.[50] Parts suppliers within this complex included firms involved in nonferrous foundry work, machining, casting, plastics molding, and electronics, among others.[51] Missile manufacturing alone involved numerous steps, many of which were carried out by smaller operators. In addition, each step of the manufacturing process required a variety of resources and generated unwanted industrial by-products.

In the Santa Ana River Basin, which spanned parts of San Bernardino and Riverside Counties, and drained through Orange County into the Pacific Ocean, the need for waste disposal was particularly urgent. In 1947 industrial activity in the Basin was second in importance to agriculture, and the total value added to products by manufacturing was $134,579,000 (see table 5.1).[52] The largest single industry in the Basin was the Kaiser steel plant near Fontana. Other important industrial activities in the area included mining, petroleum production, making of Portland cement, manufacturing of canning and packing machinery, manufacturing and repairing of transportation equipment, and the manufacturing of chemical and metallurgic products. Together, these various firms generated an unknown amount of waste, most of which went unregulated.

A 1951 study undertaken by the California Division of Water Resources,[53] at the behest of the then recently formed Regional Board, found that of the 140 known industries in the Basin, some discharged industrial waste directly into the Joint Outfall Sewer in Orange County. The California Division of Water Resources further noted an absence of "active acid sludge dumps in Orange County or in the Santa Ana River area" and concluded that oil field wastes generated from major operations in Seal Beach, Huntington Beach, Newport, West Newport, East Coyote, West Coyote, and Brea-Olinda (all in Orange County) were simply being dumped, untreated, into local sewers.[54] Meanwhile, industries in the inland counties of Riverside and San Bernardino, which had no industrial dumps, resorted to a variety of waste disposal practices (see figure 5.4). A few pursued sea dumping via already existing sewage plants. The Hunter Douglas Company in Riverside, for example, trucked their high fluorine, chromium, and arsenic wastes to the LA Hyperion Sewer, and, in Ontario, the General Electric appliance factory hauled chromic acid wastes to Anaheim for sea disposal through the Orange County Joint Outfall Sewer.[55]

Trucking wastes to distant sewage treatment plants, however, incurred additional expenses. Many manufacturing operations thus saved on costs by simply dumping industrial effluent into local sewers or rivers. A survey of pollution in the lower San Gabriel River by the California Department of Public Health, for instance, found three out of fifteen industries in that part of Los Angeles County to be discharging wastes into the tidal waters of the San Gabriel River. In 1950 the Dow Chemical Company operation to recover iodine from oil field brine was discovered to be releasing "wastes with a population equivalent of 27,700 and the Santa Fe Springs Waste Water Disposal Company was discharging wastes with the population equivalent of 83,500."[56] Further inland, a number of firms in Riverside and San Bernardino Counties resorted to neutralizing acid waste prior to disposal in percolating beds.[57] The hope, perhaps, was that these ostensibly neutralized wastes no longer posed a threat to groundwater supplies. Yet, at least one firm was cognizant of the possibility of groundwater contamination by industrial waste. The Aluminum and Magnesium Company in Corona hauled their acid wastes to Amboy in the remote Salton Sea region, where groundwater was already saline.[58]

Although there were few documented instances of industrial pollution of groundwater during this period, officials did express caution. The

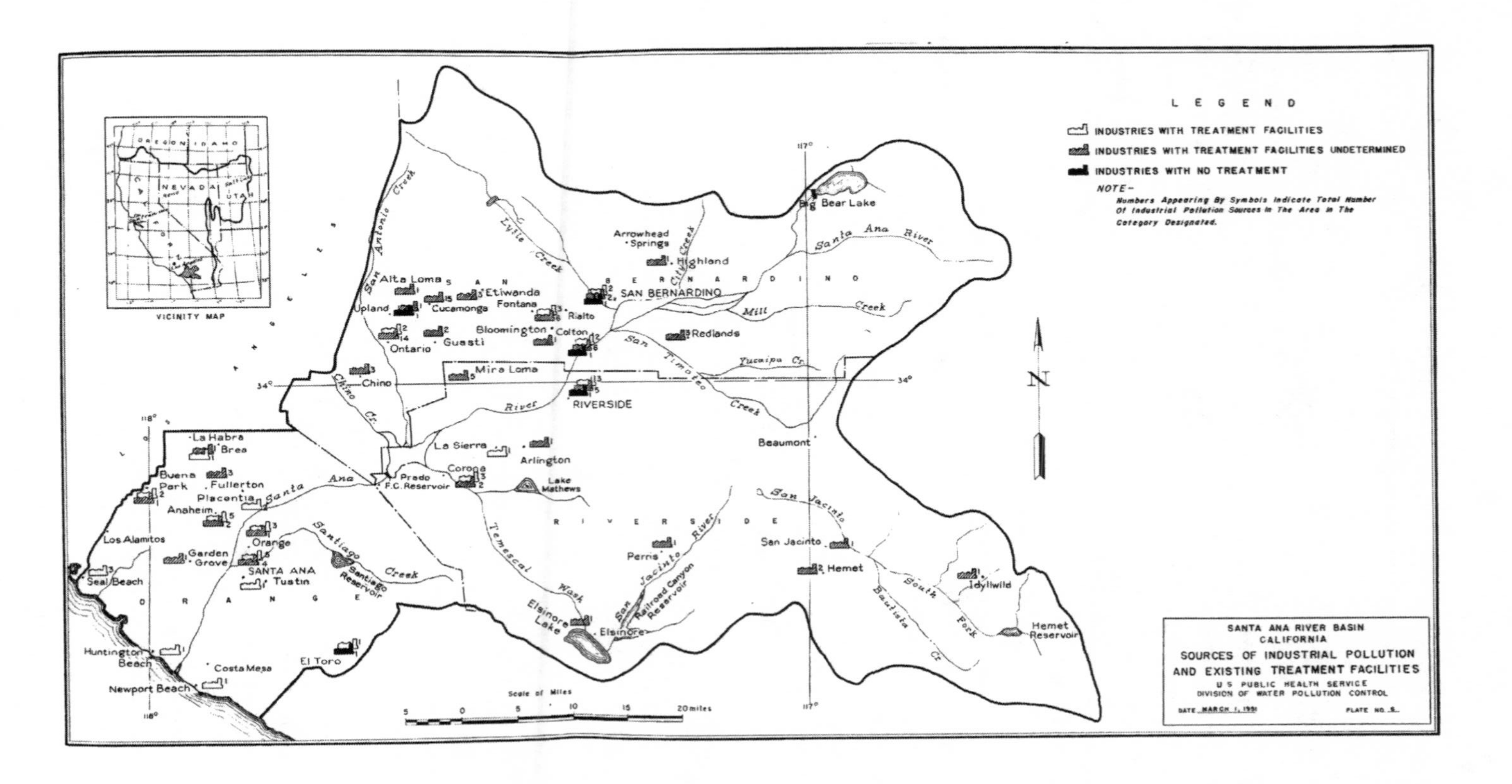

Figure 5.4. Sources of industrial pollution and existing treatment facilities in the Santa Ana River Basin, 1951. Federal Security Agency, *Report on Water Pollution Control: Santa Ana River Basin*, Water Pollution Series, Public Health Service, March 1, 1951. Plate 6.

Culligan Zeolite Company in San Bernardino, for example, was cause for considerable worry among agency officials and water users in the area. In their pollution control investigation, the state Division of Water Resources described the sodium carbonate wastes generated by Culligan Zeolite and advised that test wells be constructed to monitor for potential contamination of drinking water wells by these salt wastes.[59] Similarly, officials acknowledged the potential impacts to well water quality from aboveground dumps in Riverside County:

> There have been few cases of wells in San Bernardino and Riverside Counties having to be abandoned due to pollution, but records show significant increases in the mineral content of some well waters in recent years. Samples collected from shallow test wells downstream from the Riverside City Dump show total solids 4–5 times and sodium tests 10 times as great as in samples collected upstream from the dump.
>
> Although State and local agencies have much general information on the pollution prevention measures that have been placed in effect by industries, there is not enough factual data on the design capacities and population equivalents to give the complete picture.[60]

Riverside city and county officials anticipated industrial development in the areas of West Riverside, Corona, Elsinore, and Hemet but worried that the absence of an approved Class I dump for industrial wastes in their region would thwart growth.[61] As already noted, industries in the inland counties of San Bernardino and Riverside had limited options for waste disposal: they either hauled their wastes to authorized dumps in Orange and Los Angeles Counties—a practice that drove up their costs—or resorted to illegal means of waste disposal, a clearly risky strategy.[62] For officials in San Bernardino County, the need to find an appropriate Class I dump was further exacerbated by implicit worries about groundwater contamination from industries in Fontana, Colton, and San Bernardino, all of which were situated on "permeable alluvial fill and underlaid by ground water basins."[63] Industrial and business interests in the inland counties thus sought out the newly formed Regional Board to take the lead in finding a dump. In turn, the Regional Board requested that the California Division of Water Resources explore possible regional dump sites for industrial waste.[64]

In June 1954, state engineer A. D. Edmonston delivered the "Region-wide Municipal Industrial Dump Site Study, Santa Ana Region" to Paul G.

Brown, executive officer of the Regional Board. Overseen by Max Bookman, the principal engineer for Southern California, experts in hydraulic, civil, and geological engineering worked on the investigation. Following extensive meetings with interested county and city officials, and representatives of major industries, nine possible dump sites for toxic liquid waste were explored by Bookman's team: one in Orange County, four in San Bernardino County, and four in Riverside County. Otherwise known as the Bookman Report, the initial selection of sites was guided by geological expertise. Geologists theoretically favored areas above "non-water-bearing" or impermeable rock formations, which they considered to serve as a protective barrier between industrial wastes and underground supplies of water.[65] In contrast, sites located over water-bearing (or porous) formations were deemed inappropriate for the depositing of industrial wastes. Geologists clearly prioritized the integrity of water quality and water supply in their investigations and noted that the water-bearing series[66] "absorbs, transmits, and yields water readily to wells . . . The nature and importance of these pervious sediments as a source of water supply make them unsafe for the disposal of liquid or soluble pollutants. This is particularly true of the alluvial cone area of Upper Santa Ana Valley and the Santa Ana Forebay."[67] These geological realities ultimately ruled out most of Orange County and the Upper Santa Ana Basin from consideration for a Class I industrial dump.

Among the sites identified as feasible was an "abandoned crystalline limestone quarry excavated near the peak of a hill" in the southwestern portion of the Jurupa Mountains of unincorporated Riverside County.[68] First geologically surveyed in 1951 by Edward MacKevett of the California Institute of Technology, the Jurupa Mountains consisted of widespread siliceous metamorphic rocks with an abundance of gneiss and schist in particular.[69] Although MacKevett's study only made passing mention of existing hydrogeology, his observations about jointing (or fissures) in rocks were stated in no uncertain terms: "Any lines of rupture that had been mapped as questionable faults in earlier years were 'probably essentially joints along which a very small amount of movement took place, and which provided channel ways for local hydrothermal solutions.'"[70] The Bookman team's assessment of the Jurupa Mountains location a few years later, however, made no mention of jointing as a possible pathway for hydrothermal flows. By contrast, the brief site description was particularly striking for its key assumptions, some of which contradicted possibilities noted in the earlier MacKevett study. For example, even though the Bookman geologists observed limestone formations with joints and fissures along which water could travel, they assumed that other impermeable rocks, such as gneiss and schist, would prevent flows from reaching the groundwater below. Yet, it is unclear whether these impermeable rocks (schist and gneiss) were actually present. The reference to belief versus observed fact is evident in the Bookman Report's description of the site:

> It is not known to what depth the joints extend. It is believed that the limestone lenses out against schist or gneiss. Therefore, it is not believed that fluids could move from excavation into the permeable settlements of upper Santa Ana Valley even though they followed the joints in the limestone. . . . If an impermeable barrier were placed across the opening of the western side of the excavation and the dumping operations were controlled to prevent overflow, it is believed that waste disposed at the site could not reach waters of the State and, therefore, would have no deleterious effect.[71]

Despite incomplete hydrogeological knowledge about the former quarry site, state and county institutions moved forward in pursuit of the Jurupa Mountains option. On March 8, 1955, the Boards of Trade of San Bernardino and Riverside Counties pressed the Regional Board to further investigate the feasibility of an industrial dump at the privately owned Stringfellow Quarry.[72] A similar request followed a few weeks later from quarry owner J. B. Stringfellow himself.[73] Faced with growing economic pressure to have a Class I site[74] inland, the Regional Board asked engineers and geologists at the state Division of Water Resources to conduct further site reconnaissance. In early June, state engineer A. D. Edmonston responded to the Regional Board and merely repeated the Bookman Report's assertions about site impermeability. Addressing the Regional Board's executive officer Mr. Paul Brown, Edmonston wrote:

> The area proposed as the dump site is particularly well adapted for this use because of a natural barrier, or dike, located about one-quarter mile south of the head of the canyon. . . . It is composed of well-cemented gravels, cobbles, and boulders and appears to be impermeable. . . . Since the proposed site is underlain by impermeable bedrock, it is believed that if the aforementioned barrier is extended to the west wall of the canyon in the manner previously described, and is constructed so as to be completely watertight and thereby confine liquid wastes, the operation of the site for industrial wastes will not constitute

> a threat of pollution to waters of the State provided that surface runoff is not allowed to carry wastes into natural watercourses or drainage channels.[75]

Partly because technocrats minimized risks of groundwater contamination, they failed to conduct hydrogeological studies to verify the integrity of bedrock below the Stringfellow Quarry.[76] Faced with potentially competing sets of interests (to protect waters of the state on the one hand and to enable economic growth by making way for industrial waste disposal on the other), various government officials ultimately reached a compromise of sorts; they reinforced the protection of surface waters but relegated the fate of groundwater to the margins and recommended the siting of a Class I dump site.[77]

MANAGING STRINGFELLOW

In September 1955, the Riverside County Planning Commission granted J. B. Stringfellow a zoning variance (Variance 66) and approved the first Class I industrial dump in the inland region.[78] Variance 66 was not a blanket approval, however. It stipulated a number of provisions, including the establishment of stormwater diversion channels; approval of waste, prior to its dumping, by the county health department and Regional Board; and adequate posting of signs and fencing of the site.[79] Yet, dump operations favored industrial interests through a systemic lack of oversight.

On August 31, 1956, the first toxic wastes released at the site came from Norton Air Force Base.[80] The next month, Hunter Douglas followed suit with more military waste, and the pace of dumping increased from June 1957 on. The primary method of waste disposal was to release toxic liquids into a series of ponds and to combine this with "spray evaporation to accelerate volume reduction. As wastes accumulated in the lower ponds due to leakage of the upper berms, these wastes were pumped back up into the upper ponds."[81] During the first decade of dumping, Regional Board records are scanty and fail to provide detailed accounts on what was released or how much. Moreover, an examination of the waste approval process indicates that government agencies abdicated their regulatory authority and ensured continued dumping by industry at Stringfellow. The approval process was superficial at best and most often consisted of a phone call and verbal okay from officials at the Regional Board. Some startling examples from the Regional Board's files reveal the extent of the cavalier approval process and limited nature of visual inspections:

> July 17, 1957: "Eberhart called that N. American wished to dump a mixture of Cr plus 6 acid BlackOxide ph-12.0 and NaNo3. Black oxide a caustic acid used in black anodizing—gave verbal approval."
>
> September 25, 1957: "Inspected dumpsite—uppermost pit had been cut with a dozer. Swath was 3' deep and showed penetration to that depth—color of soil at 3' only faint and percolation *probably* stopped within."[82]

Eventually, the experimental, wait-and-see-what-happens character of Regional Board authorities' advice to dumpers—such as chronicled in the following entry—became the status quo. "August 22, 1963: 'Marge called re: permission to dump fuming HNO3. I told her to have them release slowly and *maybe fumes might not be too bad*.'" By 1969 even the minimal efforts of verbal approval proved too cumbersome and the Regional Board formally accelerated the approval process by amending Resolution 55-11 to include a "list of wastes for which blanket approval could be given."[83] In sum, the hands-off management approach of the Regional Board enabled the unburdened dumping of wastes and implicitly supported industrial growth in the region.

"YOU COULD DRINK IT"

Toxic wastes dumped at Stringfellow had an immediate impact on neighbors in the vicinity, but government officials—from both the Regional Board and Riverside County Department of Public Health—almost always dismissed these complaints. The state's lack of response marked the beginning of a pattern in which officials would protect toxic waste disposal from overregulation, excessive scrutiny, and additional costs. Officials further minimized the risk of industrial toxins to public health by viewing people's concerns about this waste in the category of "nuisance."

As early as July 1957, one neighbor complained of "odor and water pollution."[84] A month later, the same individual followed up with concerns about "flowers dying." When Regional Board officials called a Stringfellow employee to discuss such complaints, he simply assured them that "no dumping other than approved" had taken place.[85] In mid-December, a neighbor raised concerns about a dog being injured by chemicals from the site. An analysis of samples discovered dog hair "burned by a soluble caustic material which was in solid form in hair and flesh . . . and could have been [caused by] ashes."[86] Although this information was shared with the property owner at the time, it did little to raise officials' concerns about the dump itself. In 1963 one resident noted that fuming nitric acid fire at the pits resulted in fumes "so strong that it swept into our bedroom, causing us to choke and gasp for breath. The smothering fumes terrified us and we called the sheriff. We might have been more frightened if we had known that nitric acid causes pulmonary edema and nephritis."[87] This was not the only documented fire caused by chemicals at the dump. In February 1971, for example, heavy smoke was noted at the pits and only resulted in a phoned-in report to the Riverside County Department of Public Health, with no indication of additional follow-up.[88]

Although Glen Avon residents voiced worries about the Stringfellow dump since its inception, it would take more than a decade for local and regional government officials to heed community concerns. In the spring of 1969, heavy rains led to the brimming over of several toxic ponds, yet regulatory authorities expressed little, if any, regard for public safety and welfare. Testifying about the incident later, J. B. Stringfellow would recall that a staff engineer from the Regional Board had told him that he was "a worry wart; if you lost the whole thing [toxic water from ponds] *you could drink it.*"[89] In March, the holding ponds finally overflowed and washed out a section of the containment dike on site. Stormwater carried toxic effluent down Pyrite Creek and across Highway 60 into the community of Glen Avon to the south.[90]

The storms of 1969 irrevocably demonstrated that off-site toxic contamination could occur through surface runoff. Although Stringfellow made structural improvements to the site, small springs were observed bubbling to the surface by the dam despite below average levels of rainfall.[91] Further investigation found that the springs contained chemicals from waste disposed in the pits.[92] More alarming still, the Glen Avon school well, located almost two miles downstream from the dump, showed traces of hexavalent chrome.[93] Yet, disposal operations at the site continued in the same, largely unregulated manner.

By the summer of 1972, growing concerns about water contamination finally compelled the Riverside County Board of Supervisors to request a study of dump operations in relation to the conditions set forth by Variance 66.[94] Over the next several months, various local and county agency officials visited the dump and filed their respective observations. While many of their reports documented problems with dump operations or infrastructure, these findings did not coalesce in a move to close Stringfellow.

The observations and recommendations of Harold M. Erickson, director of the Riverside County Public Health Department, characterize the united front of local agencies in supporting the dump as well as their deference to the supervisory authority of the Regional Board: "On the many inspections made by this Department, at no time were conditions such that there would appear to be any condition which would adversely affect the public health or create a public health nuisance. . . . [The Regional Board conducted] a close supervision over the disposal of industrial wastes at this site with regard to probable ground water pollution or nuisances. We believe that the industrial disposal site has been operated in accordance with the requirements [set forth] by the Water Quality Control Board."[95]

Ultimately, it was not local government agencies but rather J. B. Stringfellow who, reluctant to keep incurring costs to maintain the site, voluntarily closed the dump at the end of November 1972.[96] Stringfellow's unprompted closure at the end of 1972 left manufacturing firms in the region scrambling to find other avenues through which to dispose of toxic wastes.[97] For the pro-industry regional and state water boards, this meant engaging in concerted efforts to reopen the dump. Yet, this period also marked the first time that some institutions (e.g., the Department of Water Resources, as noted in the following paragraph) broke rank with the unified, lockstep government support for the dump and began more urgently voicing concerns about toxic contamination and public welfare.

FINAL CLOSURE

As the Regional Board continued to advocate for Stringfellow, the California Department of Water Resources (DWR) dissented for the first time. No longer were DWR geologists content with unverified assumptions about site impermeability (e.g., those penned by their predecessors in the 1955 Bookman Report), and they were willing to call out their counterparts at the Regional Board. After his field visit to the dump, for example, district engineer John M. Haley expressed quite a different interpretation of the cracks and fissures he observed in plain sight: "Because of the fractured and jointed nature of the bedrock, we are concerned about leakage of Group 1 wastes that may have escaped, or could escape, from the ponds: through or under the dam; and/or through fractures and joints in the bedrock under the ponds and dam. Any leakage could then move southward and affect beneficial uses first in the Glen Avon Heights area."[98] Addressing James Anderson, executive director of the Regional Board, Haley critiqued the Regional Board for not following up on Stringfellow's site improvements and accepting at face value the geological and structural integrity of such work: "We have no information that the activity [sealing of cracks] was inspected and certified by a registered civil engineer or registered engineering geologist. Because of the severe consequences of possible leakage, we suggest that you [Regional Board] require the engineering certification."[99] Finally, John Haley refused to ignore the substantial documented observations of cracks and fissures at the Stringfellow dump. He asserted that these geological features constituted evidence of permeable bedrock that might become potential paths for groundwater contamination. More importantly, Haley questioned the unverified assumptions (about site impermeability) made by Regional Board officials and asked that their statements be changed in light of evidence to the contrary. Yet, the DWR could only go so far to protect public safety and welfare. Because it did not have regulatory authority over Stringfellow, Haley and the DWR were able to make suggestions, not mandate alternative actions.

In light of growing scrutiny over the Stringfellow operation, the Regional Board could no longer afford to maintain the status quo. In July 1973, executive officer James Anderson announced Order No. 73-20 and established new waste discharge requirements for the Regional Board.[100] Many of the ostensibly "new" requirements, however, simply reaffirmed what had been originally laid out but never enforced. In meeting with the County Board of Supervisors, State and Regional Water Board officials explicitly forwarded a logic of inevitability to justify the reopening of Stringfellow. No longer was there any dispute about the existence of fissures, cracks, and springs at the site. Rather, the rationale became "that Group [Class] 1 disposal sites all over the state leaked."[101] State and Regional Board regulators thus attenuated the public health alarms sounded by the DWR by maintaining that technical fixes were sufficient to address such concerns.[102]

Faced with growing local opposition to the dump and breaks in the agency ranks of support, the Riverside City Planning Council finally revoked Variance 66 on February 25, 1974. Unsurprisingly, Mr. Stringfellow, who had spent over $100,000 on site improvements requested by the Regional Board, appealed the revocation order. On April 18, 1974, the West Area Planning Council of Riverside County held a special hearing on the matter. The meeting saw residents who lived near the acid pits again testify about illness and physical unease that seemed to have let up only once the site closed. Meeting minutes capture their experiences as follows:

> Shirley Heel said she had headaches, except for about the past year, constantly. The last year the headaches decreased considerably. She said she would just feel more comfortable if the quarry was not there. She said she was aware of an odor in the air but was not aware of what caused the odor. The odors weren't pleasant.
>
> Mrs. Don Trust said she is a well owner in an area below the acid pits . . . she feared for the contamination of her own well. Also, the noxious fumes are very offensive. The fumes are not noticed on a daily basis, but particularly mostly when there is an inversion layer. Asked by Council member Emerson if she had noticed the fumes in the past year, Mrs. Trust said no, that the air was greatly improved.[103]
>
> Mrs. Nipp said that for the past year she hasn't had that sickish feeling in the afternoon that she used to get when the wind came up. She's lived there for seven years; and this past year was the first time she hasn't felt bad.[104]
>
> Mrs. Ethel Walter said that she has had on numerous occasions to get up in the middle of the night to try and find the source of gaseous odors. During the past year, however, they have not noticed the odors. Her asthmatic son's condition has also improved during the past year.[105]

In contrast to the ailments voiced by neighbors who lived close to the acid pits, officials from the Regional Board continued to assert that the dump did not pose a danger to public health and insisted that well contamination did not result from toxic seepage into groundwater but was, rather, a consequence of surface stormwater runoff. This line of reasoning was aimed, in particular, to alleviate growing worries about the overall integrity of the groundwater basin. As I have argued, it also kept in line with the Regional Board's tendencies to overlook groundwater contamination, minimize the threat of industrial wastes to public health, and support industrial and economic growth. Fortunately for the residents of Glen Avon, the Riverside County Board of Supervisors upheld the Planning Commission's denial of Stringfellow's permit to operate, and the pits remained closed.[106] Stringfellow went on to declare bankruptcy and defaulted paying taxes on the dumpsite property, which subsequently fell onto the Regional Board to manage. In what may only be considered a bittersweet victory, residents' long-standing concerns about surface flooding and groundwater contamination would soon be vindicated. The next drawn-out saga would be over who was responsible for cleaning up the poisoned site itself and how this would be done.

CODA

This origin story of the Stringfellow Acid Pits has sought to make clear how toxic waste is often hidden and embodies the externalized costs of industrial production and economic growth. This legacy of waste continues today with current modes of production and consumption, and has been facilitated by various projects of scientific, economic, and political rationality. The Stringfellow case—at its heart about the politics of industrial water pollution control in California—also reorients how the Golden State's "water story" is recounted, to make room for the role of waste in making place. The words of California legislator Randal Dickey at the start of this chapter still ring true, that "the limiting factor of water supply in California has been apparent to everyone. The far more limiting factor of waste disposal has been appreciated by few."[107] To this end, the political intransigence around (not) managing California groundwater up until very recently[108] may also be understood within the larger historical context of practices and policies that have served to facilitate economic growth—partly constituted by processes that generated toxic substances, that required places to contain these wastes, and that resulted in the ongoing contamination of groundwater supplies. The story of the Stringfellow Acid Pits thus illuminates the toxicity behind seemingly innocuous landscapes and underscores the reasons why such sites have been and might still continue to be produced.

ACKNOWLEDGMENTS

For their assistance in accessing archival resources, I would like to thank staff at the University of California–Riverside Special Collections and the Water Resources Institute at California State University–San Bernardino.

NOTES

Epigraph: Dickey, *Report of the Interim Fact-Finding Committee on Water Pollution*, 70.

1. Scott and Mattingly, "The Aircraft and Parts Industry in Southern California"; Scott, *Technopolis*.

2. H. K. Hatayama, B. P. Simmons, and R. D. Stephens, "The Stringfellow Industrial Waste Disposal Site: A Technical Assessment of Environmental Impact." Berkeley, CA: Hazardous Materials Management Section, California Department of Health Services, March 1979, box 11, folder 9 and folder 10. Stringfellow Hazardous Waste Site collection, MS 148. University of California–Riverside Libraries, Special Collections and Archives.

3. "Superfund" relates to the Comprehensive Environmental Response, Compensation, and Liability Act of 1980, passed in response to toxic waste disasters such as that in Love Canal, located in Niagara Falls, New York, in 1978.

4. "Acid Pits Drained Again as Part of Dam Gives Way," *Press Enterprise*, March 10, 1978, Jurupa Edition, sec. B, box 1, folder 1. Stringfellow Hazardous Waste Site collection, MS 148. University of California–Riverside Libraries, Special Collections and Archives.

5. Newman, "'It's the Pits.'"

6. For a basic site history, see California Department of Toxic Substances Control EnviroStor, "Stringfellow Hazardous Waste Site-Plume Characterization and Monitoring," accessed April 27, 2018, http://www.envirostor.dtsc.ca.gov/publicprofile_report?global_id=33490001.

7. Donald Stierman, "Chromium in Cracks: Stringfellow Fracture Study, Quantitative Contamination Analysis," March 1, 1983, box 11, folder 4. Stringfellow Hazardous Waste Site collection, MS 148. University of California–Riverside Libraries, Special Collections and Archives.

8. Dickey, *Report of the Interim Fact-Finding Committee on Water Pollution*, 13.

9. Hines, "Nor Any Drop to Drink," 201–4.

10. Layzer, *The Environmental Case*, 30; Tarr, "Industrial Wastes and Public Health"; Andreen, "The Evolution of Water Pollution Control in the United States—State, Local, and Federal Efforts, 1789–1972."

11. Dickey, *Report of the Interim Fact-Finding Committee on Water Pollution*, 33.

12. Prior to 1949, there were several administrative agencies in California overseeing water pollution, with the chief among them being the Department of Public Health. For a variety of reasons, however, this system was regarded as an admitted failure. "California's Water Pollution Problem."

13. Starr, *Golden Dreams*; Lotchin, *Fortress California, 1910–1961*.

14. Dickey, *Report of the Interim Fact-Finding Committee on Water Pollution*, 9–10.

15. Federal Security Agency, *Report on Water Pollution Control*, 13.

16. Dickey, *Report of the Interim Fact-Finding Committee on Water Pollution*, 7.

17. Dickey, *Report of the Interim Fact-Finding Committee on Water Pollution*, 71–72.

18. Hines, "Nor Any Drop to Drink."

19. Dickey, *Report of the Interim Fact-Finding Committee on Water Pollution*, 7.

20. The seven proposed regions included the North Coast, San Francisco Bay, Central Coast, South Coast—including Inland Empire regions—Central Valley Basin, Lahontan

Basin, and Colorado River Basin. Colten, "A Historical Perspective on Industrial Wastes and Groundwater Contamination," 36.

21. The permits issued by the Department of Public Health (a requirement that the Dickey Act dispensed of) "had to specify in detail both the result which the disposal device would accomplish, and how that result should be accomplished." "California's Water Pollution Problem," 650–51.

22. Moskovitz, "Quality Control and Re-Use of Water in California."

23. Hines, "Nor Any Drop to Drink."

24. Melosi, *The Sanitary City*, 96.

25. Dickey, *Report of the Interim Fact-Finding Committee on Water Pollution*, 77–78.

26. Dickey, *Report of the Interim Fact-Finding Committee on Water Pollution*, 77; emphasis original.

27. Dickey, *Report of the Interim Fact-Finding Committee on Water Pollution*, 26.

28. Dickey, *Report of the Interim Fact-Finding Committee on Water Pollution*, 170.

29. Dickey, *Report of the Interim Fact-Finding Committee on Water Pollution*, 74.

30. Dickey, *Report of the Interim Fact-Finding Committee on Water Pollution*, 8.

31. Although Dickey's pro-industry proposals eventually went through the legislature, they did not go uncontested. Opposition to the deregulation of water pollution control during this period was led by the California League of Cities, "upset by provisions overturning existing local government controls on industrial waste disposal, and by state and county health departments. They were joined by an alliance that included the Farm Bureau, water utilities, irrigation districts, and county governments" (Ross and Amter, "Deregulation, Chemical Waste, and Groundwater," 59–60). Ultimately, some requirements on water pollution control were strengthened, but they were not easy to enforce by the regional boards formed via the Dickey Act.

32. Dickey, *Report of the Interim Fact-Finding Committee on Water Pollution*, 8.

33. The committee attempted to be as inclusive as possible in gathering information on water pollution. According to the report, "every group and every governmental agency, and every person who is in any way whatever affected by or concerned with these problems, have all been invited to attend the hearings, and have been given the opportunity to present their views to the committee" (Dickey, *Report of the Interim Fact-Finding Committee on Water Pollution*, 8–9, 71). That said, business interests, including chemical producers, set up the California Association of Production Industries and proved particularly effective in getting their voices represented in the water pollution debate. Ross and Amter, "Deregulation, Chemical Waste, and Groundwater," 55.

34. Dickey, *Report of the Interim Fact-Finding Committee on Water Pollution*, 75.

35. Dickey, *Report of the Interim Fact-Finding Committee on Water Pollution*, 8.

36. Dickey, *Report of the Interim Fact-Finding Committee on Water Pollution*, 69.

37. Dickey, *Report of the Interim Fact-Finding Committee on Water Pollution*, 77.

38. Dickey, *Report of the Interim Fact-Finding Committee on Water Pollution*, 108; emphasis added.

39. Dickey, *Report of the Interim Fact-Finding Committee on Water Pollution*, 108.

40. Dickey, *Report of the Interim Fact-Finding Committee on Water Pollution*, 77.

41. Robie, "Water Pollution," 10.

42. Krieger and Banks, "Ground Water Basin Management."

43. Dickey, *Report of the Interim Fact-Finding Committee on Water Pollution*, 142.

44. As early as the 1880s, government employees such as California state engineer William Hammond Hall attempted to classify the flow of water in Southern California. Hall noted, for example, that the Santa Ana River basin was characterized by abundant groundwater, and that the river itself alternated between underground flow and aboveground flow. In this sense, Bailey was building on the work of his engineering predecessors. Hall, "Irrigation in California (Southern)." For an overview of evolving understanding about contaminant movement in groundwater, see Colten, "Groundwater Contamination: Reconstructing Historical Knowledge for the Courts."

45. Dickey, *Report of the Interim Fact-Finding Committee on Water Pollution*, 15.

46. Dickey, *Report of the Interim Fact-Finding Committee on Water Pollution*, 71.

47. Dickey, *Report of the Interim Fact-Finding Committee on Water Pollution*, 13.

48. The comprehensive management of California's groundwater only began with the passage of the Sustainable Groundwater Management Act of 2014. Blomquist, "Implementing the Sustainable Groundwater Management Act with Lessons from Existing Groundwater Management Institutions."

49. Scott, *Technopolis*, 444; Isard and Ganschow, "Awards of Prime Military Contracts by County."

50. Scott, *Technopolis*.

51. Scott, *Technopolis*, 448; Scott and Mattingly, "The Aircraft and Parts Industry in Southern California," 59–60.

52. Federal Security Agency, *Report on Water Pollution Control*, 13.

53. Up until 1955, the Division of Water Resources was housed under the California Department of Public Works. In 1956 Governor Goodwin Knight called a special session of the legislature to recommend reorganization of the state's administration of its water resources. The subsequent passage of Senate Bill 61, in July 1956, served as authorizing legislation for the California State executive branch to restructure its governance of water. Part of this restructuring saw the incorporation of the Division of Water Resources into a newly formed Department of Water Resources. For more information, see Assembly Interim Committee on Government Reorganization, "The Organization of California's New Department of Water Resources: A Report of the Assembly Interim Committee on Government Organization to the California Legislature."

54. Federal Security Agency, *Report on Water Pollution Control*, 29.

55. Federal Security Agency, *Report on Water Pollution Control*, 43.

56. Federal Security Agency, *Report on Water Pollution Control*, 29–30. "Population equivalent in waste-water treatment is the number expressing the ratio of the sum of the pollution load produced during 24 hours by industrial facilities and services to the individual pollution load in household sewage produced by one person in the same time." Wikipedia, s.v. "Population equivalent," last modified March 28, 2018, 15:36, https://en.wikipedia.org/wiki/Population_equivalent.

57. Among those industries that neutralized acid wastes before land disposal to percolation beds included the following companies: Western Tank Car Company, Portland Cement, Revere Ware, Converse Rubber Company, Culligan Zeolite Company and Norton

Air Force Base. Federal Security Agency, *Report on Water Pollution Control*, 43. Percolation beds were layered with gravel and other material to filter effluent prior to its seepage into the water table below.

58. Federal Security Agency, *Report on Water Pollution Control*, 49.

59. Federal Security Agency, *Report on Water Pollution Control*, 49–50.

60. Federal Security Agency, *Report on Water Pollution Control*, 41; Appendix plate 6.

61. "Region-wide Municipal and Industrial Dump Site Study, Santa Ana Region," 23–24.

62. In contrast to Orange County having two out of its seventeen official dumps devoted to industrial waste disposal, for example, the inland counties had not a single Class I facility. State of California, "Region-wide Municipal and Industrial Dump Site Study, Santa Ana Region," 22.

63. State of California, "Region-wide Municipal and Industrial Dump Site Study, Santa Ana Region," 24.

64. The investigation defined industrial waste as "any and all liquid or solid waste substances not sewage from any producing, manufacturing, or processing operation of whatever nature." State of California, "Region-wide Municipal and Industrial Dump Site Study, Santa Ana Region," 5.

65. Harmon, "Contamination of Ground-Water Resources"; Pickett, "Disposal of Industrial Wastes in Los Angeles County."

66. Water-bearing series refers to types of rocks that are permeable and allow for the percolation of water, often to an underground aquifer.

67. State of California, "Region-wide Municipal and Industrial Dump Site Study, Santa Ana Region," 9.

68. State of California, "Region-wide Municipal and Industrial Dump Site Study, Santa Ana Region," 29.

69. Edward MacKevett, "Geology of the Jurupa Mountains, San Bernardino and Riverside Counties, California," 5. San Francisco: Division of Mines, February 1951, box 15, folder 3. Stringfellow Hazardous Waste Site collection, MS 148. University of California–Riverside Libraries, Special Collections and Archives.

70. MacKevett, "Geology of the Jurupa Mountains, San Bernardino and Riverside Counties, California," 11.

71. State of California, "Region-wide Municipal and Industrial Dump Site Study, Santa Ana Region," 30. Groundwater basin maps created for the region in 1980 would later confirm that the Jurupa Mountain site actually rests above the Chino Basin aquifer, and its underlying bedrock was both highly fractured and perfused with underground springs. Trezek, "Engineering Case Study of the Stringfellow Superfund Site."

72. Trezek, "Engineering Case Study of the Stringfellow Superfund Site," 24.

73. Paul Brown, "Tentative Requirements for Disposal of Industrial Waste," June 14, 1955, box 16, folder 17. Stringfellow Hazardous Waste Site collection, University of California–Riverside Libraries, Special Collections and Archives.

74. Class I sites refer to landfills and dumps that are permitted to accept hazardous and nonhazardous wastes.

75. Division of Water Resources to Mr. Paul G. Brown, Executive Officer, Santa Ana Regional Water Pollution Control Board, June 2, 1955, box 16, folder 17. Stringfellow Haz-

ardous Waste Site collection, MS 148. University of California–Riverside Libraries, Special Collections and Archives.

76. Geologists and engineers at the Division of Water Resources thus deferred to untested assumptions about site impermeability despite existing knowledge in their field about the potential groundwater contamination by industrial waste. Harmon, "Contamination of Ground-Water Resources"; Pickett, "Disposal of Industrial Wastes in Los Angeles County." While it is arguable that officials did not have technologies during this period that would have made for better knowledge of hydrogeology, access to historic groundwater records dating back to 1922 indicate that geologists knew the Jurupa Mountains overlaid high groundwater supplies, and, indeed, they openly stated this. Moreover, by the turn of the century, United States Geological Survey investigators had determined that the practice of using fractured channels in limestone as sewage sinks "courted disaster," and that toxic wastes could also contaminate groundwater. Colten, "A Historical Perspective on Industrial Wastes and Groundwater Contamination." The importance of having truly impermeable barriers—and the social and economic costs of not—also had precedents in California, especially among geologists and engineers. Rogers, "A Man, a Dam and a Disaster."

77. Brown, "Tentative Requirements for Disposal of Industrial Waste."

78. Riverside County Planning Commission to J. B. Stringfellow Company, September 26, 1955, box 13, folder 6, Stringfellow Hazardous Waste Site collection, MS 148. University of California–Riverside Libraries, Special Collections and Archives.

79. Brown, "Tentative Requirements for Disposal of Industrial Waste."

80. "Handwritten Notes on Stringfellow from Santa Ana Regional Board Files," n.d., box 16, folder 9, Stringfellow Hazardous Waste Site collection, MS 148. University of California–Riverside Libraries, Special Collections and Archives.

81. Hatayama, Simmons, and Stephens, "The Stringfellow Industrial Waste Disposal Site," 6.

82. "Handwritten Notes on Stringfellow from Santa Ana Regional Board Files"; emphasis added.

83. Hatayama, Simmons, and Stephens, "The Stringfellow Industrial Waste Disposal Site," 6, emphasis added; see appendix for full list of wastes.

84. "Handwritten Notes on Stringfellow from Santa Ana Regional Board Files."

85. "Handwritten Notes on Stringfellow from Santa Ana Regional Board Files." Additional concerns voiced by Mrs. Durette in the following years were with the same dismissing attitude. In December 1960, Regional Board officials assured her that "no pollution had been found in her well, which was found to have the same water quality as the wells analyzed on the Glen Avon school in 1954 and 1957."

86. "Handwritten Notes on Stringfellow from Santa Ana Regional Board Files."

87. Kirkby, "Report to California Water Quality Control Board."

88. Stuart C. Taylor, "Complaint: Burning Trash and Dumping," February 5, 1971, box 16, folder 18, Stringfellow Hazardous Waste Site collection, MS 148. University of California–Riverside Libraries, Special Collections and Archives.

89. "J. B. Stringfellow, Jr. State Deposition," May 7, 1990; emphasis added. Hardcopy of this document is located among the files of the Center for Community Action and Environmental Justice in Jurupa Valley, California. Contact www.ccaej.org for more information.

90. "Analysis of water samples taken from Pyrite Creek by the Riverside County Health Department indicated contaminated wastewater with low pH (low pH indicates high acidity in the water). Higher levels of sulfate, chloride and total chromium were also detected in the water" ("Stringfellow Update, Issue 1: Fact Sheet, Stringfellow Hazardous Waste Site," 2). Office of Public Information and Participation, Toxic Substances Control Division, State Department of Health Services, September 1984, folder 9, box 11, Stringfellow Hazardous Waste Site collection, MS 148. University of California–Riverside Libraries, Special Collections and Archives.

91. Stephanie Sellers Cramer, "Stringfellow Report," USC Master's Program in Safety Management, n.d., box 16, folder 17, Stringfellow Hazardous Waste Site collection, MS 148. University of California–Riverside Libraries, Special Collections and Archives.

92. Cramer, "Stringfellow Report," 5.

93. "The maximum reported chromate was 0.07 mg/l (milligrams per liter) on December 4, 1972. The Glen Avon School well, located one and one-half miles downstream, showed a trace of hexavalent chrome (0.013 mg/l) in May of 1972. Subsequent samples of water from the well did not indicate any chrome." Alvin Franks, State Water Resources Control Board Division of Water Quality, to James W. Anderson, Executive Officer, Santa Ana Regional Water Quality Control Board Re: Stringfellow Quarry, February 13, 1973, Stringfellow Hazardous Waste Site collection, MS 148. University of California–Riverside Libraries, Special Collections and Archives.

94. Stuart C. Taylor of Department of Building and Safety, Land Use Division, to Ray C. Smith, Director of Building and Safety, Re: Stringfellow Chemical Dump, August 18, 1972, box 13, folder 9, Stringfellow Hazardous Waste Site collection, MS 148. University of California–Riverside Libraries, Special Collections and Archives.

95. Harold M. Erickson, Director of Public Health, County of Riverside, to Ray Smith, Director of Department of Building and Safety, County of Riverside Re: Stringfellow Dump, September 25, 1972, box 14, folder: Riverside County Department of Health, Stringfellow Hazardous Waste Site collection, MS 148. University of California–Riverside Libraries, Special Collections and Archives.

96. Richard J. Stokes, County of Riverside Department of Building and Safety, to Health Department, Planning Department, Flood Control District, Road Department (Al Fleming), State Water Quality Control Board #8, Air Pollution Control District, November 26, 1973, box 13, Stringfellow Hazardous Waste Site collection, MS 148. University of California–Riverside Libraries, Special Collections and Archives.

97. The regional effects of Stringfellow's closure were felt immediately. In but one instance, G. A. Hanke, president of the steel producing division of Ameron Steel in Etiwanda, was left to deal with the five thousand gallons per week of acid pickling fluid used in the wire mill. "Numerous Firms Affected by Disposal Pits Closure," *Press Enterprise*, December 1, 1972.

98. John M. Haley, Department of Water Resources, Los Angeles, CA, to Santa Ana Regional Water Board, Attention Mr. James Anderson, April 6, 1973, Stringfellow Hazardous Waste Site collection, MS 148. University of California–Riverside Libraries, Special Collections and Archives.

99. Haley, "Letter from Department of Water Resources, Los Angeles, CA, to Santa Ana Regional Water Board, Attention Mr. James Anderson."

100. James Anderson, "Order No. 73-20: Waste Discharge Requirements for Stringfellow Quarry Co.," July 15, 1973, box 16, folder 18, Stringfellow Hazardous Waste Site collection, MS 148. University of California–Riverside Libraries, Special Collections and Archives.

101. James Anderson, "Minutes of Meeting in Board of Supervisors Chambers, County of Riverside," June 14, 1973, box 14, Riverside County Board of Supervisors.

102. Simon Sykes, "Control Board Okays Reopening of Acid Pits," *Press Enterprise*, June 18, 1973, sec. B, box 4, folder 1, Stringfellow Hazardous Waste Site collection, MS 148. University of California–Riverside Libraries, Special Collections and Archives.

103. "Special Adjourned Meeting-Riverside West Area Planning Council," 92, April 18, 1974, box 13, folder 6, Stringfellow Hazardous Waste Site collection, MS 148. University of California–Riverside Libraries, Special Collections and Archives.

104. "Special Adjourned Meeting—Riverside West Area Planning Council," 93.

105. "Special Adjourned Meeting—Riverside West Area Planning Council," 94.

106. R. C. Smith, Building Director, to West Area Planning Council, July 2, 1974, Stringfellow Hazardous Waste Site collection, MS 148. University of California–Riverside Libraries, Special Collections and Archives; Paul, "County Upholds Denial of Permit for Stringfellow Toxic Waste Dump."

107. Dickey, *Report of the Interim Fact-Finding Committee on Water Pollution*, 70. Historical writing on California water has primarily been about agriculture, flood control, and various water supply projects, with vanishingly little on how water supplies were threatened in quality by population growth and industrial expansion. For prominent examples, see Worster, *Rivers of Empire*; Reisner, *Cadillac Desert*; Hundley, *The Great Thirst*.

108. Up until recently, California was the only state in the western United States to not manage its groundwater. This only changed in 2014, with the historic passage of the Sustainable Groundwater Management Act by the state legislature. For more, see "Sustainable Groundwater Management," accessed June 20, 2017, http://www.water.ca.gov/groundwater/sgm/.

BIBLIOGRAPHY

Archival Sources

Stringfellow Hazardous Waste Site collection, MS 148. University of California, Riverside Libraries, Special Collections & Archives, University of California, Riverside.

Published Sources

Andreen, William L. "The Evolution of Water Pollution Control in the United States—State, Local, and Federal Efforts, 1789–1972: Part II." June 7, 2004. http://papers.ssrn.com/abstract=554122.

Assembly Interim Committee on Government Reorganization. "The Organization of California's New Department of Water Resources: A Report of the Assembly Interim Committee on Government Organization to the California Legislature." Assembly Interim Committee Reports. 1955–1957. Sacramento, CA: Assembly of the State of California, February 8, 1957.

Blomquist, William. "Implementing the Sustainable Groundwater Management Act with Lessons from Existing Groundwater Management Institutions." Paper presented at the 38th Annual Fall Research Conference, Washington, DC, November 3–5, 2016. https://appam.confex.com/appam/2016/webprogram/Paper16672.html.

"California's Water Pollution Problem." *Stanford Law Review* 3, no. 4 (July 1951): 649–66. https://doi.org/10.2307/1226475.

Colten, Craig E. "Groundwater Contamination: Reconstructing Historical Knowledge for the Courts." *Applied Geography* 18, no. 3 (July 1998): 259–73. https://doi.org/10.1016/S0143-6228(98)00017-4.

Colten, Craig E. "A Historical Perspective on Industrial Wastes and Groundwater Contamination." *Geographical Review* 81, no. 2 (April 1, 1991): 215–28. https://doi.org/10.2307/215985.

Dickey, Randal. *Report of the Interim Fact-Finding Committee on Water Pollution*. Sacramento, CA: Assembly of the State of California, 1949. https://catalog.hathitrust.org/Record/007161623.

Federal Security Agency. *Report on Water Pollution Control: Santa Ana River Basin*. Water Pollution Series. Federal Security Agency, Public Health Service, March 1, 1951.

Hall, William H. "Irrigation in California (Southern)." *Second Part Report of the State Engineer of California on Irrigation and the Irrigation Question*. Sacramento: State Printing Office, 1888.

Harmon, Burt. "Contamination of Ground-Water Resources." *Civil Engineering* 11, no. 6 (1941): 345–47.

Hines, N. William. "Nor Any Drop to Drink: Public Regulation of Water Quality Part II: Interstate Arrangements for Pollution Control." *Iowa Law Review* 52, no. 3 (1966): 432–57.

Hundley, Norris. *The Great Thirst: Californians and Water, 1770s–1990s*. Berkeley: University of California Press, 1992.

Isard, Walter, and James Ganschow. "Awards of Prime Military Contracts by County, State and Metropolitan Area of the United States, Fiscal Year 1960." Regional Science Research Institute, 1961. https://catalog.hathitrust.org/Record/001896101.

"J. B. Stringfellow, Jr. State Deposition." May 7, 1990. Hard copy of this document is located among files of the Center of Community Action and Environmental Justice, Jurupa Valley, California. www.ccaej.org.

Krieger, James H., and Harvey O. Banks. "Ground Water Basin Management." *California Law Review* 50, no. 1 (1962): 56–77.

Layzer, Judith. *The Environmental Case*. 3rd ed. Washington, DC: CQ Press, 2012.

Lotchin, Roger W. *Fortress California, 1910–1961: From Warfare to Welfare*. New York: Oxford University Press, 1992.

Melosi, Martin V. *The Sanitary City: Environmental Services in Urban America from Colonial Times to the Present.* Abridged ed. Pittsburgh, PA: University of Pittsburgh Press, 2008.

Moskovitz, Adolphus. "Quality Control and Re-Use of Water in California." *California Law Review* 45, no. 5 (1957): 586–603. https://doi.org/10.2307/3478503.

Newman, Penny. "'It's the Pits': Remembering Stringfellow." Glen Avon, CA: Center for Community Action and Environmental Justice, October 2, 2004. A hard copy of this booklet is available for purchase from the Center of Community Action and Environmental Justice, Jurupa Valley, California. www.ccaej.org.

"Numerous Firms Affected by Disposal Pits Closure." *Press Enterprise,* December 1, 1972.

Paul, Kenneth. "County Upholds Denial of Permit for Stringfellow Toxic Waste Dump." *Press Enterprise,* November 15, 1974.

Pickett, A. "Disposal of Industrial Wastes in Los Angeles County." *Water and Sewage Works* 95, no. 1 (1948): 33.

Reisner, Marc. *Cadillac Desert: The American West and Its Disappearing Water.* New York: Viking, 1986.

Robie, Ronald. "Water Pollution: An Affirmative Response by the California Legislature." *Pacific Law Journal* 1 (1970): 2–35.

Rogers, J. David. "A Man, a Dam and a Disaster: Mulholland and the St. Francis Dam." *Southern California Quarterly* 77, no. 1/2 (April 1, 1995): 1–109. https://doi.org/10.2307/41171757.

Ross, Benjamin, and Steven Amter. "Deregulation, Chemical Waste, and Groundwater: A 1949 Debate." *Ambix: The Journal of the Society for the Study of Alchemy and Early Chemistry* 49, no. 1 (2002): 51–66.

Scott, Allen J. *Technopolis: High-Technology Industry and Regional Development in Southern California.* Berkeley: University of California Press, 1993.

Scott, Allen J., and Doreen J. Mattingly. "The Aircraft and Parts Industry in Southern California: Continuity and Change from the Inter-War Years to the 1990s." *Economic Geography* 65, no. 1 (1989): 48–71.

Starr, Kevin. *Golden Dreams: California in an Age of Abundance, 1950–1963.* Oxford: Oxford University Press, 2009.

State of California, Department of Public Works, Division of Water Resources. "Region-wide Municipal and Industrial Dump Site Study, Santa Ana Region." June 1954.

"Sustainable Groundwater Management." Accessed June 20, 2017. http://www.water.ca.gov/groundwater/sgm/.

Tarr, J. A. "Industrial Wastes and Public Health: Some Historical Notes, Part I, 1876–1932." *American Journal of Public Health* 75, no. 9 (September 1985): 1059–67. https://doi.org/10.2105/AJPH.75.9.1059.

Trezek, G. J. "Engineering Case Study of the Stringfellow Superfund Site." Department of Mechanical Engineering, University of California–Berkeley, August 1984.

Worster, Donald. *Rivers of Empire: Water, Aridity, and the Growth of the American West.* 1st edition. New York: Pantheon Books, 1985.

6

Processing the Past into Your Future

UNCOVERING THE HIDDEN CONSEQUENCES OF INDUSTRIAL DEVELOPMENT IN THE WEST TEXAS PETROCHEMICAL INDUSTRY

Sarah Stanford-McIntyre

The Permian Basin makes up approximately 68,000 square miles in Southwest Texas and Eastern New Mexico, stretching from Lubbock County in the Texas Panhandle south to Jeff Davis County on the Mexican border. The twin cities of Odessa and Midland, located east of El Paso on Interstate 20 and 230 miles from the Mexican border, remain the geographic and economic heart of the region.

Between 1923 and 1990, oil prospectors extracted fourteen billion barrels of crude oil from reservoirs beneath the Permian Basin, making it one of the most prolific energy-producing regions in the world. During the post–World War II period, new industrial processes led to economic growth as petroleum by-products were converted into finished, sellable goods.[1] In particular, the development of plastics and rubbers from natural gas offered new possibilities for the region's mature oil fields. For workers in the Permian Basin's network of oil refineries, drilling operations, and oil transport and storage companies, these developments meant more jobs, better pay, and a chance to claim regional importance in a technically sophisticated, globally influential industry.[2]

In contrast to other domestic oil processing communities along the Louisiana Gulf Coast, in Houston's oil refining centers, or at Sunoco's vast laboratories in Marcus Hook, Pennsylvania, oil workers in Odessa, Midland,

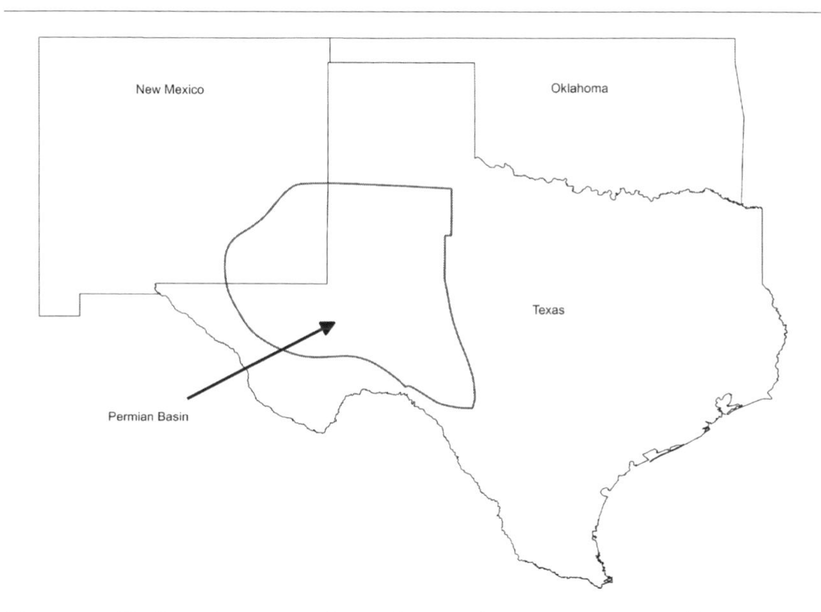

Figure 6.1. The Permian Basin is located in southwest Texas. Created by Sarah Stanford-McIntyre.

and the surrounding Permian Basin remained largely immune to broader changes in politics and industrial relations during the postwar era.[3] Nationally significant events such as the controversy caused by the 1969 Santa Barbara oil spill, the oil embargo of 1973, and increasingly public discussions about the consequences of air and water pollution did not produce critiques of industry in the Permian Basin. Efforts by the Oil, Chemical and Atomic Workers' Union (OCAWU) to establish a national oil union were met with indifference, and the push to establish the Occupational Safety and Health Administration (OSHA) was largely ignored.[4] The region's refineries and petrochemical plants experienced almost no strikes or work stoppages despite nationwide disruptions in the 1950s and 1960s.[5]

In this chapter, I argue that the Permian Basin's geographic isolation and conservative political culture helped oil industry elites create a hegemonic discourse around economic growth and shared prosperity. This discourse effectively worked to obfuscate the consequences of unregulated petrochemical production and oil processing on environmental and social well-being for over thirty years.[6]

To tell this story, I examine the popular discourse surrounding the creation and expansion of the Odessa petrochemical industry. I focus on the

Odessa Petrochemical Complex, a network of ten petrochemical facilities located in central Odessa, Texas, that was, during the 1960s and 1970s, the largest inland petrochemical plant in the world. I demonstrate that the regional political culture, helped along by racial and economic segregation, dampened acknowledgment of and action against workplace hazards and environmental contamination. More specifically, I show how industry leaders and city officials consciously positioned themselves as minor celebrities and expert bearers of industrial knowledge to advocate for unfettered industrial growth to benefit a racially white community.

ODESSA PETROCHEMICAL: GLOBAL CONNECTIONS

Plans for the facility that would become the Odessa Petrochemical Complex began during the early days of the postwar oil boom. The project was conceived and pushed heavily by the Odessa Chamber of Commerce.[7] In 1952 the chamber enlisted local banker and oil financier William D. Noel to lead a multitiered effort to facilitate petrochemical development in the area. Citing the small regional population and distance from large industrial centers, Paul Kayser, then president of El Paso Natural Gas Corporation, agreed to fund the construction of an initial butadiene and styrene plant in Odessa only if the chamber could demonstrate a regional market for at least 75 percent of the products produced.[8]

Confident that industry would breed further industry, Noel and his committee convinced the Akron, Ohio–based General Tire and Rubber Company to build a new plant in Odessa, which became the main buyer of the El Paso Corporation's petrochemicals. In 1955 El Paso Corporation began excavating the foundation for the Odessa Petrochemical Complex, and Noel was elected president of the newly formed El Paso Natural Gas Products company. Upon its completion, the facility made El Paso Corporation one of the largest single employers in Odessa, and the massive industrial facility became the crown jewel in the El Paso Corporation's growing natural gas empire.[9] Skilled at vertical and horizontal integration, during the postwar period the El Paso Corporation expanded beyond pipe-laying to build oil refineries and petrochemical facilities throughout Texas, perfecting the art of trapping and reusing residual gas—natural gas found accidentally during oil exploration—for industrial production. Through a mix of physical and ideological tools such as the laying of pipe, the building of refineries and chemical plants, the development of labor networks, and corporate philanthropy, advertising, and signage, El Paso's natural gas monopoly visually, culturally, and economically linked the sparsely populated Permian Basin to an international system of speculative commerce and wealth extraction.[10] By convincing El Paso Corporation to build a new plant in Odessa, the Odessa Chamber of Commerce thus positioned its urban labor and industrial market within El Paso's growing corporate and commercial web.[11]

Permian Basin crude oil is high in sulfur, making the refining process more complex and the production of gasoline more expensive than elsewhere. The first twenty years of oil extraction in the region had been marked by the inability of local oil producers to pay for transport of low-grade crude to far-flung refineries for processing and sale. With the region dominated by small producers having little capital, the lack of reliable pipelines stunted oil exploration in the region. Although oil speculators knew that the Permian Basin was rich in natural gas, its limited consumer use as a heating alternative and the lack of efficient transport and storage facilities meant that it was most often burned off into the atmosphere upon discovery.[12] Improvements to production technology in the 1950s, however, opened the door for what was once largely considered a petroleum industry waste by-product; these improvements revolutionized the Permian Basin oil industry.[13] The Odessa Petrochemical Complex represented an opportunity to capitalize on newfound uses for natural gas, bypassing the need for far-flung transport to Gulf or East Coast petrochemical facilities.

Noel oversaw plant expansions throughout the 1960s, with El Paso Products absorbing Brooklyn-based Beaunit Corporation in 1967 and the local Odessa Natural in 1968.[14] By 1968 the Odessa Petrochemical Complex housed six companies, including Noel's own El Paso Products and the General Tire and Rubber Company, as well as Rexall Drug and Chemical Company, Shell Oil Company, Big Three Industries, and a Morton salt brine purification plant. By 1970 the plant encompassed ten separate production facilities on two separate campuses, spread over 640 acres, and was valued at over $300 million dollars.[15] Conflicting reports from this period record anywhere from 800 to 1,400 people employed at the plant.[16]

The plant's steady expansion coincided with a larger oil industry boom in the region during the mid-1950s. Waves of oil prospectors merged with a growing manufacturing and white-collar workforce in the cities of Odessa and Midland. The populations of both cities exploded between 1950 and 1960, increasing approximately 172 percent over a ten-year period.[17] Most of these people were oil workers from other parts of Texas and the United States, either sent by large oil companies or looking for work in the new extraction boom. Closer to the oil fields, Odessa became the destination for transient, working-class oilfield workers. Some of these workers expanded an existing neighborhood of trailers and hastily built houses just southeast of the Odessa city limits and near the plant.

Even after oil extraction slowed again in the 1960s, new workers continued to be attracted by the oil industry's high wages, which, at approximately $6,279 per year in 1965, were well above the state average of $4,058.[18] In 1965 Odessa's ten petrochemical plants represented the third largest concentration of petrochemical producers in the state, only lagging behind Houston's industrial corridor.[19] When oil prices skyrocketed in the 1970s, the region experienced another, even larger boom. Median income rose to $9,251 in 1970, and the Odessa Petrochemical Complex continued to provide the city of Odessa with high tax revenue and above-average wages, making chamber officials view their industrial gamble as a unilateral success.

With household wealth steadily increasing for over thirty years, it is not surprising that many Odessa residents saw a direct connection between new uses for natural gas, the expansion of petrochemical production, and local community development. During this period, the local press took the link between industry and prosperity a step further, connecting petrochemical expansion to national narratives about the power of technology to improve upon nature as well as Cold War efforts to build an American technological empire.

FUSING TECHNOLOGICAL AND COMMUNITY PROGRESS

For its first twenty years, the Odessa Petrochemical Complex was advertised in the pages of regional newspapers such as the *Odessa American*, the *Abilene Reporter News*, and the *Midland Reporter Telegram* as an engineering marvel symbolizing Odessa's new status as part of a modern, technological world. Simultaneously, tourist souvenirs such as picture postcards, shot glasses, commemorative plates, as well as industry journals and periodicals focused on the plant's size, emphasizing its height and metallic enormity in relation to the surrounding treeless landscape.[20] The *Odessa American*, the main news periodical in Odessa and surrounding Ector County, spearheaded this effort, regularly reminding readers that the plant was "the largest inland

Petrochemical Plant in North America."[21] A typical *Odessa American* article from 1962 describes a visit from El Paso Corporation president Paul Kayser, Rexall's president Justin Dart, and other industrial elites. Along with assurances from Dart that Rexall had definite plans to expand production in Odessa, Kayser describes Odessa as "one of the most important operations in the chemical industry."[22] A similar article from 1974 narrates a plant visit by French industrialists. The article emphasizes the plant's technological sophistication and geographic improbability, stating that "strangers are amazed at such a development in an area so devoid of readily available water."[23] Petrochemical production and oil refining requires a tremendous amount of water.[24] Because of regional aridity and the high salt and mineral content of most local groundwater, petrochemical producers in the Permian Basin had to develop new water-saving techniques and cultivate reliable water filtration and transport networks to supply the plant. According to the article, the French industrialists were interested in building a similar plant in the deserts of Algeria.

By emphasizing the French visitors' desire to use the plant as a model, the *Odessa American*, the chamber, and El Paso Corporation executives reminded readers of the plant's global significance. These publications depict the plant as a symbol of Odessa, a city that used the power of science to triumph over its isolated location and recalcitrant geography. Through the plant, Odessa was no longer simply a source of extractable raw goods, no longer just a transfer point for the interregional and international flow of wealth within El Paso Corporation's vast pipeline network. Instead, the city and its inhabitants had become *creators*—builders of innovative industrial and consumer products.

Other news articles during this period focus on the plant's steady waves of expansion. They describe the increase in plant employees and rise in the number of companies that occupied facilities at the plant as clear evidence of progress and indicative of sustainable economic and infrastructural growth. In particular, the local press latched on to El Paso Corporation's national connections. Daily newspaper coverage was careful to mention El Paso's sister plants in New Mexico, Louisiana, and Tennessee when describing the company. Further texts regularly describe El Paso Corporation, Rexall, and other companies' networks of production facilities throughout the nation, again situating Odessa within a nationally and globally interconnected industrial network.[25] These articles rhetorically frame the Odessa plant as part of a larger corporate body, with all parts vital to the whole. Rather than signaling absentee corporate exploitation, such connections were designed to indicate Odessa's growing importance and demonstrate that multi-industry connectivity would help shield the community from the oil industry's notorious fluctuations in prices and the inevitability of regional oil depletion.[26]

During the 1950s and 1960s, oil company officials came to dominate every part of the regional and municipal government, replacing an established ranching elite in most elected positions.[27] The *Odessa American* supported this shift in political power, praising the Permian Basin's growing population of college-educated engineers and managers who were to lead the region in the creation of a technologically integrated, industrial West Texas. Seen as effectively blending academic knowledge and modern science with a practical understanding of the importance of constant economic growth, engineers were painted as ideal leaders and experts in community development.[28] These engineers-cum–local officials believed that municipal government existed to facilitate commerce, which was the central touchstone of civic life. Voting records demonstrate that they were even more pro–business and small government than local officials in other parts of highly conservative Texas.[29] They generally had the support of their constituents. For example, Ector and Midland Counties were two of only sixteen Texas counties (out of 254) that voted for Barry Goldwater in the 1964 general election. Throughout the 1950s and 1960s, many of the city's leaders were vocal members of the John Birch Society.[30]

Exemplifying a desire to blend business management and community leadership, the local press deified President Noel of El Paso Products as a kind of local father figure and folk hero. A 1963 article in the *Odessa American* described Noel as a patron of the arts, with the Odessa Chamber of Commerce naming Noel "Outstanding Citizen of 1963." In 1965 the *Odessa American* reminded readers of Noel's civic awards, taking care to remark that he spearheaded the establishment of the Odessa Petrochemical Complex "for the good of the people." The article further celebrated Noel's achievements "as a member of the Texas Liquor Control Board; a member of the Advisory Council of the Texas Business Administration Foundation; and a director of the Texas Tech Foundation."[31] In this context, economic boosterism, support for education, and the policing of public morality were understood as deeply intertwined components of effective civic stewardship.[32]

Despite the assurances of those who supported the oil industry's increasing social and political power, and despite the economic prosperity brought by petrochemical production, the creation and expansion of the Odessa Petrochemical Complex also meant steadily rising—and increasingly visible—levels of environmental contamination within the city limits. However, the local press's unceasing boosterism—aligning industrial growth with community progress—seems to have been shared even by those most exposed to industrial contamination and unsafe working conditions. Indeed, shop floor workers and communities neighboring the petrochemical complex did not actively mobilize for improved labor and environmental conditions, unlike those in other regions of the country. As I argue in the next section, this lack of organizing was in part influenced by the region's more conservative political culture and exacerbated by residential and labor force segregation.

WHITE WORKERS, HAZARDOUS WORK

More so than any other region of Texas, the Permian Basin oil industry was largely whites only, and the Odessa chemical industries were almost completely segregated.[33] An all-white elite managerial class of engineers and executives reigned at the top, while an almost-all-white class of skilled and semiskilled technicians made up the majority of the Odessa Petrochemical Complex workforce and were at the front lines of workplace exposure to toxins.

On a daily basis, these workers measured chemical reactions, watched temperature gauges, operated heavy machinery, and maintained or replaced machinery.[34] In the process, employees came into contact with large quantities of hot-to-molten raw liquids, alternately heating or cooling them through the manipulation of steam valves or by adding various catalysts to create specific chemical combinations.[35] In particular, the plant refined butadiene and styrene from natural gas. These two petrochemicals were—and still are—the two most common petrochemicals used in the creation of synthetic rubber.[36] Styrene resin, a further processed version of styrene, was also used as a base product in a diverse array of other plastic consumer goods, such as synthetic flooring, scrubbing brushes, and broom handles.[37] By the mid-1960s, the Odessa Petrochemical Complex began using early computers to measure these chemical combinations and to monitor equipment gauges.[38] However, few parts of the plant benefited from automation, and work remained intensely physical and often hazardous.

Prior to 1970, there were few standardized regulations governing employee exposure to petrochemical substances. Refinery and petrochemical workers often wore little to no breathing or fire protection when in close proximity to chemical fumes, distillation equipment, or raw materials.[39] According to worker testimonies filed with the OCAWU during the 1960s, a lack of individual or atmospheric protective equipment at Texas petrochemical plants produced an array of immediate health hazards for workers.[40] Crude oil and natural gas are corrosive materials and damaging to human skin if touched for an extended period of time. Constant exposure resulted in surface burns. Methane, sulfur, and other gases released when storage tanks were opened were highly toxic, causing a loss of consciousness and death after only short-term exposure.[41] Without fire-protective suits, burns from superheated liquids, overheated pipes, or pressurized steam releases were common.[42] In short, petrochemical technical workers in the Permian Basin faced real dangers on the job.

In response to hazards within the petrochemical and refinery sector, several oil worker unions, including the OCAWU, held rallies and union drives during the 1950s and 1960s to raise awareness and advocate for improved workplace conditions. The OCAWU met with success in California and the Midwest. The heavily unionized network of refineries along the Texas Gulf also held several strikes during this period and agitated for better pay and injury compensation programs.[43] While such efforts by the OCAWU and other unions ultimately led to the establishment of the OSHA in 1970, this movement held little sway with Permian Basin oil workers. On multiple occasions, the white workforce at the Odessa Petrochemical Complex voted not to join the OCAWU, and thus bucked the trend of union efforts more nationally.[44] What accounted for this? I argue that racial and residential segregation as well as a "progress" narrative that only included and benefited an all-white petrochemical community played a role.

A look at the history of Odessa residential development during the 1960s and 1970s suggests that while residents and employees publically supported oil industry expansion efforts, some were at least tacitly aware of the plant's industrial hazards. Although the 800 to 1,400 people who worked at the plant made up a relatively small portion of Odessa's total working population, the plant's sheer size coupled with the geographic realities of Odessa's population boom meant that pollution affected an increasing percentage of regional residents. When initial construction began on the plant in 1956, El Paso Corporation and the Odessa Chamber of Commerce located the complex on undeveloped, unincorporated ranchland five miles southwest of Odessa's traditionally Black and Latinx neighborhood.[45] The land was cheap and relatively isolated, with an average price of only $25 per acre. In the 1950s, the city's residential expansion was largely confined to the north and east sections of town, on the opposite side of Interstate 20 and farthest from the petrochemical complex.[46] However, a large industrial corridor quickly grew up near the plant, with the majority of the city's oil industry employees working within one or two miles of the vast complex. As non-white populations moved into the city looking for work, they expanded the existing barrio southward, toward the plant.

Suggesting that local populations were at least nominally aware of pollution from the plant, during the 1960s, the city's elite—white-collar plant managers, engineers, and salesmen—chose to build new homes increasingly farther north and east, with older, south-central Odessa neighborhoods closest to the plant becoming dominated by lower-income and minority families.[47] The map in figure 6.2 incorporates data from two historic maps, one drawn in 1940 and the other in 1975. It illustrates residential growth on the north and east sides of town as well as the expansion of industrial operations south of the interstate.[48]

While the movement of wealth out of central Odessa was connected to a national postwar suburban housing boom and the desire to conspicuously display newfound wealth via real estate, it is also an example of the broader history of urban decay and white flight that plagued American industrial centers during the second half of the twentieth century.[49] As land closest to the plant became increasingly undesirable, those with the economic means were quick to abandon a recognizably hazardous neighborhood. As Black and Latinx families migrated to Odessa looking to escape low-paying agricultural work, they moved into the increasingly inexpensive homes nearest to the plant.[50] Excluded from lucrative plant jobs, these residents would have been most immediately affected by the plant's air and water pollution.[51] In contrast, white city officials, engineers, middle-class oil workers, and petrochemical industry executives who had left for the surrounding suburbs were shielded from constant contact with plant contamination and had little direct incentive to enact or agitate for industrial ordinances against a major source of tax revenue and regional employment. The starkly segregated racial environment of twentieth-century Texas thus rendered non-white residents invisible and gave white elites the freedom to claim that the plant unilaterally benefited the "community."

The El Paso Corporation also went to great lengths to shore up its reputation with white oil workers and further draw a line between the oil-

Figure 6.2. The Odessa Petrochemical Plant was located south of both the Texas and Pacific Railroad and Interstate 20. Created by Sarah Stanford-McIntyre.

producing community and everyone else. Building upon the local press's sunny booster narrative of regional importance and sophistication through oil, El Paso officials used pageantry and civic displays to forge an undeniable connection between the Odessa plant, technological progress, and the city's mostly white population of middle-class oil workers. Beginning in the 1950s and continuing throughout the 1960s and 1970s, El Paso Corporation opened the Odessa Petrochemical Complex to educational tours, with local schools, the Odessa Study Club, the local women's Desk and Derrick Club, and other organizations regularly visiting the plant to ostensibly learn about synthetic production.[52] Such tours reinforced awe at the plant's size, complexity, and scientific sophistication even as it integrated the plant into the community processes of youth education and recreation. These tours were paired with an array of industry booster events such as the annual Odessa Oil Show, industry athletic competitions, and barbecue dinners.[53] In the early 1950s, the American Petroleum Institute Oil Industry Information Committee began a nationwide promotional tour, using theater, song, and dance to exalt oil industry economic growth and demonstrate the industry's connection to small-town America. People in the Permian Basin were quick to demonstrate their support for the campaign. In 1952 the city of Odessa sponsored the largest rally of its kind in the country to welcome the American Petroleum Institute boosters. Seeking to demonstrate that the oil industry was a fundamental part of Permian Basin culture and community, the event culminated in a citywide barbeque facilitated by "local high school bathing beauties" and the presentation of the "largest vat of baked beans in the world," served from a miniaturized oil storage tank constructed at an Odessa oil industry machine shop.[54]

Such programs were reinforced through *Odessa American* coverage of the Odessa Oil Show, which featured an array of detailed, technical op-ed pieces written by local industry leaders and engineers. Each year, this series of articles followed single products through the plant's assembly-line process and emphasized in exhaustive detail the array of uses for the plastics and rubbers produced at the plant. Reminiscent of DuPont's consumer product–focused "Wonderful World of Chemistry" presentation at the 1964 World's Fair, these media profiles sought to valorize the daily labor of oil workers as part of the globalization of American industry and the expansion of twentieth-century automated technology.[55]

El Paso's efforts tapped into a wider Cold War context in which oil companies used narratives of military and technological triumph to gain public support for industry expansion.[56] Well into the 1960s, oil producers were joined by companies such as the Houston-based Union Carbide Corporation and the Firestone Chemical Company to fund a cottage industry of pamphlets, posters, books, and documents trumpeting the chemical origins of new synthetic products and their role in elevating the American standard of living.[57]

The El Paso Corporation was no exception. Its propaganda coupled the power of science with corporate expansion and sought to usher Odessa, and its people, into a new technological future. A period-typical El Paso Products ad from 1967 features a single image. A well-dressed white woman sits amid a pastel cornucopia of household goods, presumably manufactured from El Paso petrochemicals. Below the image a caption reads: "Processing the past into your future." Similar to petrochemical ads found in industry and employee periodicals such as *Petroleum Week*, *Petroleum Refiner*, or *The Lamp*, this ad focused on the final result instead of the industrial process, implying that through the miraculous processes of science, raw nature (crude oil and natural gas) could be brought—along with the rapidly industrializing Permian Basin—into a modern, synthetic age.[58]

Such propaganda effectively linked the oil and petrochemical industries to local civic pride, which was enacted in the Permian Basin via plant tours, industry fairs, and corporate philanthropy programs. Importantly, these events also reinforced the social marginalization of the region's Black and Latinx populations, who were excluded from oil industry jobs. Finally, the Cold War nationalism and scientific utopianism applied in such campaigns also blurred the lines between (white) workers, corporations, and human communities. In essence, full community citizenship was synonymous with oil industry employment, and the oil industry's corporate interests aligned with the needs of the local community. Any meaningful opposition or critique of the industry was implicitly rendered unacceptable.

CRACKS IN THE NARRATIVE

As the push for industrial health and safety reform, rising environmental awareness, and the energy crisis grew in the late 1960s and 1970s, the discourse of community progress through oil began to be challenged at the national level.[59] In the Permian Basin, however, the press, civic actors, and oil industry officials all doubled down on their narratives about growth and progress. It would take another decade and an almost total collapse of the region's oil industry to finally make space for an emerging counternarrative and concerted legal action against environmental contamination.

Soon after the establishment of the Environmental Protection Agency (EPA) in 1970, Texas state regulators began documenting contamination at the Odessa Petrochemical Complex, finding crude oil and other chemicals in the city water supply and high levels of chemicals in a nearby stream, Monahans Draw. In the late 1970s, the EPA began investigations into the particular hazards posed by exposure to petrochemicals, publishing a series of exposure recommendations for regulators and employers. By 1978 butadiene was publically known to be particularly toxic when airborne and was considered more toxic than formaldehyde when ingested.[60] A 1980 EPA report on the toxicity of styrene and ethylbenzene found that both could become easily vaporized and travel airborne long distances, and national epidemiological analysis of chemical plant workers linked styrene and ethylbenzene exposure to nervous system and reproductive disorders.[61] The public nature of these documents, coupled with the OCAWU's vocal and explicit testimonies during the OSHA campaigns of 1968 and 1969, make it clear that by the mid-1970s it was widely known within industry regulatory circles that the type of contamination documented at the Odessa plant was immediately hazardous to both workers and neighboring residents. However, almost no action was taken to mitigate pollution at the Odessa Petrochemical Complex by either state or federal officials, and almost no discussion of these events appeared in the regional press.[62]

Although state authorities and plant administrators remained indifferent, a few residents were aware of these reports and took action. The *Odessa American* noted that in 1977 a group of undisclosed parties began a $300,000 lawsuit against El Paso Products, seeking damages for health problems caused by long-term exposure to toxic chemicals at the plant.[63] However, this isolated case does not indicate popular outcry against the chemical industries in Odessa. Such evidence is tempered by the fact that while some oil workers in the Permian Basin did participate in the OCAWU's membership drives, union cards from the mid-1970s document that employees unionized hoping for better wages, not over any concern about workplace hazards.[64]

Yet, not all oil workers necessarily had blind faith in their companies' proclaimed benevolent intentions. Oral history interviews with oilfield workers in the 1970s, for example, reveal that the region's working-class white population was more pragmatic. They expressed a pride in the Permian Basin's regional importance as part of a globally significant industry and hoped that further technological developments would prevent oil depletion and save the region's oil communities.[65] Others, boosted by high wages and removed from the worst sources of regional contamination, were reluctant to entertain critiques by outsiders whom they believed had little understanding of the industry or regional politics.[66]

The reluctance to openly critique the industry was likely also connected to timing, with residents and the local press hesitant to condemn one of the few expanding sectors of the 1970s. In 1973 the Organization of Arab Petroleum Exporting Countries began an oil embargo against the United States, damaging the domestic oil industry's public image. However, it served as an economic boon to the Permian Basin. While easy-to-find oil had been largely depleted in the Permian Basin by the end of the 1960s, the desperate need for domestic crude in the mid-1970s meant that oil companies could make huge profits drilling deeper and with more expensive technologies. In order to process this jump in extraction, the Odessa Petrochemical Complex underwent a series of expansions in the early 1970s, increasing employment at a time of national economic crisis and stagflation. Oil production in the Permian Basin increased over the next ten years, reaching its peak in 1980.[67]

During this period, the popular press proudly featured the wave of plant expansions as a way to reassure locals that industry would continue to expand in the face of national economic troubles. On October 17, 1976, an ad in the *Odessa American* for the Odessa Oil Show pleaded with readers to "support the oil industry. . . . It supports all of us." Other *Odessa American* articles, while similar to those in the 1950s and 1960s, were almost hyperbolic in their shrill industry support. One article spouted, "So now just two miles south of town is a multi-million-dollar complex, considered to be one of the largest inland complexes in the United States today, all of which started as a dream back in the early 1950s and resulted from a lot of hard work in the ensuing years by the companies involved and the invaluable help and cooperation of the people of Odessa."[68] This narrative connected the petrochemical complex to community stability even as it relegated the "people of Odessa" to helpers on the plant's larger journey. Here the plant became a conscious entity, anthropomorphized and acting independently to help the community and the nation out of economic doldrums.

In 1977 another article remarked: "The complex is ever expanding, searching for new markets and new techniques. The $300 million plus complex is symbolic of new science that is changing the American way of life."[69] Again, humans were downplayed, and the life of the plant became the most important factor in community development. Especially in tough economic times, what was good for the plant was considered fundamentally good for the people of Odessa. Individual citizens—understood to be industry employees—were incomplete component parts that made up the fully realized organism of both the local body politic and the corporate body of the industrial plant.

These rhetorical connections between the industry and the community are important. In the late 1970s, the local press parroted copy from early 1950s booster literature, demonstrating an inability to see beyond technological progress based in a single industry, even in the face of public health problems and a looming economic crisis. Constant reassurance by Odessa's politicians and civic leaders, engineers and plant managers, and advertising campaigns and civic pageantry that the plant would sustain constant growth and that the city would continue to defy its isolated and arid location muffled the consequences of daily workplace exposure to toxic waste and hazardous materials. Moreover, the rigid lines of racial segregation kept residents in the most affected neighborhoods out of sight and out mind, ultimately allowing this slow-moving tragedy to evolve for over forty years.

Public silence about industrial contamination was only disrupted by the collapse of the entire Texas oil industry in the 1980s. In 1982 the Organization of Petroleum Exporting Countries began cutting oil prices. In early 1983 they dropped from $34 to $29 per barrel. The next year, prices plummeted as low as $7.[70] The crisis was felt across Texas, with the oil industry falling from one-quarter of the total Texas economy in 1981 to around 12 percent in 1991. Statewide oil and gas employment dropped by one-third between 1982 and 1994.[71] While these developments created trouble for many people in Texas, it was most painfully felt in the Permian Basin. Average income plummeted as oil companies laid off thousands. The crisis peaked with the collapse of the First National Bank of Midland, the largest independent bank in Texas, with dozens of businesses in the Permian Basin declaring bankruptcy soon after. In 1980, during the final days of the boom, Ector County residents living in poverty made up only 8.5 percent of the population (compared with a state average of 14.6 percent). In 1990, seven years after the sudden drop in oil prices, poverty levels remained at a staggering 20.4 percent. Particularly hard hit was West Odessa, a suburb of Odessa adjacent to the petrochemical plant and made up almost entirely of temporary oilfield workers. In 1985 three-quarters of the population lived in rented mobile homes, with a poverty rate of 24 percent.[72]

This economic disaster finally halted the unequivocal veneration of the oil and chemical industries by the local popular press. In the face of such a catastrophe, nothing of the local boosters' original narrative could be salvaged. One plant employee seemed to understand this break in the narrative as an opportunity to finally expose the industry's environmental violations. In 1981 *Odessa American* reporter Pat Weir contacted the Texas Water Commission after receiving several phone calls from an anonymous source claiming to work at the Odessa Petrochemical Complex. In a letter to Weir, the source expressed concern about the plant's negligent water treatment practices, saying: "Know this though, they know what they are doing but after reading my letter so will you." The source accused the plant of contaminating local groundwater with massive amounts of benzene, nonsulfur inhibitor, and other chemicals.[73] Although the letter's ominous tone suggested the discovery of a massive internal scandal, the Texas Water Commission ultimately found no violations, and Weir never published the letter. The plant would spend the next several years slowly scaling back operations, and the Texas Water Commission would document an increasing stream of public complaints about air and water contamination coming from the plant.[74]

By the mid-1990s, inhabitants of the low-income Black and Latinx neighborhoods closest to the plant were also fed up. For decades, residents had reaped only a fraction of the plant's economic benefits, and while the downturn in the oil economy probably decreased the scale of regional air and water pollution, it did not make the problem disappear. Instead, residents were faced with the dual challenge of a bad economy, few job prospects, and a string of health problems related to plant contamination. In 1995 the Odessa, Texas, branch of the National Association for the Advancement of Colored People (NAACP) took over an ongoing class action lawsuit filed in 1991 by neighborhood residents against petrochemical processors Shell, Rexene, and Dynagen.[75] The lawsuit accused the companies of gross negligence, declaring that inadequate waste disposal and air emissions equipment at the Odessa Petrochemical Complex were responsible for breathing problems and increased cancer rates among people living in the predominately Black and Latinx neighborhood.[76] The NAACP cited Dynagen's failure to fully address the over 500 safety violations—including unsafe wastewater disposal practices and unchecked air contamination from daily chemical burn-offs—that had been found by state inspectors in 1988, when, in an abrupt reversal of their past practices, the state fined Dynagen $1.4 million and ordered the company to update waste disposal equipment immediately.[77]

As late as 1994, Dynagen had installed little of the court-ordered emissions control equipment.

The NAACP lawsuit gained international attention, with commentators calling into question the safety of not just the neighborhood but of the 200,000 people living in the Odessa-Midland metro area. After rejecting an initial lump-sum settlement, the NAACP won the case in 1997. Over 850 plaintiffs received an undisclosed monetary settlement, and Dynagen was forced to install $12 million worth of pollution-control equipment.[78] Odessa gained a dubious reputation as "the most polluted city in America," and descriptions of the "Odessa syndrome"—a malady that included sore throat and eyes, headaches, nausea, and bloody noses—proliferated in the popular press.[79] The trial also revealed a final irony. The NAACP lawsuit and subsequent investigations revealed that contamination from the plant had reached much farther than assumed. Although more affluent white Odessa residents had fled the south-central neighborhoods, in part to escape the waves of toxic contaminants produced by the plant, emissions traveled much farther, with even the wealthiest residents subjected to increased liver cancer rates, respiratory problems, and contaminated groundwater. Those who had clung fastest to the narrative that the plant was fundamentally integral to their stability and health were no more immune to toxic exposure and industrial harm than the racial minorities excluded from the oil-producing "community." Even as the Odessa Petrochemical Complex thus gave the white community and its body politic a means of livelihood and power, it simultaneously harmed lives, including those most directly benefiting from it.

In sum, for almost sixty years a consistent booster narrative that promised constant growth and linked oil industry expansion to community cohesion and sustainability dominated Odessa politics, press, and mass culture. This narrative, backed by a national Cold War veneration of technology and the realities of racial segregation, helped to shut down public discussion about the petrochemical industry's environmental and public health consequences. Only the collapse of the entire Texas oil industry would allow space for some to seek restitution. As a new Permian Basin oil boom plays out in the 2010s, ongoing debates about hydraulic fracturing and horizontal drilling suggest that many in the region are less willing to assume the oil industry's benevolent intentions. Only time will tell if such public dialogue will prevent another tragedy.

ACKNOWLEDGMENTS

I would like to thank the researchers and archivists at several institutions, including the Permian Basin Petroleum Museum, the Hagley Museum and Library, the Walter P. Reuther Library, and the Lemelson Center for the Study of Invention and Innovation. I would also like to thank the Washington DC Labor History Working Group for their time and attention to this project.

NOTES

1. "War Productions Board," Sun Oil Company Collection, folder: "Mobil Corporation Subject Files, Affiliates . . . Mobil Chemical Company: Petrochemicals Division 1956–1999," ExxonMobil Historical Collection.

2. During the 1920s and 1930s, Houston became a center of global oil tool and well services operations. Permian Basin boosters wanted to be a part of this growing regional powerhouse. Childs, *The Texas Railroad Commission.*

3. Seligson, *Oil Chemical and Atomic Workers*; Eastabrook, *Labor-Environmental Coalitions*; Revathi, "African Americans' Struggle for Environmental Justice and the Case of the Shintech Plant," 777–89.

4. "Average State Membership, Aug 1957–July 1958," Otto Pragan Collection.

5. Notebook, Tony Mazzocchi Collection.

6. Harvey, *A Brief History of Neoliberalism*; Cullen and Wilkinson, eds., *The Texas Right*; Hinton and Olien, *Oil and Ideology.*

7. While members of the oil industry made up the majority of the Odessa Chamber during the early 1950s, a mix of bankers, lawyers, and real estate brokers also held seats on the executive board. The 1953 chamber president, Harold Downs, was a partner in the Downs-Clark Equipment Company, which sold oilfield tools. Jesse Owens, member of the Texas Railroad Commission and chamber manager, was instrumental in the push for a petrochemical plant, later leaving public service to work for El Paso Corporation. In 1953 Odessa mayor Fred Gage owned Gage and Cochran Real Estate.

8. According to the 1950 Census, the combined population of both Ector and Midland counties was only 67,800. U.S. Census Bureau, Census of the Population, *Characteristics of the Population, 1950*, vol. 2, 1953. https://www.census.gov/library/publications/1953/dec/population-vol-02.html.

9. A pipeline company founded by Houston-based lawyer Paul Kayser in 1928, El Paso Corporation profited heavily from West Texas's geographic isolation and the regional oil industry's chronic lack of transport infrastructure. The company expanded during the 1930s and 1940s to become the main natural gas distributor in Texas. After World War II, El Paso Corporation formed a natural gas monopoly that dominated the U.S. Southwest and northern Mexico for much of the century.

10. The creation of this administrative, technological, and economic network echoes the history of other extractive industries including coal, agriculture, and timber. Mitchell, *They Saved the* Crops; Boyd, *The Slain Wood*.

11. Already the natural gas supplier for most of the state of California, in the 1990s El Paso Corporation, by then a subsidiary of Burlington Northern, acquired the largest natural gas pipeline in the country. With this move, El Paso Corporation gained control of the East Coast to form a short-lived natural gas empire. After several years of highly publicized legal troubles, Kinder Morgan bought El Paso Corporation for $38 billion in 2012.

12. E. N. Beane, interview, 1970, Oral History Collection.

13. Meikle, *American Plastics*; "New Plant to Produce More Geon," *Goodchemco News*, September 1950, Warshaw Collection.

14. "El Paso Natural Gas Products Co. Grew with Odessa," *Odessa American*, 2B.

15. Texas Commission for Environmental Quality, "TCEQ WST IHW 30142 Reports MF307606 Hydrogeological Investigation at the Xylene Tank Area Facility 28," microfiche.

16. "El Paso Products Odessa, Tex," Oil, Chemical and Atomic Workers Union Collection (hereafter OCAWU).

17. U.S. Census Bureau, *Census of Population: Texas Volume 1 Part 45, 1960, Characteristics of the Population, 1960*. https://www.census.gov/prod/www/decennial.html.

18. "Statistical Records: 1965," Mid-Continent Oil and Gas Association Records.

19. Harris County held thirty-three plants and Jefferson County held sixteen. Both counties sit within Houston's extended metropolitan area. Folder: "Statistical Records: 1965," Mid-Continent Oil and Gas Association Records.

20. These tchotchkes were handed out at the regional industrial fair, the Permian Basin Oil Show, or sold at gas stations and local motels to passing tourists.

21. The *Odessa American* was first published in 1940 by local Texas historian, essayist, and moral crusader R. Henderson Shuffler. In 1948 Shuffler sold the paper to Freedom Communications, an independent news conglomerate that owned several newspapers in California and around the United States.

22. Sliney, "Financiers to Visit Odessa," *Odessa American*, 1.

23. The absence of potable water was a constant barrier to Permian Basin oil exploration and civic development, with towns reporting into the 1930s the need to ship drinking water from Alpine in far West Texas. "Complex is Growth Beacon," *Odessa American*, 82.

24. Petrochemical production relies on a huge amount of water. In 1978 El Paso produced 1.5 million gallons of wastewater per day. In 2008, with the majority of the plant shut down, the complex still consumed over five million gallons of groundwater. Texas Commission for Environmental Quality, "TCEQ WST IHW 30142 Correspondence MF307608 2–3," microfiche.; Texas Commission for Environmental Quality, "TCEQ WST IWH 30142 Vol 001 Correspondence 2007–2008," microfiche.

25. "Gas Firm President Speaks Here," *Odessa American*, 29.

26. Stabilizing the oil industry's cycles of boom and bust has been a preoccupation for U.S. presidents, federal regulators, and industry leaders. In particular, Harold Ickes, Franklin Roosevelt's minister of the interior, made it a personal mission to regulate oil extraction, stabilize prices, and prevent natural resource waste. He met with only limited success. Ickes, *Fightin' Oil*. In the Permian Basin, voluntary extraction quotas during the 1920s also

had only limited success. However, industry cooperation was widely praised as the best way to manage oil production. Oral history interviews from 1970 reveal only limited support for federal regulation among regional landowners but a general faith that better technologies had resulted in less waste and the revival of aging oil wells. Olien and Hinton, *Wildcatters.*

27. For example, when W. T. Edwards was elected vice president of El Paso Products Company in 1970, he was also the past president of the Odessa Chamber of Commerce, the Odessa Kiwanis Club, Permian Basin Petroleum Association, Southern Little League, and the Odessa Country Club. "Edwards Elevated to Vice President," *Odessa American*, 36.

28. This is not to say that everyone in the industry welcomed engineers with open arms. The nineteenth-century battle between engineers and working-class oil drillers is documented in Frehner, *Finding Oil.*

29. Merrill, "The Illusions of Independence."

30. In 1962 the arch-conservative, famously anti-integration general Edwin Walker won Odessa's vote for governor over ex–oil industry lawyer John Connally in the Texas Democratic primary. According to one author, during the 1960s the president of the Odessa United States Junior Chamber was a paid recruiter for the far-right, stridently anticommunist John Birch Society. King, "Who's Number One in the Permian Basin?"

31. "Fish Engineering is Pioneer in Odessa Complex," *Odessa American*, n.p.

32. Regional philanthropy was expected from those who had success in the oil industry and was considered part of good citizenship by the general population. W. M. Allman and Mrs. Allman, interview, June 19, 1970, Oral History Collection; George W. Ramer, interview, February 12, 1970, Oral History Collection.; James J. Wheat and Mr. George Clayton, interview, July 17, 1970, Oral History Collection.

33. In contrast, by 1939 approximately 1,450 African Americans and 750 self-identified Mexicans and Mexican Americans worked in refining along the Gulf. On Texas's upper coast, African American workers made up 12 percent of refinery workers. According to a 1943 Fair Employment Practices Committee survey, 59 percent of Mexican refinery workers in Texas were born outside the United States. Zamora, *Claiming Rights and Righting Wrongs in Texas*, 161. People identifying as Black, White, or Latinx made up 99 percent of Odessa's, and the Permian Basin's, population during the twentieth century. The 1950 Census documents ten people who described their race as "other" in Odessa. This number increased to fifty-six by 1960. U.S. Census Bureau, *1960 Census of Population—Preliminary Reports Population Counts for States.* https://www.census.gov/prod/www/decennial.html.

34. "General Petroleum Corporation," ExxonMobil Historical Collection.

35. Folder: "Exploration and Producing," ExxonMobil Historical Collection.

36. Spitz, *Petrochemicals.*

37. "Products and Processes," Union Carbide and Carbon Corporation, 1953, 21, Warshaw Collection.

38. Folder: "Human Resources: Education and Training Learning about Computers," ExxonMobil Historical Collection.

39. Folder: "Human Resources Employee Relations: R&D Supervisors Manual ca. 1950," ExxonMobil Historical Collection.

40. Folder: "BP 78–4109," Tony Mazzocchi Collection; Notebook, Tony Mazzocchi Collection.

41. The dangers of sulfur gas were well known even in the 1920s. E. N. Beane, interview, 1970, Oral History Collection, 21.

42. Notebook, Tony Mazzocchi Collection; Folder: "Union News 1968–69," OCAWU Collection.

43. The OCAWU organized fierce union action in the state's largest refining hubs in Houston, Texas City, and Port Arthur throughout the 1950s, with workers protesting increasing hours and shrinking pay. These efforts culminated in a wave of bitter strikes along the Texas Gulf Coast, with a watershed coming during 1962 and 1963 as striking Shell workers threatened to shut down the majority of the nation's refineries. Shell ultimately broke the strike, heavily publicizing the fact that they could run their newly automated plants without employees. Although demoralized by these losses, Texas union agitation continued on an isolated basis throughout the 1960s and 1970s. Tony Mazzocchi Collection.

44. Folder: "04–40 Rexall Chemical Company Odessa, Texas Potential: 86," OCAWU Collection; Folder: "04–66 Rexall Chem Co. Odessa, Tex Potential: 110," OCAWU Collection.

45. U.S. Census Bureau, *1960 Census of Population—Preliminary Reports Population Counts for States.* https://www.census.gov/prod/www/decennial.html.

46. The exception to this rule was West Odessa, established during the postwar boom on unincorporated land west of the city and made up of temporary, working-class oil workers and their families.

47. Weber, "A Comparison of Two Oil City Business Centers." Along with racial prejudice, there are several reasons for this migration. In the postwar era, Midland boomed as a center for regional banking and oil industry administration, so some Odessa residents probably moved north seeking better access to jobs in Midland. Further, new housing developments on larger lots and with modern conveniences such as multicar garages and multiple bathrooms probably appealed to nouveau riche employees looking to display their wealth. Jackson, *Crabgrass Frontier.*

48. U.S. Census Bureau, *16th Census of the United States: 1940, Population, Volume 1, Number of Inhabitants,* https://www.census.gov/prod/www/decennial.html; Texas Commission for Environmental Quality, "TCEQ WST IHW 30142 Reports MF307518 Hydrogeologic Investigation at the Styrene Plant Area," microfiche.

49. Sugrue, *The Origins of the Urban Crisis*; Zamora, *The World of the Mexican Worker in Texas.*

50. The Tejano population of Ector County in 1980 was 23,790, while Midland County held 11,727 Tejanos. De Leon, *Tejano West Texas,* 66.

51. Nelson, "A Service Classification of American Cities." While the League of United Latin American Citizens and other Tejano activist organizations had a marked presence in several West Texas cities, including Lubbock and San Angelo, there is little evidence of their presence in Odessa. De Leon, *Tejano West Texas*; Zamora, *The World of the Mexican Worker in Texas.*

52. Desk and Derrick was a social and educational network for secretaries and office staff working in the oil industry. Overwhelmingly female, the organizations provided technical and scientific education for administrative employees.

53. The Odessa Oil Show is an outdoor industry fair similar in personality to a state

agricultural fair. The Oil Show began in 1940 as the "Little International Oil Show" and is still held every two years.

54. "Odessa Oil Progress Week Observance Expected to Rank as Nation's Largest," *Odessa American*, 1.

55. Cowan, *A Social History of American Technology*.

56. Ekbladh, *The Great American Mission*.

57. "Chemicals," Warshaw Collection.

58. *Petroleum Weekly* was a weekly oil industry news magazine that ran from 1954 to 1962. It was published by McGraw-Hill and marketed to oil executives. *Petroleum Refiner*, first published in 1942, was a product of Gulf Oil in Houston, Texas. *The Lamp* was the employee magazine of the Standard Oil Company, first published in 1918. ExxonMobil still publishes an online version of the magazine.

59. Gottlieb, *Forcing the Spring*; Merrill, *The Oil Crisis of 1973–1974*.

60. Miller, "Ingestion of Selected Potential Environmental Contaminants: Butadiene and Its Oligomers, Final Report."

61. Santodonato et al., "Investigation of Selected Potential Environmental Contaminants."

62. Analysis of TCEQ inspection reports reveal that violations resulted in letters to the plant management. Sometimes this included a fine. Despite reporting consistent and significant water quality violations throughout the 1970s, the TCEQ issued the plant a new operation permit each year. Texas Commission for Environmental Quality, "TCEQ WST IHW 30142 Inspection Report MF307610 5–6," microfiche; Texas Commission for Environmental Quality, "TCEQ WST IHW 30142 Correspondence MF307609 1–16 (pre 1985)," microfiche. During the 1970s, there is scant reference to either OSHA regulations or EPA research in the *Odessa American*. In particular, there is no mention of local chemical exposure, environmental contamination, or the industry's long-term health effects.

63. In contrast to their scant coverage in the *Odessa American*, these efforts were heavily publicized in union periodicals and in national newspapers, ostensibly as evidence of the new law's effectiveness.

64. Texas Commission for Environmental Quality, "TCEQ WST IHW 30142 Correspondence MF307609 1–16 (pre 1985)," microfiche.

65. George R. Bentley and Judge Carl D. Estes, interview, 1970, Oral History Collection.

66. W. M. Allman and Mrs. Allman, interview, June 19, 1970, Oral History Collection.

67. "Subsidiary of Products Firm Formed," *Odessa American*, 11.

68. "Joint Venture in Complex Started Eight Years Ago," *Odessa American*, 69.

69. "Building Boom Is Continued," *Odessa American*, 119.

70. Burrough, *The Big Rich*; Yergin, *The Prize*.

71. McDonald, *Postwar Urban America*, 161.

72. U.S. Census Bureau, *Characteristics of the Population, Number of Inhabitants, Texas, 1980, Vol. 1*, https://www.census.gov/prod/www/decennial.html.

73. A 1983 memo to Robert Fleming, director of enforcement, from environmental quality specialist Gary Raven reveals that the Water Commission only reluctantly followed up on these accusations. Raven had been assigned to the case in 1982 after the abrupt resignation of Robert Bradshaw, the agent assigned to the case in 1981. After conducting tests of the

water table near the plant, two years after the original complaint, the Water Commission found no hazardous levels in local groundwater and only recommended further monitoring. Texas Commission for Environmental Quality, "TCEQ WST IHW Correspondence MF307608 8 of 16_0000," microfiche.

74. Brown, "Residents Suing Rubber Plant over Health Problems from Air Emissions."

75. In the initial lawsuit against Dynagen, residents asked for $27 million in damages. "Rubber Plant Emissions Unhealthy, Suit Claims," *Orlando Sentinel*, November 18, 1991. http://articles.orlandosentinel.com/1991-11-18/news/9111180426_1_rubber-plant-pollution-emissions.

76. Rexene Corporation, "Rexene Corporation 1995 Annual Report on Form 10-K Part 1"; Rose, "NAACP Fumin' over Rubber Plant Emissions in West Texas."

77. The initial fine of $4.4 million was later reduced to $1.4 million. It represents the largest fine by the State of Texas for environmental contamination since the Texas Clean Air Act of 1965. The five hundred violations still represent the largest number of citations ever given to a Texas industrial site. "75th Anniversary," *Odessa American*.

78. AP, "Officials Say Feud at an End," *Odessa American*, 9B.

79. Rexene Corporation, "Letter to Stockholders"; Buncombe, "Odessa Texas: Is It the Most Polluted City in America?"

BIBLIOGRAPHY

Boyd, William. *The Slain Wood: Papermaking and Its Environmental Consequences in the American South*. Baltimore: Johns Hopkins University Press, 2015.

Brown, Chip. "Residents Suing Rubber Plant over Health Problems from Air Emissions." AP News Archive, November 17, 1991. http://www.apnewsarchive.com/1991/Residents-Suing-Rubber-Plant-Over-Health-Problems-From-Air-Emissions/id-c13b8632d3ff814e42ea7da094db5ead.

Buncombe, Andrew. "Odessa, Texas: Is It the Most Polluted City in America?" *Independent*, April 1, 2001. https://www.independent.co.uk/environment/odessa-texas-is-it-the-most-polluted-city-in-america-5365449.html.

Burrough, Bryan. *The Big Rich: The Rise and Fall of the Greatest Texas Oil Fortunes*. New York: Penguin, 2009.

Childs, William. *The Texas Railroad Commission: Understanding Regulation in America to the Mid-Twentieth Century*. College Station: Texas A&M University Press, 2005.

Cowan, Ruth Schwartz. *A Social History of American Technology*. Oxford: Oxford University Press, 1997.

Cullen, David O'Donald, and Kyle Wilkinson, eds. *The Texas Right: The Radical Roots of Lone Star Conservatism*. College Station: Texas A&M University Press, 2014.

De Leon, Arnoldo. *Tejano West Texas*. College Station: Texas A&M University Press, 2015.

Eastabrook, Thomas. *Labor-Environmental Coalitions: Lessons from a Louisiana Petrochemical Region*. Amityville, NY: Baywood Publishing, 2007.

Ekbladh, David. *The Great American Mission: Modernization and the Construction of an American World Order*. Princeton, NJ: Princeton University Press, 2011.

ExxonMobil Historical Collection. Briscoe Center for American History. University of Texas–Austin.

Frehner, Brian. *Finding Oil: The Nature of Petroleum Geology, 1859–1920*. Lincoln: University of Nebraska Press, 2012.

Gottlieb, Robert. *Forcing the Spring: The Transformation of the American Environmental Movement*. New York: Island Press, 2005.

Harvey, David. *A Brief History of Neoliberalism*. Oxford: Oxford University of Press, 2007.

Hinton, Diana Davids, and Roger Olien. *Oil and Ideology: The Cultural Creation of the American Petroleum Industry*. Chapel Hill: University of North Carolina Press, 2000.

Ickes, Harold. *Fightin' Oil: The History and Politics of Oil*. New York: Hyperion Press, 1976.

Industry Periodical Collection. Archives Center. National Museum for American History.

Industry Periodicals Collection. Hagley Museum and Library.

Jackson, Kenneth. *Crabgrass Frontier: The Suburbanization of the United States*. Oxford: Oxford University Press, 1985.

King, Larry L. "Who's Number One in the Permian Basin?" *Texas Monthly*, June 1975. https://www.texasmonthly.com/articles/whos-number-one-in-the-permian-basin/.

McDonald, John F. *Postwar Urban America: Demography, Economics, and Social Policies*. New York: Routledge, 2014.

Meikle, Jeff. *American Plastics: A Cultural History*. New Brunswick, NJ: Rutgers University Press, 1997.

Merrill, Karen R. "The Illusions of Independence: Texas Oilmen and the Politics of Postwar Petroleum." In *The Political Culture of the New West*, edited by Jeff Roche, 74–96. Lawrence: University Press of Kansas, 2008.

Merrill, Karen R. *The Oil Crisis of 1973–1974: A Brief History with Documents*. Boston: St. Martin's, 2007.

Mid-Continent Oil and Gas Association Records. Briscoe Center for American History. University of Texas at Austin.

Miller, Lynne M. *Ingestion of Selected Potential Environmental Contaminants: Butadiene and Its Oligomers, Final Report*. Office of Toxic Substances. U.S. Environmental Protection Agency, December 1978.

Mitchell, Don. *They Saved the Crops: Labor, Landscape, and the Struggle over Industrial Farming in Bracero-Era California*. Athens: University of Georgia Press, 2012.

Mitchell, Timothy. *Carbon Democracy: Political Power in the Age of Oil*. New York: Verso, 2011.

Nelson, Howard J. "A Service Classification of American Cities." *Economic Geography* 33, no. 3 (July 1955): 189–210.

Odessa American. 1950–1990. Historical Newspapers Online. Accessed October 15, 2016. Newspapers.com.

Oil, Chemical and Atomic Workers Union Collection. Special Collections. University of Colorado–Boulder.

Olien, Roger M., and Diana Davids Hinton. *Wildcatters: Texas Independent Oilmen*. College Station: Texas A&M University Press, 2007.

Oral History Collection. Special Collections. Permian Basin Petroleum Museum.

Pragan, Otto Collection. Oil Chemical and Atomic Workers Union. Walter Reuther Library. Wayne State University.

Revathi, I. "African Americans' Struggle for Environmental Justice and the Case of the Shintech Plant: Lessons Learned from a War Waged." *Journal of Black Studies* 31, no. 6 (July 2001): 777–89.

Rexene Corporation. "Letter to Stockholders." United States Security and Exchange Commission, July 28, 1997. https://www.sec.gov/Archives/edgar/data/829218/0000950134-97-005523.txt.

Rexene Corporation. "Rexene Corporation 1995 Annual Report on Form 10-K Part 1." United States Security and Exchange Commission, 1995. https://www.sec.gov/about/annual_report/1995.pdf.

Rose, Mark. "NAACP Fumin' over Rubber Plant Emissions in West Texas." *Crisis*, February-101, no. 2 (1994): 24.

"Rubber Plant Emissions Unhealthy, Suit Claims." *Orlando Sentinel*, November 18, 1991. http://articles.orlandosentinel.com/1991-11-18/news/9111180426_1_rubber-plant-pollution-emissions.

Santodonato, Joseph, et al. *Investigation of Selected Potential Environmental Contaminants: Styrene, Ethylbenzene, and Related Compounds*. Final Report. Office of Toxic Substances. U.S. Environmental Protection Agency, May 1980.

Seligson, Harry. *Oil Chemical and Atomic Workers: A Labor Union in Action*. Denver: University of Denver Press, 1960.

Spitz, Peter H. *Petrochemicals: The Rise of an Industry*. New York: John Wiley and Sons, 1988.

Sugrue, Thomas. *The Origins of the Urban Crisis: Race and Inequality in Postwar Detroit*. Princeton, NJ: Princeton University Press, 1997.

Sun Oil Company Collection. Archives Center. Hagley Museum and Library.

Texas Commission for Environmental Quality. Austin, Texas. Microfiche Archive.

Tony Mazzocchi Collection. Special Collections. University of Colorado–Boulder.

Warshaw Collection. Archives Center. National Museum of American History.

Weber, Dickinson. "A Comparison of Two Oil City Business Centers (Odessa-Midland, Texas)." PhD diss., University of Chicago, 1958.

Yergin, Daniel. *The Prize: The Epic Quest for Oil, Money, and Power*. New York: Simon and Schuster, 1991.

Zamora, Emilio. *Claiming Rights and Righting Wrongs in Texas: Mexican Workers and Job Politics during World War II*. College Station: Texas A&M University Press, 2009.

Zamora, Emilio. *The World of the Mexican Worker in Texas*. College Station: Texas A&M University Press, 2000.

7

Vast, Incredible Damage

HERBICIDES AND THE U.S. FOREST SERVICE

James G. Lewis and Char Miller

The Douglas-fir tussock moth (*Orgyia pseudotsugata*) caterpillar is small in size, with brightly colored tufts of black hair projecting from the head and rear of its body. However diminutive and decorative, this caterpillar's fierce appetite—especially during outbreaks in the late spring and early summer—can quickly defoliate individual trees and collectively damage large swaths of that arboreal species whose name it bears. Its capacity to chew through forests gained notoriety in the 1960s and 1970s, so much so that in 1965 the U.S. Forest Service sprayed DDT mixed with fuel oil over 66,000 infected acres in the Pacific Northwest. After conducting posttreatment analysis, agency scientists proclaimed the aerial assault a complete success, achieving "a tussock moth kill ranging from ninety to one hundred percent, with an overall average of ninety-eight per cent."[1]

Less than a decade later, an even larger outbreak blew up along the Washington, Oregon, and Idaho borders, which overflights estimated had damaged upward of 500,000 acres. Although the Forest Service, along with state, tribal, and private landowners, wanted to replicate the successful control-and-eradication operations that had occurred in the mid-1960s, there was a catch. In the interim, Rachel Carson's *Silent Spring* (1962) had appeared, and her revelations of the devastating impact that indiscriminate use of DDT—what she decried as "a bright new toy"—was having on wood-

Figure 7.1. Adult male of Douglas-fir tussock moth (*Orgyia pseudotsugata*). This species of moth was responsible for numerous outbreaks in the Pacific Northwest in the postwar years, which the U.S. Forest Service and other federal and state agencies attempted to control with aerial spraying. Photo courtesy of Ladd Livingston, Idaho Department of Lands.

ed, riparian, and marine habitats, and the animals that inhabited them, had led to closer scrutiny of the insecticide and related chemicals.[2] Indeed, DDT had been banned in the United States, complicating the Forest Service's managerial response to the 1973 outbreak. As then regional forester Ted Schlapfer later recalled: "We were really caught between a rock and a hard place, knowing that the only way we could positively control [the tussock moth] was to use DDT."[3]

A legal loophole opened up just such an opportunity. DDT could still be deployed if the relevant agencies and entities determined that its use constituted a national emergency. Together, the Forest Service, the Bureau of Indian Affairs, and the Bureau of Land Management joined the Oregon State Forester's Office and the Oregon State University School of Forestry in petitioning the U.S. Environmental Protection Agency (EPA) for an exemption. Although the EPA initially denied their request, in 1974, after evidence that the tussock moth defoliation now sprawled across 1.2 million acres and in response to what historian Harold K. Steen has described as "unprecedented political pressure," a reluctant EPA granted the petitioners one-time use of the chemical that summer. The operation was only partly successful, as the outbreak may have already run its course. Yet the massive scale of the operation—it was the largest aerial spraying of DDT ever undertaken in the United States—also caused considerable concern within the Forest Service. "One of the real positive things that came out of [it]," remembered Schlapfer, was the conclusion that agency leaders reached: "We don't want to do this again. We have got to find alternative solutions to controlling [the] tussock moth." Shortly thereafter, researchers identified a nontoxic way to disseminate *Bacillus thuringiensis*, a biological agent that infects the tussock moth with a virus. The 1974 aerial spraying was the last time that DDT was applied in American forests.[4]

That happy outcome and the implication that policymaking, and the scientific expertise on which it depends, could come to know its limits; that postmortem analyses could lead to better science more carefully applied; and that its better application could lead to less environmentally damaging results is only part of the story surrounding the Forest Service's overdependence on herbicides, pesticides, and insecticides in the post–World War II era. In addition to the internal debates surrounding the use of DDT, outside forces exerted considerable pressure on the agency to halt its use of these toxic chemicals. Communities in and around national forests—particularly in Northern California and the Pacific Northwest—pushed back against the Forest Service's aerial campaigns. So did workers' organizations seeking to protect their members' health and laboring conditions, who did so by challenging agency science. An emboldened, post–*Silent Spring* environmental movement went to court, filing lawsuits in defense of endangered species, biodiversity, and water quality. Rather than simply demonstrating the limitations of the technological fix to land management dilemmas, then, the tussock moth incidents of the 1960s and 1970s are a reminder of the degree to which the Forest Service—the single largest agency in the U.S. Department of Agriculture, at the time employing upward of 35,000 people—dominated land management decisions at the federal level. That dominance helps explain why the Forest Service, and its peer agencies, routinely utilized chemicals whose impact on environmental and public health had not been fully assessed or completely understood—a process that has continued into the twenty-first century.

LIMITATIONS OF A CAN-DO AGENCY

The Forest Service's ready use of chemicals in the mid-twentieth century depended in good part on its leadership's firm belief that empirical science and rational planning had been the keys to its ability to resolve many of conservation's gravest problems. Chief Ferdinand Silcox's apparent success in suppressing forest fires in the 1930s gave an inkling of what could be accomplished if the Forest Service applied the right mix of personnel, technology, research, and budget. The same outcome seemed to have been true during the post–World War II era. As the Eisenhower administration came to a close, the Forest Service was reaching its peak in power and prestige, and was the undisputed leader of American conservation.[5] Its centrality was largely attributable to the agency's robust timber program. Because private industry had logged out most of its holdings by the end of the war, it turned to the national forests for timber to meet the burgeoning peacetime demand for lumber. The Forest Service willingly obliged in what it perceived to be a win-win situation. Through its ever-increasing timber yields, the Forest Service was making tangible contributions to the growing U.S. economy amid the Cold War—no small incentive for ambitious employees of the goal-oriented agency. In the age of Sputnik, scientific achievement mixed with a can-do attitude made the Forest Service a model agency. Its managerial strategy, former chief Michael Dombeck (1996–2001) declared, was reactive: "If commercially valuable timber was inaccessible, build a road. If a harvested forest on south-facing slopes resisted regeneration, terrace the mountainside. If soil fertility was lacking, fertilize the area. If pests or fire threatened forest stands, apply pesticides and marshal all hands to combat fire. If people grew unhappy with the site of large clearcuts, leave 'beauty strips' of trees along roadways to block timber harvest units from view."[6]

But when it came to timber, the postwar agency was also proactive. The Forest Service had dispelled the long-standing fear of a timber famine. *Timber Resources for America's Future,* which the agency had published in 1958, revealed that for the first time, timber growth on all lands—public and private—was exceeding the annual cut.[7] One reason for this was the agency's fire-suppression campaign: since the 1930s, the amount of acreage lost each year to fire had steadily dropped. In the 1960s, the average annual acreage burned was 4.6 million acres, down from a high of 39.1 million acres thirty years earlier. More importantly, the national forests had become the nation's lumberyard. The amount of timber sold from national forest lands nearly quadrupled from 1950 to 1960 (3,434,114 MBF to 12,167,180 MBF), and the harvest rate in that same time span nearly tripled (3,501,568 MBF to 9,366,897 MBF).[8] Higher sale and harvest rates meant more money coming in to the federal treasury's coffers. And that was good for Forest Service careers; those who made their timber targets could expect to be promoted. But the higher rates also gave birth to what historian Paul W. Hirt has dubbed a "conspiracy of optimism"—the belief that the Forest Service could deliver timber at the levels Congress demanded now and well into the future.[9] The goals could be achieved through intensive timber management techniques that included clearcutting and artificial regeneration supplemented and complemented by an accelerating application of herbicides.

The agency's continued emphasis on timber management, however, left it blind to ecological considerations and social concerns; it turned a deaf ear to rising public criticism on issues such as the impact of chemicals on human and animal populations as well as the public's changing values that favored recreation over resource extraction. One source of the agency's problems was that it suffered from groupthink. In 1960 Herbert Kaufman published a probing study of the Forest Service employees' administrative behavior. He sought to understand how field personnel operating within the agency's decentralized system, which allowed the lowest-ranking officers to make decisions without consulting superior officers, functioned at such a high level. Kaufman found that the agency recruited men with similar technical knowledge and practical skills who also had the will to conform and carry out what he called "the preformed decisions" of their superiors, which could be found in the ranger's bible, the *Forest Service Manual*. The agency designed the manual to do most of the thinking for its line officers, and the text laid out in full detail how to reach decisions on everything from "free-use permits to huge sales of timber, from burning permits to fighting large fires, from requisitioning office supplies to maintaining discipline."[10] The manual and the agency culture it nurtured and legitimized ensured a standard way of handling most situations or problems.

Adding to the self-scrutiny was the requirement that each ranger had to keep a diary and file multiple reports each year that would eventually reveal any deviation from accepted policy. Because personnel were rotated every two to three years, supervisors would be able to spot any inconsistencies in staff behavior or action that might be noted in their personnel record. In such an atmosphere, a forester who questioned operations might be labeled a troublemaker and place his career at risk. By handling personnel this way,

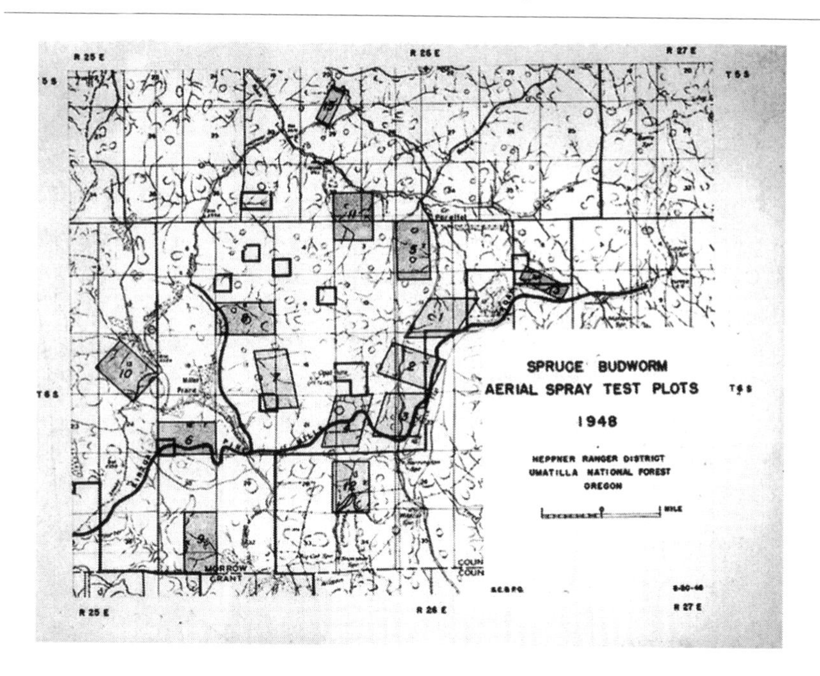

Figure 7.2. To assess the effectiveness of aerial spraying of herbicides over extensive forested acreage, the U.S. Forest Service established a series of test plots on the Umatilla National Forest in Oregon and Washington. Photo courtesy of U.S. Forest Service, Pacific Northwest Region, State and Private Forestry, Forest Health Protection.

the Forest Service, Kaufman asserted, "enjoyed a substantial degree of success in producing field behavior consistent with headquarters' directives and suggestions."[11]

This insularity was one reason why the agency proved particularly prickly about external criticism. In the mid-1960s, a seasoned forester told newly hired foresters: "We must have enough guts to stand up and tell the public how their land should be managed. As professional foresters, we know what's best for the land."[12] This assertion of expertise—which, as Kaufman indicated, was the result of the professionalizing nature of these employees' education and their adherence to the agency's internal mind-set—proved problematic. For as Rachel Carson had demonstrated, what the federal government had assured the public was better for humanity was not necessarily better for the environment or the species that depended on it.

In 1962 Carson, a former U.S. Fish and Wildlife Service biologist, published *Silent Spring*. Her book was a powerful indictment of the use of toxic chemical pesticides, primarily DDT, due to their poisonous impact on the food chain and the magnified threat this posed for human populations. She was highly critical of governmental agencies such as the Forest Service for their failure to test chemicals in biotic settings. In 1958, in Wyoming's Bridger National Forest, the Forest Service had taken a "shotgun approach to nature," she wrote, spraying upward of ten thousand acres of sage in response to "the pressure from cattlemen for more grasslands."[13] The intended target died, but so did "the green, life-giving ribbon of willows that traced its way across these plains" and the trout, beaver, and moose that had lived within this ecosystem's embrace.[14] A shocked Supreme Court justice-cum-conservationist, William O. Douglas, visited the area one year later and was appalled by what he saw, writing that "the damage is vast, incredible, awful." Visiting again a year after that, Douglas saw "more depredation by government."[15] Although the agency justified its decision based on the "improvement" it would bring to the range, Carson countered that its actions here and elsewhere were ripping apart "the whole closely knit fabric of life."[16] Her arguments, observes historian Stephen Fox, drew on the insights of ecologist Charles Elton, and with him she "argued that diversity was the key to biological health. It was imperiled by the human conceit that sorted out wild species according to their human uses and eliminated the 'bad' ones."[17]

Carson's book triggered a national controversy. Pesticide manufacturers and large agricultural organizations threatened lawsuits and attacked Carson's credibility; so did the U.S. Department of Agriculture.[18] Overlooked in the furor was Carson's call for research to determine how to use pesticides safely and to find alternate techniques for pest control; she had not urged the abandonment of pest control. Instead of following her suggestions, however, many in the timber and agricultural industries, along with the Forest Service, spent the next twenty years and countless resources arguing that they could not carry out their work without the chemicals they had on hand.[19]

By the early 1960s, for example, the Forest Service was annually conducting aerial spraying of DDT on more than one million acres of national forest lands to generate ever-higher timber yields. By the end of that decade, scientists inside and outside the agency determined that herbicides were adversely affecting wildlife and habitat. Instead of changing course or exploring alternative herbicides, however, Forest Service leaders responded to the perceived threat of Carson's work by launching an "information and education" program engineered out of its Washington office. Its publication, "The Forest Service in a Changing Conservation Climate" (1965), attempted to counter *Silent Spring* on several different fronts. The booklet's goal was to educate the public: "We need the understanding and support that comes from an informed public," Chief Ed Cliff declares in the text's epigraph. "[Our story] must be told and retold so that people everywhere will recognize and comprehend the forest patterns they see in America today."[20]

Among the twenty-nine "problems" listed in the booklet, was one titled "Use of Pesticides in Forestry." Strikingly, the word "herbicides" appears only once in the relevant text:

> Judicious use of pesticides and herbicides is necessary to control several important forest pests. In fact, pesticides are the only known effective method of control for several destructive forest insects and diseases. Many persons and several organized groups, believing that all pesticides are dangerous to wildlife and to people, oppose their use under any circumstances. The Forest Service, working in close cooperation with several other agencies . . . , is engaged in a widespread program to insure safe and effective use of pesticides. This program includes intensified research, detailed screening, controlled field testing, careful planning of action programs, and critical evaluation of the results and consequences.

The "objective," the agency declared, was to "develop public confidence in Forest Service decisions to use pesticides, emphasizing our equal concern that pesticides will always be used under safe, scientific, and carefully controlled conditions."[21]

The agency's literary efforts did not match *Silent Spring*'s reach, but in retrospect that mismatch in influence is less important than the Forest Service's effort to blunt criticism of its default use of pesticides and herbicides. Those who continued to oppose its actions were dismissed as being "ignorant or acting on 'misinformation,'" a dismissiveness demonstrating the agency's (almost willful) remaining out of touch.[22] Indeed, many agency foresters even advocated managing the land more intensively to achieve what they called "full utilization." Hoping to pull his colleagues back from this high-stakes gamble of defying the public interest, Charles Connaughton, who served as the regional forester for three regions from 1951 to 1971, urged his peers to take seriously the growing gap between what foresters did and public perceptions of why they did what they did when managing the national forests. In a 1966 article in the *Journal of Forestry*, he noted that the "toughest problem facing the forestry profession today results from a major

Figure 7.3. Rachel Carson's *Silent Spring* (1962), which closely analyzed the deleterious impact that DDT and other chemicals had on all life, profoundly influenced American environmental culture and politics and disrupted the once unquestioned deference to scientific expertise. Courtesy of U.S. Fish and Wildlife Service.

segment of the public not realizing commercial forest lands can be managed without destroying their utility and appearance. Consequently, much of the public lacks confidence in foresters as stewards of the land." He encouraged his fellow professionals to adopt management objectives and techniques that "result in acceptable conditions on the land that the public can and should be shown."[23] Four years later, fellow regional forester Neal Rahm, in a letter to the journal's editor, reinforced Connaughton's point: although confident in foresters' ability to do the job, he too wondered why they failed to prioritize educating the public, suggesting that this failure was because "we lack the will!"[24] Their urgings were too little, too late.

That their pleadings fell on deaf ears is a reminder, environmental historian Thomas Dunlap has noted, that "*Silent Spring* marked a watershed, as the private, scientific debate became a public, political issue."[25] In short order, Congress passed the Clean Air Act (1963) and the Water Quality Act (1965), which since have been amended, updated, and extended. Along with the Wilderness Act (1964) and a host of other new environmental regulations protecting endangered species and requiring public participation in land management planning, these legislative initiatives, and related concerns over quality-of-life issues, helped usher in the modern environmental movement.[26] In one sense, the movement argued that the human species no longer stood apart from the rest of the natural world. Yet, paradoxically, human survival was of growing concern. The threat of nuclear war, along with the use of chemicals to control nature domestically and abroad, when combined with photographs of Earth taken from space were reminders that despite humanity's impressive technological achievements, life on this blue planet seemed increasingly fragile.

This sense of fragility came coupled with a growing disillusionment with government policies that deepened as a result of the Vietnam War and the Watergate scandal. Swept up in this culture of distrust was the concept of scientific land management and managerial expertise—the once-unquestioned foundation of the Forest Service's mission. Historian Paul W. Hirt observed: "The same deference for scientists that contributed to public acceptance of intensive management for maximum production in the 1950s now contributed to widespread questioning of the faith in technological fixes and a growing skepticism" toward the Forest Service.[27] An agency that long had thought of itself as heroic now was perceived to be villainous.

VIETNAM AND THE HERBICIDE WARS

This perception was bound up with the Forest Service's ready deployment of herbicides. In limited use before World War II, chemical pesticide usage on the national forests accelerated in 1947, when Congress passed the Forest Pest Control Act. This legislation charged the Forest Service with preventing, controlling, or eradicating destructive pests on private and public forests. Industrial foresters and the Forest Service considered insecticides necessary to protect timber and range animals from harmful insects. Herbicides provided an efficient way to foster regeneration of economically desirable trees by killing undesirable ones, maintaining fuel breaks, and destroying noxious weeds. The agency's confidence in the findings of its researchers underscored its faith that it could effectively handle land management problems and control outcomes. Although Rachel Carson's *Silent Spring* had inspired further studies that showed how insect populations adapted to the chemicals and how pesticides killed beneficial parasites and predators along with the targeted insects, that research persuaded many Forest Service entomologists that "one hundred percent control or eradication of an insect was neither necessary nor practical to prevent economic loss." That finding notwithstanding, the agency persisted in its use of chemicals.[28] The continued reliance on such chemicals troubled some of its field scientists and also the EPA; the latter accused the Forest Service of conducting inadequate research on the impact of herbicidal spraying.[29]

After the EPA banned DDT in 1972, the Forest Service turned to other toxins—Malathion, Zectran, Sevin-4-Oil, and Orthene—that had not specifically been banned. Another herbicide of choice was 2,4,5-Trichlorophenol (or 2,4,5-T). The U.S. Army had developed it during World War II and then afterward released the formula for domestic use as a weed and brush killer. Beginning in the late 1940s, the Forest Service began using 2,4,5-T on American hardwoods to clear weeds from around shade-intolerant softwood stands. Twenty years later, the military launched widespread, aerial application of a mixture of 2,4,5-T and 2,4-Dichlorophenoxyacetic acid (or 2,4-D), called Agent Orange, over Vietnam to defoliate the hardwood jungle canopy and deny the enemy safe haven.[30] The levels of its application were extreme: the U.S. Air Force saturated the land, using twenty-seven times more herbicide per area unit than the Forest Service was spraying stateside for weed control. By 1966 studies had revealed that Agent Orange's primary

Figure 7.4. This B-18, one of many military surplus airplanes that federal agencies such as the U.S. Forest Service deployed in their aerial spraying campaigns after World War II, is laying down herbicides to control a Spruce Budworm outbreak on the Boise National Forest (Idaho), July 22, 1955. Photo courtesy of U.S. Forest Service/Forest History Society.

active ingredient, TCDD, or dioxin, caused birth defects in laboratory animals and was suspected of causing illnesses, birth defects, and miscarriages in humans. Domestic scientists protested the use of these poisons in Vietnam as early as 1964, and their challenge accelerated across that decade.[31] The federal government soon imposed restrictions on its use at home, such as banning it for household use and on food crops intended for human consumption.

Curiously, public and private foresters were exempt from these restrictions, so although antiwar protesters succeeded in 1970 in getting the military to stop using Agent Orange in Vietnam, aerial spraying of toxic herbicides in national forests continued. The Forest Service operated under the assumption that a chemical registered for use with the federal government did not have any significant adverse effects on the environment. But it had not conducted any risk analysis on the health effects of these and related chemicals, and therefore had not considered the need for alternatives, including manual or mechanical brush removal or hand spraying.[32]

Debate over the continued use of 2,4,5-T and Silvex (2,4,5-TP), another dioxin-containing herbicide, quickly became a national issue. A teacher in Alsea, Oregon, for example, did preliminary research that seemed to link the Forest Service's aerial spraying of 2,4,5-T and 2,4,5-TP on the nearby Siuslaw National Forest with local women's miscarriages. A pair of EPA-sponsored studies appeared to confirm that significantly higher percentages of miscarriages had occurred following the spraying of these toxins.[33] On April 27, 1978, at a public forum on the use of herbicides on public lands, assistant agriculture secretary M. Rupert Cutler announced that until EPA finished its latest study, he would oversee all Forest Service decisions to use these sprays. That same day, Forest Service chief John McGuire issued a directive authorizing herbicide use only after all other alternatives had been considered. His failure to ban the chemicals sparked what one historian has called the "herbicide wars."[34]

Although Assistant Secretary Cutler cast doubt on the causal connection between the use of defoliants in the Vietnam War and their domestic application—"because of the more concentrated and volatile ingredients used in 'Agent Orange,' the Vietnam experience is not comparable to the current use of herbicides in the United States"—he knew the connection was on people's minds. As he noted at a joint EPA/Forest Service symposium: "Part of today's concerns about the use of herbicides on the environment and human health grew out of a 1969 charge that an increase in human birth defects in Vietnam was caused by 'Agent Orange,'" which was ramified when complaints "at home from people who lived near treated forest areas began to receive wide attention in the news."[35]

This issue would have received a lot more attention had Americans known the extent to which the Forest Service was involved in the war in Southeast Asia. What is known is that the agency, as historians Ronald B. Hartzer and David A. Clary observe, "conducted important programs in support of both civilian and military interests in forest management, fire control and employment, and defoliation."[36] Its personnel were also involved in tactical, strategic, and logistical decisions that they carried out on their own or in coordination with the Central Intelligence Agency, the Department of Defense, and the U.S. Agency for International Development; even years later, then chief Ed Cliff refused to speak on the record about these aspects of the Forest Service in Vietnam, because he believed the missions were still classified.[37]

One such cooperative venture involved testing whether the armed forces could integrate the use of herbicides and forest fires to degrade the environment and thus erode the enemy's capabilities. From 1965 to 1967, Forest Service scientists from the Montana and California fire research laboratories were in Vietnam advising on various projects, including Operation Ranch Hand. This operation, which began in 1962 and ended in 1971, involved the aerial spraying of Agent Orange and other defoliants to open up the hardwood jungle canopy to expose enemy movements. Poor initial test results were no deterrent. The Military Assistance Command, Vietnam ordered additional spraying using formulas with increased levels of dioxin. The military command then expanded its list of targets to include food crops, both to starve the enemy and to drive the South Vietnamese off the land and into internment camps.[38] The deleterious impact led Ranch Hand team members in Vietnam to modify Smokey Bear's motto on a Forest Service poster to read: "Only you can prevent a forest." That the Forest Service became involved in efforts to destroy forests is one of several ironies, not the least of which was that the Forest Service had for several years advised and assisted the South Vietnamese in the development of their lumber industry.[39]

Forest Service fire researchers also worked on Operations Sherwood Forest and Pink Rose, which involved chemically defoliating the jungle to create dry fuel and then dropping incendiary weapons, such as magnesium firebombs, to ignite an inferno. Sherwood Forest launched in January 1965 with the intensive bombing of Boi Loi Woods, a dense forest twenty-six miles northwest of Saigon that the U.S. military believed served as an enemy stronghold. Over a two-day period, military aircraft dropped eight hundred tons of bombs before a squadron of C-123s began dispensing 78,800 gallons of herbicide over the next twenty-nine days. Forty days later, after the canopy had fallen and the vegetation had dried, bombers dropped diesel fuel and incendiaries. The rising heat from the fires, however, triggered a rainstorm that doused the flames. The quick return of the Viet Cong—the South Vietnamese communists fighting the South Vietnamese government forces and U.S. forces—to the area soon thereafter indicated that chemical agents alone would not deny them permanent use of the Boi Loi Woods.[40] The official U.S. Air Force historian of Operation Ranch Hand, of which the Sherwood Forest was a part, noted that the ecological conditions made it "almost impossible to set a self-sustaining forest fire in the jungles of South Vietnam."[41]

Figure 7.5. The challenges and controversies surrounding Operation Ranch Hand defoliation missions in Vietnam impelled team members to alter Smokey Bear's motto on a Forest Service poster. Photo courtesy of Ranch Hand Association Vietnam Collection, the Vietnam Center and Archive, Texas Tech University.

That failure did not stop the Defense Advanced Research Projects Agency from contracting with the Forest Service again to explore additional ways that forest fires could become part of the military's arsenal. Enter Operation

Pink Rose, which began in May 1966 and ended a year later. Its planners decided to defoliate the targeted areas in War Zones C, D, and the so-called Iron Zone—Viet Cong strongholds in and around Saigon—and do so three times over the course of a year before attempting to ignite the desiccated vegetation with incendiary bombs.[42] The military had high hopes for Pink Rose and even sent up a planeload of journalists to watch the burn experiments. Results, however, were similar to Sherwood Forest—the heat created rain clouds that extinguished the fires. The military discontinued the firestorm experiments, which one government official later admitted was a "nutty" idea to begin with.[43] Yet defoliation operations to expose communication and travel routes the Viet Cong employed continued in South Vietnam and then expanded into Laos in December 1965 before spreading into North Vietnam in the summer of 1966.[44]

Even as it defoliated and burned forests, the U.S. Forest Service participated in projects designed to regenerate Vietnam's forested domain. In January 1967, as fighting in Vietnam escalated, the Forest Service dispatched a seven-person team of foresters to help the U.S. Agency for International Development conduct forestry operations in South Vietnam. The loan of the foresters came after Chief Ed Cliff and other forestry experts visited Vietnam in 1966 at the request of the secretary of agriculture to study the lumber supply situation. After examining the situation, Cliff agreed to supply Forest Service personnel to help increase local production of lumber and plywood, and tapped Jay H. Cravens, a forester with nearly twenty years of experience, to lead the Forest Service team.[45] Planners hoped that locals would become economically self-sufficient and not side with the Viet Cong—yet another attempt to win the hearts and minds of this embattled people. Whatever the results of that effort, when Cravens arrived in Vietnam in late February 1967, Operation Pink Rose was in full swing, and immediately he was called on to provide technical expertise to the military's deployment of toxic chemicals. Their use was so pervasive throughout the country that Cravens, who visited all forty-four provinces of South Vietnam, later recalled that the country reeked of herbicide.[46]

Not everyone in the Forest Service supported the strategy of using chemicals to incinerate the jungle. William "Bud" Moore, who had grown up in and spent most of his career in western Montana, was serving as national deputy director of fire control at the time of the Sherwood Forest and Pink Rose operations. He was in the process of reevaluating the Forest Service's overall approach to land management, a reevaluation that founds its source in his witnessing the deadly downstream consequences of a 1956 Forest Service DDT spraying operation in the Bitterroot Mountains. A decade later, he was privately questioning the agency's use of herbicides and clearcutting to meet required harvest levels; he also privately questioned and the negative impacts of its fire-suppression policies.[47] In the midst of this reflective process, Moore was offered the opportunity to go to Vietnam to contribute to the agency's fire research experiments in Southeast Asia. He declined. "I didn't have any heart for blowing up the forest, you know, [or] the people over there," Moore told an interviewer. "I just didn't have it. So I told them, no, I'm not going to put my name in . . . I can fight a war if I'm cornered but I don't want to ruin a country or a lot of their important places."[48]

Environmental activists, labor organizers, community officials, and scientists did not want the Forest Service to ruin the Pacific Northwest, either. Many of them felt that that would be the end result of the agency's repeated use of herbicides and pesticides in its land management operations in the region during the 1970s and 1980s, anxiety that was compounded by the fact that the Bureau of Land Management and state forestry departments were following suit. That these agencies were spraying many of the same chemical agents that had been used in Vietnam—such as 2,4,5-T, 2,4-D, and 2,4,5-TP—in Washington, Oregon, and Northern California lent credence to the fears that the Forest Service was bringing the war home.

Certainly, the Pacific Northwest seemed like a battleground, given that environmental groups such as the Sierra Club, workers organizations like the collaborative known as the Hoedads (a progressive reforestation-workers cooperative), and at-risk mothers, among others, protested nearly every announcement of an upcoming aerial-spraying project. With reason. A number of serious, if accidental, incidents of wind drift of aerial-sprayed herbicides destroyed agricultural crops, settled over human habitations, and damaged riparian habitats. One of the most egregious was an early 1970s herbicide spill in the Alsea River watershed in the Siuslaw National Forest in Oregon; it may have resulted from Forest Service contractors brush-spraying 2,4,5-T upstream of the Alsea State Fish Hatchery. Whatever the cause, hatchery officials there suddenly noticed a dramatic fish die-off. Water sampling indicated that the "percentage of 2,4,5-T in the hatchery was the equivalent of about a 55-gallon drum of the stuff being dumped directly in the hatchery."[49] This incident, combined with wind-drift killing of resident-owned gardens as well as flocks of domesticated ducks, geese, and chickens, led Carol Van Strum, who lived within the Alsea watershed, to form Citizens Against Toxic Sprays, a grassroots organization devoted to banning the use of herbicides on public lands. It joined with the Northwest Coalition for Alternatives to Pesticides, the Oregon Environmental Council, and the Hoedads to sue to halt federal and state land management agencies' use of dioxin-laced herbicides and pesticides.

To build these legal cases, the Hoedads in particular developed alternative scientific evidence—what is now called citizen science—to challenge the (usually) uncritical acceptance of chemical applications that the Forest Service, the Bureau of Land Management, and state forestry officials favored. The Hoedads established the Herbicide Study Committee, which boned up on herbicide research, assessed the economic value of the use of chemicals, and conducted field analyses of the efficacy of aerial spraying versus manual clearing of brush. The results led the committee to "question the whole array of confident statistics which are the very underpinnings of justification for aerial herbicide use."[50] The pressure grew to such an extent that Wendell Jones, former timber manager of Forest Service's Region 6, pushed back against the regional office's decision to start an herbicide program on the Mount Hood National Forest, writing later: "My position is that herbicides were very necessary in the management of the Siuslaw, and to a lesser extent the Siskiyou and Umpqua" National Forests. But he was convinced that an herbicide program on the Mount Hood National Forest "would be met with rigorous opposition by the Portland area enviros, and we didn't need to do that and jeopardize the use of herbicides on these other forests. I was able to get the [regional office] folks to back off using that argument."[51]

However politically savvy his plea, Jones admitted, too, that by not deploying herbicides on the Mount Hood National Forest, the agency learned an important lesson in local ecology. "In later years the Ceanothus brush, that we were being pushed to treat to release young DF [Douglas-fir] trees, turned out to be protective cover for the increased elk herds who were turning to DF as a source of food in the winter."[52] Had land managers been willing to incorporate a wider managerial focus that included not just the production of timber but also the maintenance of biodiversity and other nonextractive resources—as their many critics had pressed for in the court of public opinion and the court of law—they might have avoided a number of bitter battles that ended in defeat. In 1983, responding to scientific research and public pressure, the EPA issued its final decision to stop the use of herbicides containing dioxin. In the Forest Service's Region 5 (Oregon and Washington) and Region 6 (California), where debate over pesticides had raged the loudest, the agency and the Bureau of Land Management continued to use other herbicides until March 1984. Then, at that time, an Oregon judicial ruling stated that a government body that used herbicides must fully consider potential human health problems associated with its operations, and that all potential risks associated with their use must be incorporated in the planning process under the National Environmental Policy Act.[53] The two federal land management agencies immediately suspended the use of herbicides on federal lands in Oregon; shortly thereafter, regional forester Zane Grey Smith Jr. also issued a moratorium in California on herbicides. Within five years, those who once had been eager to spray had reached a different conclusion about their once default position. "I don't foresee ever having to use chemical herbicides on this forest again," Siuslaw reforestation expert Tom Turpin observed in 1989. "We have proved that we can manage without chemicals, and we've seen that what we are doing now works better and is less expensive," a realization that Region 6 spokesman Michael Ferris seconded: "We don't want to go back to doing business as usual . . . We were wrong to use chemicals the way we did."[54]

The fights over herbicide use in the 1970s and 1980s forced the Forest Service and others to reconsider their approach to plant and pest control. Where once cost-effectiveness and resource extraction were the main criteria the Forest Service employed for whether to use such chemicals, internal disagreements and external pressure had forced it to weigh and evaluate herbicide-free strategies up front. Manual and mechanical cutting, along with controlled burns, are among the tools that land managers began to adopt as part of an integrated pest-management approach. Currently, where pesticides—whether sprayed from the ground or air—appear to be the best option, their use is tightly regulated and monitored, and must cover a much smaller area, a sharp contrast to the one-time, indiscriminate application of these toxins over tens of thousands of acres.[55]

This shift in the Forest Service's approach became the anvil on which the Skykomish Valley Environmental and Economic Alliance (SVENA) hammered state and private timberland managers in Washington State. In late December 2015, and in language reminiscent of that which Forest Service critics employed four decades earlier, SVENA, which represents an array of grassroots organizations, residents, and businesses in the Sultan-Startup area forty miles east of Seattle, decried a troubling incident of aerial spraying on private forestlands. "There was no advance warning to the residents of this area for this huge spraying operation in our watershed," SVENA alleged. "Local citizens were horrified and upset when they observed for hours a helicopter with toxic clouds around it. This neighborhood has numerous homes, families, children, businesses, farms and organic farms, gardens and orchards. The residents are very concerned about their well water. There are many private wells in this area and most of them are shallow."[56] Because, as the organization observed, no "company or government agency performed follow-up testing or monitoring for possible drift or contamination of non-targeted properties and resources, such as air, surface water and well water," there could be no accountability for any damage to life or property.[57] This need not have happened. After all, SVENA observed, the "United States Forest Service has managed to conduct successful commercial forestry in the Mount Baker–Snoqualmie National Forest for a great many years without using chemical pesticides (herbicides or insecticides). We strongly suggest that private and state timberlands in WA could be managed in the same way, without the aerial application of chemical pesticides."[58] What SVENA demanded was that the state's Forest Practices Board align its practices with federal land managers to "protect Washington's citizens from pesticides applied to forestlands, and monitor the effects of these chemicals on the ground."[59]

Fittingly, SVENA's analysis suggests just how far the Forest Service had come since 1965, when it unleashed chemical warfare on the Douglas-fir tussock moth. Yet SVENA's concerns and those of its peers in the West also indicate that the struggle "to be free from chemical trespass" persists.[60]

ACKNOWLEDGMENTS

We are grateful to the Forest History Society and the Texas Tech University Archives for permission to publish images from their collections.

NOTES

1. Williams, *The U.S. Forest Service in the Pacific Northwest*, 234.
2. Carson, *Silent Spring*, 69.
3. Williams, *The U.S. Forest Service in the Pacific Northwest*, 234.
4. Harold K. Steen, *The Chiefs Remember*, 35–36.
5. Hirt, *A Conspiracy of Optimism*, 82–131; Lewis, *The Forest Service and the Greatest Good*, 136–62; Steen, *The Forest Service*, 246–77.

6. Dombeck, Wood, and Williams, *From Conquest to Conservation*, 29–30.

7. *Timber Resources for America's Future*, 26–88.

8. "FY 1905–2015 National Summary Cut and Sold Data and Graphs, USDA Forest Service," 1. Lumber volume is measured in board feet, which is a board twelve inches wide, twelve inches long, and one inch thick. MBF is 1,000 board feet. The M is the Roman numeral for one thousand.

9. Hirt, *A Conspiracy of Optimism*, xxxii.

10. Kaufman, *The Forest Ranger*, 153, 96, 161; for an analysis of how groupthinking shaped other federal policymaking, see Eden, *Whole World on Fire*, and Hamblin, *Arming Mother Nature*.

11. Kaufman, *The Forest Ranger*, x.

12. Behan, "The Myth of the Omnipotent Forester," 398.

13. Carson, *Silent Spring*, 67–68.

14. Carson, *Silent Spring*, 67–68.

15. Douglas, *My Wilderness*, 53. He recounts his visit to the area and the impact of the spraying program under the Forest Service's "multiple use" policy in chapter 2, "Wind River Mountains." Carson misquotes him as saying the "vast, incredible damage." Carson, *Silent Spring*, 68.

16. Carson, *Silent Spring*, 67–68.

17. Fox, *The American Conservation Movement*, 297.

18. Lear, "Rachel Carson's *Silent Spring*," 23–48.

19. Lewis, *The Forest Service and the Greatest Good*, 148.

20. "The Forest Service in a Changing Conservation Climate," 1.

21. "The Forest Service in a Changing Conservation Climate," 27.

22. Clary, *Timber and the Forest Service*, 185.

23. Connaughton, "Forestry's Toughest Problem," 446.

24. Rahm, "Education and the Environment," 257.

25. Dunlap, *DDT*, 97.

26. Hays, *Beauty, Health, and Permanence*, 1–40.

27. Hirt, *A Conspiracy of Optimism*, 219.

28. Paananen, Fowler, and Wilson, "The Aerial War against Eastern Region Forest Insects, 1921–86," 185.

29. Johnson, "Keynote Address," 13–17, passim. The symposium was held by the Forest Service and the EPA in response to the criticism and lawsuits over the Forest Service's use of herbicides, particularly those containing 2,4,5-T and 2,4,5-TP. By the late 1980s, some field scientists in the Forest Service, known as "ologists," were disenchanted enough with the Forest Service's timber program that they formed the Association of Forest Service Employees for Environmental Ethics (later shortened to FSEEE). As an internal counter to agency groupthink and whistleblowers, FSEEE challenged agency leadership's continued emphasis on timber production at the expense of other ecological values and launched lawsuits to stop timber sales and other management decisions that it believed violated federal laws, threatened biodiversity, or despoiled wilderness. Hirt, *A Conspiracy of Optimism*, 283–85; and Lewis, *The Forest Service and the Greatest Good*, 204–5.

30. Lewis, *The Forest Service and the Greatest Good*, 188.

31. Dunlap, *DDT*, 177–231.

32. Fedkiw, *Managing Multiple Uses on National Forests*, 165.

33. Robbins, *Landscapes of Conflict*, 199–200.

34. Robbins, *Landscapes of Conflict*, 199–200.

35. Cutler, "Keynote Address," 11.

36. Hartzer and Clary, *Half a Century in Forest Conservation*, 29.

37. Hartzer and Clary, *Half a Century in Forest Conservation*, 29. For more on the Forest Service and international forestry, see Terry West, "USDA Forest Service Involvement in Post–World War II International Forestry," 277–91, and Williams, *The USDA Forest Service*, 138–41.

38. Cecil, *Herbicidal Warfare*, 29–35. The chemicals Agent Orange and Agent White received their names from the markings on their barrels. They had been developed to retard the growth of broad-leaved weeds and for the defoliation of crops such as cotton so that mechanical pickers could work more efficiently, but the concentration levels used in Southeast Asia ensured maximum and prolonged effect on a broad range of jungle vegetation. Agent Blue was more effective on food crops but only in dry weather because it was water-soluble. Other agents included Green, Pink, and Purple, which were used between 1962 and 1964 before being replaced by Orange, Blue, and White. For more on this, see Cecil, *Herbicidal Warfare*, 225–32, and Buckingham, *Operation Ranch Hand*, 195–202.

39. Lewis, *The Forest Service and the Greatest Good*, 193–94; and Martini, *Agent Orange*, 47. Two assistance programs had been proposed in 1954. One was a loan program for purchasing equipment and the other was a reforestation and soil erosion control to stabilize sand dunes on the coast. Page, *Forest Industries of the Republic of Vietnam*, 42.

40. Cecil, *Herbicidal Warfare*, 57–58; Buckingham, *Operation Ranch Hand*, 109–12.

41. Buckingham, *Operation Ranch Hand*, 112; Martini, *Agent Orange*, 46.

42. Cecil, *Herbicidal Warfare*, 77–78.

43. "U.S. Admits Move to Burn Forests," *New York Times*, July 22, 1972, 5.

44. Cravens, *A Well Worn Path*, 324; Buckingham, *Operation Ranch Hand*, 127; Robert Reinhold, "A Modest Proposal—'Sherwood Forest,'" *New York Times*, July 23, 1972, E2.

45. Hartzer, *Half a Century in Forest Conservation*, 263.

46. Flamm and Cravens, "Effects of War Damage on the Forest Resources of South Vietnam," 789; Jay H. Cravens, interview with James G. Lewis, June 23, 2005, Forest History Society.

47. At the time he was asked to go to Vietnam, Moore was thinking about land management in holistic terms and called for "ecologically enlightened change" in the Forest Service's wildfire policy. In 1972 he led a Forest Service team that conducted a pilot study on wildfires in federally designated wilderness areas. See Smith, "From Research to Policy," 6. The National Park Service changed its wildfire policy in 1968.

48. Lewis, "Thinking Like a File Cabinet," 244. See also Moore's memoir, *The Lochsa Story*, 341–52.

49. Williams, *The U.S. Forest Service in the Pacific Northwest*, 277–80.

50. Loomis, *Empire of Timber*, 177–78.

51. Loomis, *Empire of Timber*, 170–80; Jones quoted in Williams, *The U.S. Forest Service in the Pacific Northwest*, 280.

52. Jones quoted in Williams, *The U.S. Forest Service in the Pacific Northwest*, 280.

53. Larsen, "Herbicides, the Forest Service, and NEPA," 38.

54. Stein, "U.S. to End Ban on Spraying Herbicides in State's Forests," *Los Angeles Times*, February 28, 1989, 3, 21.

55. Fedkiw, *Managing Multiple Uses on National Forests*, 167. For an update and overview on early twenty-first-century use of herbicides on public lands, see Wagner, Antunes, Irvine, and Nelson, "Herbicide Usage for Invasive Non-native Plant Management in Wildland Areas of North America."

56. "Letter of Protest for Toxic Spray."

57. "Letter of Protest for Toxic Spray."

58. "Letter of Protest for Toxic Spray."

59. "Letter of Protest for Toxic Spray."

60. "Letter of Protest for Toxic Spray." Similar protests erupted in Oregon in 2014: Clarren, "Timberland Herbicide Spraying Sickens a Community," *High Country News*, November 10, 2014, https://www.hcn.org/issues/46.19/timberland-herbicide-spraying-sickens-a-community.

BIBLIOGRAPHY

Behan, R. W. "The Myth of the Omnipotent Forester." *Journal of Forestry* 64, no. 6 (June 1966): 398.

Buckingham, William A., Jr. *Operation Ranch Hand: The Air Force and Herbicides in Southeast Asia, 1961–1971*. Washington, DC: Office of Air Force History, U.S. Air Force.

Carson, Rachel. *Silent Spring*. Boston: Houghton Mifflin, 1962.

Cecil, Paul Frederick. *Herbicidal Warfare: The Ranch Hand Project in Vietnam*. New York: Praeger, 1986.

Clarren, Rebecca. "Timberland Herbicide Spraying Sickens a Community." *High Country News*, November 10, 2014. https://www.hcn.org/issues/46.19timberland-herbicide-spraying-sickens-a-community.

Clary, David A. *Timber and the Forest Service*. Lawrence: University of Kansas Press, 1986.

Connaughton, Charles. "Forestry's Toughest Problem." *Journal of Forestry* 64, no. 7 (July 1966): 446–48.

Cravens, Jay H. Interview with James G. Lewis, June 23, 2005, Forest History Society.

Cravens, Jay H. *A Well Worn Path*. Huntington, WV: University Editions, 1994.

Cutler, M. Rupert. "Keynote Address." In *Symposium on the Use of Herbicides in Forestry [Proceedings]*, 10–13. Washington, DC: Department of Agriculture, Office of the Secretary, 1978.

Dombeck, Michael P., Christopher A. Wood, and Jack E. Williams. *From Conquest to Conservation: Our Public Lands Legacy*. Washington, DC: Island Press, 2003.

Douglas, William O. *My Wilderness: East to Katahdin*. Garden City, NY: Doubleday, 1961.

Dunlap, Thomas R. *DDT: Scientists, Citizens, and Public Policy*. Princeton, NJ: Princeton University Press, 1981.

Eden, Lynn. *Whole World on Fire: Organizations, Knowledge, and Nuclear Weapons Devastation*. Ithaca, NY: Cornell University Press, 2006.

Fedkiw, John. *Managing Multiple Uses on National Forests*. Washington, DC: U.S. Forest Service, 1998.

Flamm, Barry R., and Jay H. Cravens. "Effects of War Damage on the Forest Resources of South Vietnam." *Journal of Forestry* 69, no. 11 (November 1971): 784–89.

"Forest Service in a Changing Conservation Climate: I and E Approaches to Organizational and Public Needs, 1966." Washington, DC: U.S. Forest Service, 1966.

Fox, Stephen. *The American Conservation Movement: John Muir and His Legacy*. Madison: University of Wisconsin Press, 1986.

"FY1905–2015NationalSummaryCutandSoldDataandGraphs,USDAForestService."http://www.fs.fed.us/forestmanagement/documents/sold-harvest/documents/1905-2015_Natl_Summary_Graph.pdf.

Hamblin, Jacob Darwin. *Arming Mother Nature: The Birth of Catastrophic Environmentalism*. New York: Oxford University Press, 2013.

Hartzer, Ronald B., and David A. Clary. *Half a Century in Forest Conservation: A Biography and Oral History of Edward P. Cliff*. Washington, DC: U.S. Forest Service, 1981.

Hays, Samuel P. *Beauty, Health, and Permanence: Environmental Politics in the United States, 1955–1985*. New York: Cambridge University Press, 1989.

Hirt, Paul W. *A Conspiracy of Optimism: Management of National Forests since World War Two*. Lincoln: University of Nebraska Press, 1996.

Johnson, Edwin L. "Keynote Address: The Regulation of Herbicides for Use in Forest Management." In *Symposium on the Use of Herbicides in Forestry [proceedings]*, 13–17. Washington, DC: Department of Agriculture, Office of the Secretary, 1978.

Kaufman, Herbert. *The Forest Ranger: A Study in Administrative Behavior*. Baltimore: Johns Hopkins Press for Resources for the Future, 1960.

Larsen, Gary L. *EPA Journal* 14, no. 1 (January 1988): 38–39.

Lear, Linda J. "Rachel Carson's *Silent Spring*." *Environmental History Review* 17 (Summer 1993): 23–48.

"Letter of Protest for Toxic Spray." *SVENA*, December 29, 2015. https://svena.org/letter-of-protest-for-toxic-spray/.

Lewis, James G. *The Forest Service and the Greatest Good: A Centennial History*. Durham, NC: Forest History Society, 2005.

Lewis, James G. "Thinking Like a File Cabinet: Eco-cruising in the Bitterroots with Bud Moore." In *The Land Speaks: New Voices at the Intersection of Oral and Environmental History*, edited by Debbie Lee and Kathryn Newfont, 233–47. New York: Oxford University Press, 2017.

Loomis, Erik. *Empire of Timber: Labor Unions and the Pacific Northwest Forests*. New York: Cambridge University Press, 2016.

Martini, Edwin A. *Agent Orange: History, Science, and the Politics of Uncertainty*. Boston: University of Massachusetts Press, 2012.

Moore, Bud. *The Lochsa Story: Land Ethics in the Bitterroot Mountains*. Missoula, MT: Mountain Press Publishing, 1996.

Paananen, Donna M., Richard F. Fowler, and Louis F. Wilson. "The Aerial War against Eastern Region Forest Insects, 1921–86." *Journal of Forest History* 85, no. 10 (October 1987): 173–86.

Page, Rufus H. *Forest Industries of the Republic of Vietnam*. Washington, DC: U.S. Forest Service, 1967.

Rahm, Neal M. "Education and the Environment." *Journal of Forestry* 68, no. 5 (May 1970), 257.

Reinhold, Robert. "A Modest Proposal—'Sherwood Forest.'" *New York Times*, July 23, 1972, E2.

Robbins, William. *Landscapes of Conflict: The Oregon Story, 1940–2000*. Seattle: University of Washington Press, 2010.

Smith, Diane. "From Research to Policy: The White Cap Wilderness Fires Study." *Forest History Today* (Spring/Fall 2014): 4–12.

Steen, Harold K. *The Chiefs Remember: The Forest Service, 1952–2001*. Durham, NC: Forest History Society, 2004.

Steen, Harold K. *The Forest Service: A History*. 2nd edition. Durham, NC: Forest History Society, 1976, 2005.

Stein, Mark A. "U.S. to End Ban on Spraying Herbicides in State's Forests." *Los Angeles Times*, February 28, 1989, 3, 21.

Timber Resources for America's Future. Forest Resource Report No. 14. Washington, DC: U.S. Forest Service, 1958.

"U.S. Admits Move to Burn Forests." *New York Times*, July 22, 1972, 5.

Wagner, Viktoria, Pedro M. Antunes, Michael Irvine, and Cara R. Nelson. "Herbicide Usage for Invasive Non-native Plant Management in Wildland Areas of North America." *Journal of Applied Ecology*, June 29, 2016. https://doi.org/10.1111/1365-2664.12711.

West, Terry. "USDA Forest Service Involvement in Post–World War II International Forestry." In *Changing Tropical Forests: Historical Perspectives on Today's Challenges in Central and South America*, edited by Harold K. Steen and Richard P. Tucker, 277–91 Durham, NC: Forest History Society, 1992.

Williams, Gerald W. *The USDA Forest Service—The First Century*. Washington, DC: U.S. Forest Service, 2005.

Williams, Gerald W. *The U.S. Forest Service in the Pacific Northwest: A History*. Corvallis: Oregon State University Press, 2009.

8

Neighborhood Oil Drilling and Environmental Justice in Los Angeles

Bhavna Shamasunder

The Los Angeles Basin contains one of the highest concentrations of crude oil in the world, with over five thousand active oil wells in Los Angeles County.[1] Oil development in this region began in the 1890s and reached its peak in the 1930s, making up nearly half of California's oil production at the time and nearly one-quarter of the world's oil output.[2] Urban and suburban development grew alongside active oil production. Oil development is both ubiquitous and invisible in the city of Los Angeles, which has the largest urban oil field in the country and over one thousand oil wells. While many Angelenos are surprised today to learn of the extent of oil development in their region, this obfuscation is the result of specific and often uneasy historical and regulatory compromises. Since the 1890s, after oil was first discovered, residents fought for a share in oil-driven prosperity but bristled at the pollution and environmental destruction caused by oil drilling. Efforts to assert local control over drilling faltered against state, federal, and corporate measures to maintain economic and political control over oil resources.[3] The legacy of decision making over oil drilling is that thousands of active wells in the greater Los Angeles area today are located among a dense population of more than ten million people. Moreover, 70 percent of active oil wells in the city are located within 1,500 feet of a home or sensitive land use such as a school, playground, or hospital, places where people live, work, and play.[4]

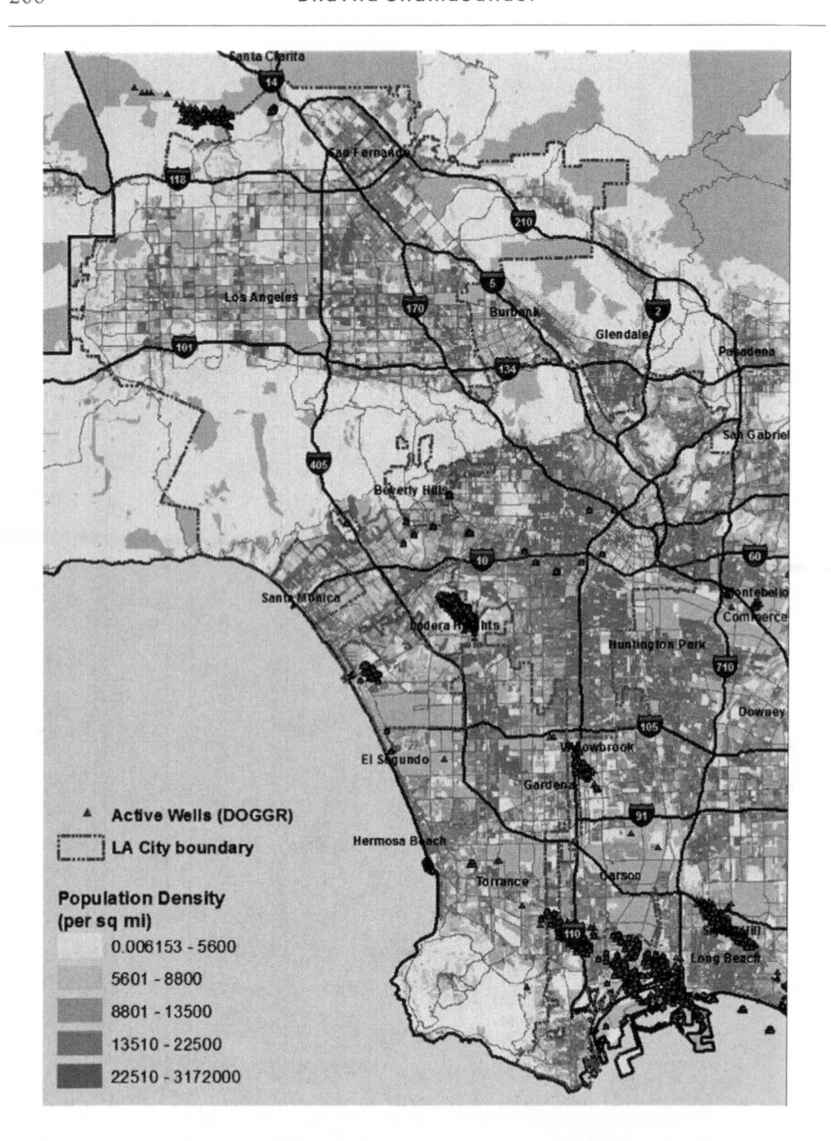

Figure 8.1. Active oil wells in Los Angeles County and their relationship to population density. Map courtesy of Dr. James Sadd and Liberty Hill Foundation.

Oil wells are scattered across the city and county, but in poor communities and communities of color, the distance between wells and their neighbors is closer than in wealthier and whiter neighborhoods. These communities are further exposed to contamination via outdated emissions equipment and uncovered rather than enclosed fields.[5] Narrow definitions of environmental injustice as distributional harm distort struggles between frontline environmental justice communities and government and regulatory officials over oil drilling.

In 2015 the California Council on Science and Technology conducted an independent scientific assessment of well stimulation in California. They concluded that "while it is clear that oil and gas is being developed in low-income communities and communities of color, there does not appear to be a disproportionate burden of oil and gas development on any one demographic in the Los Angeles Basin. In other words, oil and gas wells are not located disproportionately near the rich, the poor, or any race/ethnicity more than any other."[6] With this framing by scientific and regulatory authorities, race and poverty do not explain the presence of oil wells, the operations of oil fields, or the long-standing government and regulatory neglect of oil-adjacent communities of color.[7]

The South Los Angeles neighborhood of West Adams is a case in point. There, the Jefferson oil field wall is located just three feet away from the nearest home, and the field itself constitutes a complex of more than sixty active oil wells. For members of this primarily Black and Latinx South Los Angeles neighborhood, where over 60 percent of residents live below the poverty line, information about the oil field, its ongoing operations, and on-site hazards are hard to obtain. Chemicals trucked into the closed compounds are shrouded in claims of trade secrecy, and it takes constant effort by community residents to learn about plans and activities at the field.[8] Despite hard-won changes to regulation that require notice by oil companies to the local air district seventy-two hours in advance of any type of oil development activity, such as well drilling or maintenance, residents most often learn of toxic activity when their health is directly impacted and they suffer nosebleeds, headaches, or worsened asthma.[9] West Adams residents have filed a petition for nuisance abatement to enclose the oil field and afford them the same protections found in wealthier neighborhoods.[10] Such disparities in the enforcement and regulation of oil industry operations have prompted oil field–adjacent communities such as those in South Los Angeles to raise a number of questions about systemic environmental injustice by city and county agencies.[11] Certainly, the proximate distances of oil development to dense urban populations make Los Angeles a prime case for examining contested expertise in a toxic environment.

The more recent introduction of hydraulic fracturing and other unconventional technologies exacerbates ongoing issues from conventional oil drilling. During the second Bush administration, Congress gave fracking and related technologies special exemptions from federal laws such as the Clean Water Act and the Safe Drinking Water Act, and devolved authority over these protections to the states. The 2005 Energy Policy Act affirmed that states, rather than the federal government, had jurisdiction over the oil and gas industry.[12] Local decision-making authority can be preempted by state laws that can hinder local action. Further, communities are often unclear as to where authority lies. For example, noise or smell pollution is reportable to the South Coast Air Quality Management District, while the Los Angeles County Department of Public Health can intervene at the time when a community experiences health impacts from pollution. But these agencies do not often work in coordination. This fragmented regulatory landscape further dampens progress on environmental justice. As a result, and despite acknowledgment by state and local governments that poor communities of color are disproportionately and cumulatively burdened by hazardous industries, patterns of racial injustice remain in place.

In this chapter, I situate oil drilling in Los Angeles in the context of long-standing fights for environmental justice, both as a social movement and a policy concern, and examine how environmental justice communities in Los Angeles have responded to increases in urban drilling since 2010. I first contextualize oil production in the Los Angeles region during the early twentieth century, during which discoveries of several oil-rich fields led to rampant overproduction. The resulting environmental degradation from uncontrolled oil development prompted early social movements in the 1920s that contested unregulated drilling and sought to restrict drilling practices.[13] I then turn to the more contemporary landscape of oil development with the arrival of unconventional technologies in Los Angeles (e.g., hydraulic fracturing, i.e., "fracking" or acidizing, the related technology more commonly used in Los Angeles). While conventional drilling is a one-hundred-year-old practice in Los Angeles, the use of new technologies combined with the national wave of anti-fracking activism in the early 2000s set the stage for the emergence in 2014 of an organized coalition of front-line environmental justice groups—constituted through STAND-L.A. (Stand Together against Neighborhood Drilling)—to respond to these developments. I argue that efforts to achieve environmentally just policies and community protections at the local level have historically been stymied by minimal regulation of the oil industry at local, state, and national levels, although local activism against neighborhood oil drilling did flourish in the 1920s and 1930s. Today's organizing by frontline communities against pollution by neighborhood oil fields can be considered a "second wave" of local opposition to oil industry practices, separated by a nearly sixty-five-year gap during which the oil industry set deep roots, built a strong lobby, and expanded in neighborhoods across the Southland.

EARLY DRILLING AND FIGHTS FOR LOCAL CONTROL

The oil industry has been central to the economic, political, and spatial development of California. The struggle over property rights and regulation of oil in the early twentieth century shaped how oil drilling is managed today. Oil extraction, transport, and sales are a mix of jurisdictional authority that was negotiated through the courts, legislatures, state policy, and economic interests. The discovery of oil across California in the 1890s prompted a rush to develop common oil pool resources and, subsequently, to fights over property rights among state, federal, and private entities. The laws governing oil followed the "right of capture," whereby individuals could only claim ownership by pulling oil from the ground, leading to a rush for rapid extraction that destabilized oil markets. State and federal governments jockeyed for jurisdiction over oil revenues alongside other economic interests, and there ensued a constant political struggle in the first half of the twentieth century over local and federal petroleum policy and property rights. From the 1920s through World War II, politicians and industry sought to contain oil overproduction, a consequence of the rush to develop huge pools of discovered oil, and in so doing to stabilize oil prices and more efficiently develop petroleum reserves.[14] In the resulting compromises, California retained latitude to influence patterns of economic development and environmental change related to oil development at the state level, but national policies such as tax incentives for oil drilling constrained the state's options.[15] At the local level, conflicts over oil pitted those who asserted property rights as central against those who sought protection from the ill effects of drilling. Meanwhile, the oil industry funded public parks and other voluntary efforts that aimed to marshal public support and deflect efforts at regulatory control. Thus, for more than one hundred years, oil drilling proceeded in Los Angeles with few regulatory constraints.[16]

A political economy of oil dictated California's regulatory landscape over a range of issues (e.g., transportation policy, property rights over subsurface resources); however, a discussion of the ways in which the state and the oil industry interfaced with national policy is beyond the scope of this chapter.[17] Local response to rampant oil production has been described in Los Angeles's beach cities, but fewer accounts exist in the interior communities where today's frontline struggles are centered.[18] Nancy Quam-Wickham describes one of the few accounts of the response to oil drilling in Los Angeles's working-class suburbs. She finds that local politics diverged from state and federal policies, particularly in the urban environment of Los Angeles, arguing that the environmental degradation that followed on the heels of oil extraction led to early working-class mobilizations to restrict drilling in neighborhoods.[19]

In Los Angeles, the first oil wells were opened in the 1890s, and their presence dictated land use patterns, encouraged side-by-side industrialization and suburbanization, and propelled rampant real estate speculation. In 1913 Los Angeles produced 14 percent of California's oil. After World War I, the discovery of numerous oil fields in the Los Angeles Basin led to further surges in oil development. By 1924 Los Angeles accounted for 80 percent of the state's oil production at 872,000 barrels per day, and the region dominated state oil production through the 1920s and 1930s.[20] As in the rest of California, Los Angeles faced overproduction, and with unplanned and widespread extraction came extensive oil pollution. Pipelines leaked, ships dumped oil-contaminated water into public waterways, waste oil clogged the Los Angeles harbor, and beachgoers routinely coped with oily tar on beaches. Terrible pollution led to intense local and largely suburban opposition from those who lived and worked near oil production sites.[21] Residents worked with local city authorities to pursue strategies for local control and to improve the surrounding environment. Local efforts were initially successful at slowing the place of oil development, particularly in blue-collar Southland communities. Led by unions, communities such as Lomita, Hawthorne, and Torrance worked to pass prohibitions. Residents had first celebrated newfound oil because of jobs, but soon concerns arose over public health, street damage from heavy equipment, water table reduction, and fire protection, issues chronicled by residents in public meetings. In the working-class suburb of Hawthorne, incorporation papers prohibited oil development within city boundaries.[22] Unions and elected leaders worked to rezone neighborhoods and held referendums to ban oil drilling within city limits. In the city of Torrance, a ban on oil drilling in the city's residential and business districts was approved in 1923.[23]

Local victories such as these turned out to be short lived and rarely resulted in longer-term reforms at the state or federal levels, particularly as the industry lobby found ways to circumvent local authority and appeal directly to state officials. However, these local political struggles did slow oil development and left a legacy of "an intensely local political culture throughout suburban Los Angeles, a culture that revolved around debates over the pace and character of industrial development in the region."[24] To push back against local regulatory victories, the oil industry at this time forged deep political ties and developed a forceful lobby that successfully moved authority out of local jurisdictions. A previously divided industry now worked to remove local opposition and suppress working-class organizing efforts. In the decades that followed, the petroleum industry emerged as a leading sector in the California economy, producing one-quarter of the world's oil by 1930 and reaching peak production of over 120,000 barrels per day in the 1980s.[25] In the aftermath of the first oil boom of the 1920s and 1930s, and following World War II, efforts to curb oil production were largely structured as a fight between competing economic interests—the tourism sector and the oil industry—rather than being linked to grassroots or working-class mobilizations.[26] "Growth machine" politics, a coalition of local business and local government, pursued a well-documented joint agenda of land use that dominated Los Angeles political culture after World War II. This political consensus largely took the place of local organizing efforts in curbing the landscape of oil production and played an important role in how the oil industry in Los Angeles looks today.[27]

In the early part of the 1900s, Southern California beaches were largely privately held but undeveloped and open for recreational uses. Angelenos came to think of the beaches as their own. The drilling boom of the 1920s alarmed beachgoing Angelenos, and in the 1940s, conflict arose over the incompatible land uses of the shoreline. Advocates who fought to keep the shoreline free of drilling asserted beaches as public, noncommercial, and nonindustrial spaces. These efforts were championed primarily by planning associations, state governments that moved to purchase beach land, tourism business interests, and chambers of commerce, rather than the popular and grassroots working-class protests of the 1920s. Cities such as Santa Monica and Huntington Beach were generally successful in purchasing lands from private holders and reclaimed their shorelines for public access. But efforts to curb oil production had little traction outside of recreational policy.[28] The flip side of the success of recreation- and tourism-related victories over oil drilling was the inability to curb oil production in neighborhoods farther

Figure 8.2. The THUMS Islands are a set of four artificial islands in San Pedro Bay off the coast of Long Beach, California. They were built in 1965 to tap into the East Wilmington Oil Field. The landscaping and sound walls were designed to camouflage the operation and reduce noise, and they are the only decorated oil islands in the United States. Photo courtesy of Adam Benwell.

from the shore. In the midst of World War II, federal authorities and the military effectively pressured Los Angeles to open the interior of the city and residential neighborhoods to drilling as a show of patriotism for the war effort.[29] Working-class neighborhoods that had fought the oil booms of the 1920s and that were increasingly home to low-income communities of color were opened to largely unregulated drilling, neighborhoods that would spur the environmental justice movement of the 1980s.[30]

In the continued effort to curb regulatory oversight, the oil industry made massive investments in camouflage and beautification in both offshore and interior oil fields. These efforts were aimed at keeping oil activities from the attention of neighboring residents and government regulators as Los Angeles's population grew.[31] Companies built tall hedges around wells and planted landscaped gardens in front of "no trespassing" gates. In Long Beach in 1965, oil companies hired a famed Disney theme park architect, Joseph H. Linesch, to design the $10 million THUMS Islands. Named for the consortium of companies Texaco, Humble (now Exxon), Union, Mobil, and Shell, THUMS embodied the "aesthetic mitigation of technology."[32] The complex is a set of four artificial islands built to camouflage drill rigs with landscaping, waterfalls, and tall structures, and its purpose is to hide from view forty-two acres of oil fields and 1,100 wells in a vast underground oil field.[33] By the late 1980s and 1990s, as the price of oil dropped and local property values rose, many oil wells around Los Angeles were capped and oil production fell.[34]

UNCONVENTIONAL DRILLING TECHNOLOGIES IN THE EARLY TWENTY-FIRST CENTURY

Through the 1990s and the early 2000s, though oil prices and production decreased, drilling continued. Over this time period, and in the absence of any unified oppositional social movement, sustained public attention to the oil industry was generally lacking. The *Los Angeles Times* reports on oil drilling as occasional, accidental, and isolated local incidents rather than as part of the structural fabric on which the city was built. Accidents from oil drilling came to be seen as unanticipated or surprising. One such event was the hydrogen sulfide leak underneath Belmont High School, built atop an oil field in downtown Los Angeles, in 1998. The leak was discovered before the school opened, but the district had already invested in the construction, seemingly taken off guard by the presence of oil-related gases. The issue dragged out for ten years as the school district was forced to close the hazardous site, reopening it in 2008 after installing a $17 million gas mitigation system.[35] This period in Los Angeles history marks a departure from the earlier postwar period. Public activism shifted away from oil drilling as a focus, and oil companies worked toward a politics of invisibility.

While conventional oil-drilling practices in Los Angeles faced limited large-scale social resistance, a new unconventional drilling technology—hydraulic fracturing—began to draw attention nationally. Over the past two decades, United States energy policy encouraged the development of domestic energy sources to reduce America's dependence on foreign oil and advance energy independence.[36] As oil was depleted from conventional and more accessible geologic formations, the fossil fuel industry aggressively pursued "unconventional oil" deposits that were located in deeper, more difficult to recover deposits.[37] By the early 2000s, widespread use of hydraulic fracturing and related technologies such as acid fracturing made the recovery of hard-to-access deposits possible. "Fracking" shatters subterranean rocks by introducing liquids at high pressures to extract oil or gas. Most unconventional drilling aims to access natural gas within shale rock, which is typically located at significant depths several miles below the surface. Large volumes of water or acid, sand, and chemical additives are injected at high pressure into horizontally drilled boreholes to break apart shale formations and release trapped gas.[38]

The 2005 "Halliburton loophole" (named after the Texas energy company that invented the technology) exempts unconventional fossil fuel extraction from the nation's major environmental protection laws, including the Clean Air Act, Clean Water Act, and the Safe Drinking Water Act, that were passed in the 1970s.[39] Since 2014, more than six hundred peer-reviewed studies have been published that investigate the environmental and public health consequences of natural gas development, with 84 percent of them finding public health hazards, elevated risks, or adverse health outcomes. In sum, unconventional oil-drilling practices are connected with exposure to benzene and other volatile organic compounds that are known carcinogens and reproductive toxins. Hydraulic fracturing is also significantly linked with groundwater contamination, elevated air pollution, and associated risks to human health.[40] Federal regulation of these practices is thus both lacking and lax, and has left impacted communities at the local level, including those in Los Angeles, to largely fend for themselves.

By 2009 the use of horizontal drilling technologies (i.e., acid fracking) in Los Angeles neighborhoods revived local concerns over oil drilling. The national wave of anti-fracking activism arrived in Los Angeles by 2010 and with it the presence of national environmental groups that had earlier fought fracking in New York. Groups such as Food and Water Watch formed initial connections with local neighborhoods over drilling, but these alliances quickly frayed as their interests diverged. Neighborhood groups, particularly environmental justice communities, bristled with what they saw as large national environmental organizations deprioritizing local concerns over public health, lax regulatory enforcement, and entrenched environmental injustice related to oil drilling over cries to "ban fracking now." The following section chronicles what can be considered a "second wave" of local, working-class, and neighborhood activism in Los Angeles that has not existed since the 1920s. This new movement is informed by and integrated with the environmental justice movement that emerged in the 1980s, a movement that advanced an understanding of racial patterns of injustice and disproportionate exposures to pollution and other environmental hazards in poor communities of color.[41]

ANTI-DRILLING ACTIVISM AND ENVIRONMENTAL JUSTICE IN LOS ANGELES

In Los Angeles, unconventional oil extraction methods such as "acidizing" (where corrosive acids are used instead of water as a fracturing fluid) and "directional drilling" (where a well is vertically drilled thousands of feet below the surface and then directionally, horizontally or at an angle, for up to two miles) ushered in renewed investment into old and idle wells.[42] New wells have been drilled or redrilled under the auspices of old permits that were granted by the city of Los Angeles in the 1960s. Moreover, oil fields in South Los Angeles and many other sites across the city predate major environmental legislation enacted in the 1970s, such as the Clean Air Act, the Clean Water Act, and the California Environmental Quality Act, which requires an Environmental Impact Review.[43] Early drilling permits have thus been grandfathered, in some cases allowing the field owner to drill scores of additional wells under the original oil permits without any type of environmental review.

Horizontal or directional underground drilling also enables companies to reach underneath urban residences and buildings and hides the magnitude of oil development in Los Angeles. On the street level, the oil company is hidden behind closed and often landscaped fences. The drilling begins behind this fence at a well pad where drilling spiders horizontally below the ground beneath homes, businesses, playgrounds, schools, and streets. For example, the Jefferson and Murphy oil fields in South Los Angeles are connected underground through a web of pipes that are invisible at the surface. As a result, residents are often unaware they live near or on top of an active drill site. The use of corrosive acids such as hydrochloric and hydrofluoric acid in acid fracturing further threatens public health. Oil companies bring corrosive acids through neighborhoods in trucks marked with cautionary warnings for hazardous chemicals and chemicals that have been linked with birth defects.[44] Chemical additives used in unconventional oil development and regular well maintenance have been identified as a toxicity risk with a lack of adequate evaluation of the hazard posed by their use.[45] In the West Adams neighborhood of South Los Angeles, residents can look over their backyard walls into the oil compound and see workers wearing hazmat suits, while they themselves have no protection.

By 2010 residents of South Los Angeles started to notice and report odors, nosebleeds, and headaches, and began to loosely connect with other communities in Los Angeles also reporting similar issues.[46] During this period, national anti-fracking activism brought mainstream environmental organizations into the fray of local and state politics. In December 2014, Governor Andrew Cuomo signed legislation to ban fracking in New York State.[47] In the wake of this victory, Food and Water Watch brought their campaign to Los Angeles, with slogans such as "Don't Frack California." However, this framing slighted working-class neighborhoods and environmental justice communities that had differences in strategy, ideology, and proposed solutions.

The clash between mainstream environmental organizations and environmental justice groups is long-standing. The largest mainstream environmental organizations known as the Big 10 (e.g., the Sierra Club, Natural Resources Defense Council, and Environmental Defense Fund) were founded with goals of wilderness preservation and conservation of natural resources. Following World War II, groups that grew as a response to the rise of industrialization, such as the Sierra Club, turned to preserving places that were increasingly under threat from resource use, such as Yosemite Valley.[48] In the 1960s, Rachel Carson's book *Silent Spring*, written about the harms of DDT to wildlife, ushered in the modern-day environmentalism that was concerned with the widespread use of pesticides and the impacts of chemicals on the environment.[49] The mainstream environmental movement was largely spearheaded by white activists. By contrast, the environmental justice movement has its roots in the civil rights organizing of the 1960s and is characterized by grassroots organizing by communities of color. Starting in the 1980s, environmental justice communities publicized fights against hazardous waste landfills and toxic waste dumping in their urban neighborhoods, and forwarded a theoretical framework describing distributional injustice—that environmental harms were disproportionately located in poor communities and communities of color.[50] Environmental justice communities argued that mainstream environmentalism had neglected where people "live, work, and play," and in particular had abandoned environmental problems in the urban core. Today, mainstream environmental groups continue to be better funded and often lack voices of communities of color in their ranks.[51] The friction between environmental justice and mainstream environmental groups remains, though there have been some important strides to patch tensions and work together to find common ground.[52]

In the mid-1980s, Los Angeles became an epicenter for environmental justice organizing. The Mothers of East Los Angeles formed to fight the proposed construction of a state prison, and residents of Wilmington organized against the pollution emitted by the Phillips 66 oil refinery, with support from Communities for a Better Environment.[53] These early struggles were important in highlighting health and safety hazards of living near oil refineries, and they framed oil development as an environmental justice problem, with communities of color facing a disproportionate burden of harm from a widely used resource. In subsequent decades, the environmental justice movement deepened and expanded the framework of environmental injustice to include ongoing issues such as enforcement disparities and cumulative burdens of harm.[54] Through the various efforts of environmental justice leaders, in 2002 California became one of the first states to acknowledge the role of government agencies in alleviating environmental injustice. After a decade of research, the California Environmental Protection Agency published CalEnviroScreen, which ranks communities according to economic, social, and environmental vulnerability.[55] Low-income neighborhoods with active oil development, such as South Los Angeles and Wilmington, are thus identified as communities that face some of the highest burdens of pollution and are recognized by state agencies as "environmental justice communities." In practice, this identification supports efforts to deter new hazardous industries from entering communities and aims to help target investment into these neighborhoods.[56] Yet, as already noted, oil drilling and existing oil fields continue to be exempted from many major environmental laws, making the effort of environmental justice communities to seek public health protections an ongoing challenge.

The tensions between more recent mainstream anti-fracking activism and the decades-long struggle of low-income and minority communities suffering from exposure to undesirable land uses in Los Angeles were epitomized in the effort of city officials to oppose unconventional drilling practices. In February 2014, Los Angeles City Council members Mike Bonin and Paul Koretz, who represent districts on the wealthier west side and beach cities of Los Angeles, took up a fracking moratorium championed by Food and Water Watch and Los Angeles Waterkeeper, two national mainstream environmental organizations. Food and Water Watch brought the campaign "Ban Fracking in California" to Los Angeles, with the main problems described as water pollution, air pollution, and earthquake and property damage.[57] It was a broad agenda that lacked local specificity. In championing the ban with city leaders, Food and Water Watch and Waterkeeper made no mention of health impacts to communities living near oil wells, well proximity to houses, lack of regulation, or the incompatible land uses in neighborhoods already burdened by many other sources of toxic pollution—the core concerns of frontline environmental justice groups. The city council voted to send the moratorium to the city attorney's office and the city planning department to be written as a zoning ordinance. Following the vote, Bonin tweeted: "Los Angeles just became the largest city in the nation to support a moratorium on fracking & other dangerous drilling."[58] But by November 2014, the city planning department released a sixty-seven-page report in which the deputy director of planning in the city of Los Angeles, Alan Bell, stated that there were no qualified city staff with expertise in petroleum and natural gas engineering and geology who could work with the planning department to implement the moratorium.[59] The planning department was hesitant to move forward and suggested the city take a different route. Bell argued that federal, state, and county rules would supersede any city restrictions on unconventional drilling. He noted that "the oil industry has argued that local jurisdictions have no authority to regulate how oil and gas extraction occurs. This issue is the subject of pending litigation in Western States Petroleum Association v. City of Compton et al. (LASC Case No. BC552272). The lawsuit is in response to an ordinance passed in Compton prohibiting hydraulic fracturing, acidizing, and other well stimulation activities. It is unlikely that this legal issue will be resolved anytime soon. On September 24, 2014, the City Council of Compton rescinded the moratorium at the City Attorney's recommendation."[60]

Indeed, in April 2014, the predominantly African American and Latinx city of Compton in Los Angeles County had passed an ordinance to ban new drilling within their city borders. City officials sought to prevent underground drilling operations that could drill into the city from outside the city limits. These drills could access oil deposits located beneath the city's boundaries. The Western States Petroleum Association sued Compton and rightly claimed that the local ban was unconstitutional and preempted by state regulation. The association argued that Compton could not control drilling into their city that originated outside of its borders. Under the threat of an extended legal battle, the Compton city government was forced to drop the ordinance.[61] Given the historic loss of local jurisdictions to regulate oil field development, the neighboring city of Carson soon backed out of a similar moratorium effort. Ultimately, Los Angeles, too, dropped its fracking moratorium.[62]

In this scenario, environmental justice organizations viewed national environmental groups as alienating frontline communities by promoting an agenda that did not include long-standing local concerns over public health from undesirable and incompatible land uses. Moreover, many frontline neighborhood groups understood the attempt of the Los Angeles city fracking moratorium as shortsighted, vulnerable to oil industry pressure, and limited in terms of securing long-term public health protections for environmental justice communities because it did not cover existing conventional drilling operations that continue unabated in neighborhoods.[63] Communities in South Los Angeles and Wilmington, which house the majority of active wells in Los Angeles, instead sought reforms that would ensure public health protections from all oil development activities, address proximity of drilling to neighborhoods, improve lax regulatory enforcement, incorporate recognition of cumulative harm, and afford communities basic public health protections through zoning changes, protective buffers, and setbacks from active oil fields. The importance of framing oil drilling in low-income communities of color as a public health issue pushed already-connected frontline groups to form the coalition STAND-L.A., which had until then been a set of loosely organized groups.

SETTING BOUNDARIES AND BUILDING ALLIANCES

In early 2014, frontline communities began meeting to share strategies. Local groups sought to clearly define their organizing agenda by creating a joint mission and vision, and collectively defining the path forward. Since April 2014, Redeemer Community Partnership, Esperanza Community Housing, Communities for a Better Environment, Physicians for Social Responsibility–Los Angeles, and Liberty Hill Foundation (which served as convener) had been meeting as the Oil Extraction Working Group. By January 2015, the working group had created a name (STAND-L.A.) and an organizational structure. The coalition created a concentric circle system. STAND-L.A. made decisions internally, and organizations with common concerns (e.g., the Sierra Club and other mainstream environmental groups) were invited to offer their support. This strategy engaged larger environmental organizations on broader issues related to oil development, such as climate change, but kept the coalition focused on frontline and neighborhood issues. In this way, STAND-L.A. sought to control the policy agenda and media framing

Figure 8.3. The Jefferson Drill Site, West Adams Neighborhood, South Los Angeles. Photo courtesy of Richard Parks.

to ensure that the issue of health and safety risks from oil development in working-class neighborhoods was at the forefront of organizing.[64]

In November 2014, STAND-L.A. organized its first effort in support of one of its members, Redeemer Community Partnership, whose neighborhood sits adjacent to the Jefferson oil field. Freeport-McMoRan Oil and Gas, a multinational mining giant based in Phoenix, Arizona, and operator of the Jefferson oil field at that time, sought to drill and redrill three new wells at this site. That month, the South Coast Air Quality Management District held a public meeting to discuss this issue. Over 150 residents and activists attended the meeting, and by January 2015, in the face of overwhelming opposition, Freeport-McMoRan withdrew its application. This was an early and important victory to protect community health. The Los Angeles planning department, however, could not impose any conditions on Freeport-McMoRan's existing operations, an issue that continues to be a struggle for communities given the aforementioned grandfathered permits.

The following year, in November 2015, Communities for a Better Environment, with a coalition of youth in member organizations of STAND-

L.A., sued the city of Los Angeles over racially discriminatory oil-drilling permitting. The suit, filed in Los Angeles County Superior Court, accused the city of systematically violating the California Environmental Quality Act by exempting new wells and other proposed changes in oil extraction sites from required environmental review. The lawsuit called out the pattern of racial injustice, with Black and Latinx communities disproportionately placed in harm's way. The suit argued that more stringent conditions, including taller walls, better sound protection, and greater pollution controls on equipment were required for drilling sites in West Los Angeles, a higher-income neighborhood, than in Wilmington or South Los Angeles. The suit sought to halt what communities see as a "pattern and practice" of approvals that result in unequal protection.[65] In response to the lawsuit and prior to settlement, the Los Angeles planning department implemented new procedures and guidelines to ensure that the city complies with the California Environmental Quality Act when permitting new oil wells. The city will not issue permits for oil drilling without public notice and a hearing, and will examine the project's health, safety, and environmental threats.[66] Following the settlement, a state-based oil and gas trade group, the California Independent Petroleum Association, sued both the youth and the city, arguing that they did not receive an appropriate opportunity to provide input into the new guidelines, which "unilaterally and improperly [override] state law."[67] This claim by the California Independent Petroleum Association draws from the long history of the oil industry seeking to remove oil regulation from local control. For environmental justice groups, this victory created an important direction for addressing future drilling in Los Angeles. STAND-L.A.'s organizing agenda has given local neighborhoods a voice in oil development for the first time since the 1920s and has highlighted how communities in noncoastal and working-class neighborhoods suffer from pollution and health consequences from living near oil and gas development operations.

The coalition's ongoing work has brought wider attention to neighborhood oil drilling and, from a grassroots and environmental justice perspective, has been able to rebuild bridges with broader social movements while keeping frontline agendas as their core struggle. They have forged partnerships with new allies such as Oscar-nominated actor Mark Ruffalo, who, after a neighborhood oil visit, penned the blog post "Why is L.A. Toxic?"[68] Neighborhood oil drilling was also a focal point of the "Break Free" fossil fuels march held in countries across the world, including the United States, the

UK, Australia, South Africa, and Indonesia over a two-week period in May 2016. Neighborhood oil activists from Los Angeles were profiled in international newspapers with a photo caption reading: "Activists oppose oil wells in immigrant neighborhoods."[69] These efforts demonstrate the connections that environmental justice organizations make with broader social movements on issues such as climate change, "keep it in the ground" organizing, 350.org, and fossil-free future efforts. But STAND-L.A. clearly frames their work as protecting community safety and public health in poor neighborhoods of color as their central concern and the driver of their policy agenda.

CONCLUSION

Residents in dense urban areas view neighborhood oil drilling as a safety hazard and a threat to their health and environment, a problem further magnified by extreme weather events and environmental catastrophes and crises. Frontline communities today, located in many of the same neighborhoods that fought against rampant extraction in the 1920s, have revived the intense culture of local political organizing that was suppressed by oil industry and growth machine politics that dominated Los Angeles after World War II. Bolstered by the legacy of environmental justice activism, interior communities near oil wells far from Los Angeles's beaches are primarily working-class communities of color that have cut their teeth on organizing against undesirable land uses since the 1980s. The tight-knit consensus of business and city leaders that dominated the past fifty years of growth in Los Angeles is also eroding, making space for local activism and environmental groups.[70] This convergence of factors creates an important space for the current upsurge in local activism against neighborhood oil drilling, epitomized by the work of STAND-L.A.

In this chapter, I argue that social movement activism and racialized histories have shaped the politics and landscape of oil extraction in the city of Los Angeles. Communities in the interior, without the protective arguments of beach recreation, became home to unregulated drilling after World War II. At the same time, these neighborhoods were increasingly racialized and segregated by land use and zoning practices of the 1950s and 1960s.[71] The ongoing struggle to create safe and healthy communities in the shadow of the long history of oil development in Los Angeles might thus provide useful insights for impacted communities elsewhere.

ACKNOWLEDGMENTS

To the activists and residents of STAND-L.A., who struggle at the frontlines every day, you have been inspiring collaborators on difficult research questions. I am grateful to my colleagues in the Urban and Environmental Policy Department at Occidental College—Robert Gottlieb, Martha Matsuoka, and Peter Dreier—who champion engaged scholarship and have supported my work in numerous ways. Thank you to Robert Gottlieb and Brinda Sarathy for helpful edits on this chapter, and to Brinda, Vivien, and Janet for shepherding this project to the finish line.

NOTES

1. Signal Hill Petroleum, "Signal Hill Petroleum Successfully Conducts 1st Aerial Gravity Gradiometry Survey of Los Angeles Basin."

2. Taylor, "The Urban Oil Fields of Los Angeles."

3. Elkind, "Black Gold and the Beach."

4. Sadd and Shamasunder, "Oil Extraction in Los Angeles."

5. Reyes, "Community Group Petitions City to Enclose South L.A. Drilling Site."

6. Shonkoff and Gautier, "A Case Study of the Petroleum Geological Potential and Potential Public Health Risks Associated with Hydraulic Fracturing and Oil and Gas Development in the Los Angeles Basin," 245.

7. Ranganathan, "Thinking with Flint."

8. Redeemer Community Partnership, "Odor Control Chemical Spotted at Jefferson Drill Site."

9. South Coast Air Quality Management District, "Oil and Gas Well Electronic Notification and Reporting"; Sahagun, "Chemical Odor, Kids' Nosebleeds, Few Answers in South L.A. Neighborhood."

10. Earthjustice, "Petition for Abatement of Public Nuisance"; Reyes, "Environmental Advocates Sue L.A., Accusing It of 'Rubber Stamping' Oil Drilling Plans."

11. Bogado, "The Sad, Sickening Truth about South L.A.'s Oil Wells."

12. Warner and Shapiro, "Fractured, Fragmented Federalism."

13. Quam-Wickham, "'Sacrificed on the Alter of Oil.'"

14. Elkind, "Oil in the City."

15. Sabin, *Crude Politics*, 206.

16. Elkind, "Oil in the City."

17. Sabin, *Crude Politics*.

18. Elkind, "Black Gold and the Beach."

19. Quam-Wickham, "'Cities Sacrificed on the Altar of Oil.'"

20. Quam-Wickham, "'Cities Sacrificed on the Altar of Oil,'" 191. The full title is "'Cit-

ies Sacrificed on the Alter of Oil"; Popular Opposition to Oil Development in 1920s Los Angeles"

21. Quam-Wickham, "'Cities Sacrificed on the Altar of Oil.'"

22. Quam-Wickham, "'Cities Sacrificed on the Altar of Oil'"; 'Cities Sacrificed on the Alter of Oil"; Popular Opposition to Oil Development in 1920s Los Angeles"; Nicolaides, *My Blue Heaven.*

23. Quam-Wickham, "'Cities Sacrificed on the Altar of Oil,'" 201.

24. Quam-Wickham, "'Cities Sacrificed on the Altar of Oil,'" 202. 'Cities Sacrificed on the Alter of Oil"; Popular Opposition to Oil Development in 1920s Los Angeles"

25. Department of Conservation, "Fact Sheet." Independent Petroleum Association of America, "The Story of California Crude."

26. Elkind, *How Local Politics Shape Federal Policy.*

27. Fulton, *The Reluctant Metropolis*; Elkind, *How Local Politics Shape Federal Policy.*

28. Elkind, *How Local Politics Shape Federal Policy*, 2011.

29. Elkind, "Oil in the City."

30. Pulido, Sidawi, and Vos, "An Archaeology of Environmental Racism in Los Angeles."

31. Momtastic WebEcoist, "Fuels Paradise"; Gougis, "THUMS Oil Islands."

32. Schoch, "Toasting Industry as Art."

33. Gougis, "THUMS Oil Islands."

34. Gamache and Frost, "Urban Development of Oil Fields in the Los Angeles Basin Area 1983–2001."

35. Blume, "New Name, New Life for Belmont School."

36. White House, "Advancing American Energy."

37. U.S. Environmental Protection Agency, "Sector Performance Report."

38. Gallegos and Varela, *Trends in Hydraulic Fracturing Distributions and Treatment Fluids, Additives, Proppants, and Water Volumes Applied to Wells Drilled in the United States from 1947 through 2010.*

39. Howarth, Ingraffea, and Engelder, "Natural Gas."

40. Hays and Shonkoff, "Toward an Understanding of the Environmental and Public Health Impacts of Unconventional Natural Gas Development."

41. Bullard, ed., *Unequal Protection.*

42. There are large stores of accessible oil under Los Angeles, which is the aim of most drilling and recovery efforts Yerkes et al., "Geology of the Los Angeles Basin, California"; Sharp and Allayaud, "California Regulators."

43. Freeport-McMoRan Oil and Gas, "Project Description and Discussion."

44. STAND-L.A., "Acid-Filled Trucks Arrive in South L.A., Residents Hold a Communal Phone-In to Report Odors."

45. Stringfellow et al., "Identifying Chemicals of Concern in Hydraulic Fracturing Fluids Used for Oil Production."

46. Sahagun, "Chemical Odor, Kids' Nosebleeds, Few Answers in South L.A. Neighborhood."

47. Gerken, "Gov. Andrew Cuomo to Ban Fracking in New York State."

48. Dunlap and Mertig, *American Environmentalism.*

49. Griswold, "How 'Silent Spring' Ignited the Environmental Movement."

50. Cole and Foster, *From the Ground Up.*

51. Taylor, "The State of Diversity in Environmental Organizations."

52. Green 2.0, "Leaders Supporting Diversity Transparency."

53. Pardo, *Mexican American Women Activists.*

54. Sze and London, "Environmental Justice at the Crossroads."

55. Office of Environmental Health Hazard Assessment, "California Communities Environmental Health Screening Tool, Version 2.0."

56. California Environmental Protection Agency, "California Climate Investments to Benefit Disadvantaged Communities."

57. Food and Water Watch, "Ban Fracking in California."

58. Braker, "Breaking."

59. Bell, "Council Files 13–1152, 13–1152-S1 Regulatory Controls over Well Stimulation."

60. Bell, "Council Files 13–1152, 13–1152-S1 Regulatory Controls over Well Stimulation," 1.

61. O'Hara, "Law to Keep Fracking Outta Compton Challenged by Oil Industry."

62. Kirby, "The Fight over Fracking Heats Up."

63. Oil Extraction Working Group, Oil Extraction Working Group Meeting, June 26, 2014. Los Angeles, California.

64. Oil Extraction Working Group, Oil Extraction Working Group Planning Meeting, January 17, 2015. Los Angeles, California.

65. Communities for a Better Environment; Gupta Wessler, PLLC; and the Center for Biological Diversity, Youth for Environmental Justice; South Central Youth Leadership Coalition; Center for Biological Diversity v. City of Los Angeles; City of Los Angeles Department of City Planning, Case No. Verified Complaint for Declaratory and Injunctive Relief and Petition for Writ of Mandate; November 6, 2015.

66. Communities for a Better Environment, "Youth Groups Settle Lawsuit Challenging L.A.'s Oil Drilling Approvals."

67. Braun, "After Los Angeles Youth Sued City for Discriminatory Drilling Practices, the Oil Industry Sued Back."

68. Ruffalo, "Why Is L.A. Toxic?"

69. Oliver Milman, "'Break Free' Fossil Fuel Protests Deemed 'Largest Ever' Global Disobedience."

70. Purcell, "The Decline of the Political Consensus for Urban Growth."

71. Davis, *City of Quartz.*

BIBLIOGRAPHY

Bell, Alan. "Council Files 13–1152, 13–1152-S1 Regulatory Controls over Well Stimulation." Los Angeles: Department of City Planning, November 5, 2014. http://clkrep.lacity.org/onlinedocs/2013/13-1152-s1_rpt_plan_11-6-14.pdf.

Blume, Howard. "New Name, New Life for Belmont School." *Los Angeles Times*, August 10, 2008. http://articles.latimes.com/2008/aug/10/local/me-belmont10.

Bogado, Aura. "The Sad, Sickening Truth about South L.A.'s Oil Wells." *Grist* (blog), November 16, 2015. http://grist.org/cities/the-sad-sickening-truth-about-south-l-a-s-oil-wells/.

Braker, Brandon. "Breaking: Los Angeles Passes Fracking Moratorium." *EcoWatch* (blog), February 28, 2014. http://ecowatch.com/2014/02/28/breaking-los-angeles-passes-fracking-moratorium/.

Braun, Ashley. "After Los Angeles Youth Sued City for Discriminatory Drilling Practices, the Oil Industry Sued Back." *DeSmogBlog*. Accessed June 21, 2017. https://www.desmogblog.com/2017/04/03/youth-color-lawsuit-los-angeles-drilling-discrimination-oil-industry.

Bullard, Robert D., ed. *Unequal Protection: Environmental Justice and Communities of Color.* San Francisco: Sierra Club Books, 1997.

California Environmental Protection Agency. "California Climate Investments to Benefit Disadvantaged Communities." 2016. http://www.calepa.ca.gov/EnvJustice/GHGInvest/.

Cole, Luke W., and Sheila Foster. *From the Ground Up: Environmental Racism and the Rise of the Environmental Justice Movement.* New York: New York University Press, 2001.

Communities for a Better Environment. "Youth Groups Settle Lawsuit Challenging L.A.'s Oil Drilling Approvals: City Implements New Procedures to Promote Compliance with State Laws in Oil Permitting." October 20, 2016. http://www.cbecal.org/wp-content/uploads/2016/11/16.10-20-Settlement.Pr_.Release.pdf.

Communities for a Better Environment; Gupta Wessler, PLLC; and the Center for Biological Diversity. Youth for Environmental Justice; South Central Youth Leadership Coalition; Center for Biological Diversity v. City of Los Angeles; City of Los Angeles Department of City Planning. Case No. Verified Complaint for Declaratory and Injunctive Relief and Petition for Writ of Mandate; November 6, 2015. http://guptawessler.com/wp-content/uploads/2015/11/Youth-EJ-v-City-Complaint-fnl.pdf.

Davis, Mike. *City of Quartz: Excavating the Future in Los Angeles.* New ed. London: Verso Books, 2006.

Department of Conservation, Division of Oil, Gas, and Geothermal Resources. "Fact Sheet." n.d. http://www.conservation.ca.gov/index/Documents/LA%20oil%20production%20history.pdf.

Dunlap, Riley E., and Angela G. Mertig. *American Environmentalism: The U.S. Environmental Movement, 1970–1990.* London: Taylor and Francis, 2014.

Earthjustice. "Petition for Abatement of Public Nuisance." June 9, 2016. http://earthjustice.org/sites/default/files/files/Petition%20to%20Department%20of%20Planning%20%E2%80%93%20Final%20(Electronic).pdf.

Elkind, Sarah S. "Black Gold and the Beach: Offshore Oil, Beaches, and Federal Power in Southern California." *Journal of the West* 44, no. 1 (Winter 2005): 8–17.

Elkind, Sarah S. *How Local Politics Shape Federal Policy: Business, Power, and the Environment in Twentieth-Century Los Angeles.* Chapel Hill: University of North Carolina Press, 2011.

Elkind, Sarah S. "Oil in the City: The Fall and Rise of Oil Drilling in Los Angeles." *Journal of American History* 99, no. 1 (June 1, 2012): 82–90. https://doi.org/10.1093/jahist/jas079.

Food and Water Watch. "Ban Fracking in California." June 2012. http://www.foodandwaterwatch.org/sites/default/files/Ban%20Fracking%20California%20FS%20June%202012.pdf.

Freeport-McMoRan Oil and Gas. "Project Description and Discussion." 2015. http://jeffersonparkunited.org/sites/jeffersonparkunited.org/files/user160/pdf_nodes/FMOG%20CEB%20ZA%20Ex%201%20Prjct%20Descrp%20Discuss%20140711.pdf.

Fulton, William. *The Reluctant Metropolis: The Politics of Urban Growth in Los Angeles.* Baltimore: Johns Hopkins University Press, 2001.

Gallegos, Tanya J., and Brian A. Varela. *Trends in Hydraulic Fracturing Distributions and Treatment Fluids, Additives, Proppants, and Water Volumes Applied to Wells Drilled in the United States from 1947 through 2010: Data Analysis and Comparison to the Literature.* USGS Numbered Series. Scientific Investigations Report 2014-5131. Reston, VA: U.S. Geological Survey, 2015. http://pubs.er.usgs.gov/publication/sir20145131.

Gamache, Mark T., and Paul L. Frost. "Urban Development of Oil Fields in the Los Angeles Basin Area 1983–2001." Sacramento: California Department of Conservation, Division of Oil, Gas, and Geothermal Resources, 2003. ftp://ftp.consrv.ca.gov/pub/oil/publications/tr52.pdf.

Gerken, James. "Gov. Andrew Cuomo to Ban Fracking in New York State." *Huffington Post*, December 18, 2014. http://www.huffingtonpost.com/2014/12/17/cuomo-fracking-new-york-state_n_6341292.html.

Gougis, Michael. "THUMS Oil Islands: Half a Century Later, Still Unique, Still Iconic." *Long Beach Business Journal*, October 26, 2015. http://www.lbbizjournal.com/#!THUMS-Oil-Islands-Half-A-Century-Later-Still-Unique-Still-Iconic/c1sbz/562e5c680cf27ade992c760f.

Green 2.0. "Leaders Supporting Diversity Transparency." *Green 2.0* (blog). Accessed December 20, 2016. http://www.diversegreen.org/talking-about-data/.

Griswold, Eliza. "How 'Silent Spring' Ignited the Environmental Movement." *New York Times*, September 21, 2012. http://www.nytimes.com/2012/09/23/magazine/how-silent-spring-ignited-the-environmental-movement.html.

Hays, Jake, and Seth B. C. Shonkoff. "Toward an Understanding of the Environmental and Public Health Impacts of Unconventional Natural Gas Development: A Categorical Assessment of the Peer-Reviewed Scientific Literature, 2009–2015." *PLOS ONE* 11, no. 4 (April 20, 2016): e0154164. https://doi.org/10.1371/journal.pone.0154164.

Howarth, Robert W., Anthony Ingraffea, and Terry Engelder. "Natural Gas: Should Fracking Stop?" *Nature* 477, no. 7364 (September 15, 2011): 271–75. https://doi.org/10.1038/477271a.

Independent Petroleum Association of America. "The Story of California Crude." Accessed December 15, 2016. http://oilindependents.org/the-story-of-california-crude/.

Kirby, David. "The Fight over Fracking Heats Up." TakePart, July 25, 2016. http://www.takepart.com/feature/2016/07/25/fight-over-fracking-heats.

Milman, Oliver. "'Break Free' Fossil Fuel Protests Deemed 'Largest Ever' Global Disobedience." *Guardian*, May 16, 2016. http://www.theguardian.com/environment/2016/may/16/break-free-protest-fossil-fuel.

Momtastic WebEcoist. "Fuels Paradise: THUMS Islands Help Big Oil Look Good." March 16, 2010. http://webecoist.momtastic.com/2010/03/16/fuels-paradise-thums-islands-help-big-oil-look-good/.

Nicolaides, Becky M. *My Blue Heaven: Life and Politics in the Working-Class Suburbs of Los Angeles, 1920–1965*. 1st ed. Chicago: University of Chicago Press, 2002.

Office of Environmental Health Hazard Assessment. "California Communities Environmental Health Screening Tool, Version 2.0." 2014. http://oehha.ca.gov/media/CES20FinalReportUpdateOct2014.pdf.

O'Hara, Mary Emily. "Law to Keep Fracking Outta Compton Challenged by Oil Industry." *VICE News*, July 25, 2014. https://news.vice.com/article/law-to-keep-fracking-outta-compton-challenged-by-oil-industry.

Oil Extraction Working Group. Oil Extraction Working Group Meeting, June 26, 2014.

Oil Extraction Working Group. Oil Extraction Working Group Planning Meeting, January 17, 2015.

Pardo, Mary. *Mexican American Women Activists.* Philadelphia: Temple University Press, 1998.

Pulido, Laura, Steve Sidawi, and Robert O. Vos. "An Archaeology of Environmental Racism in Los Angeles." *Urban Geography* 17, no. 5 (July 1, 1996): 419–39.

Purcell, Mark. "The Decline of the Political Consensus for Urban Growth: Evidence from Los Angeles." *Journal of Urban Affairs* 22, no. 1 (January 1, 2000): 85–100.

Quam-Wickham, Nancy. "'Cities Sacrificed on the Altar of Oil': Popular Opposition to Oil Development, in 1920s Los Angeles." *Environmental History* 3, no. 2 (April 1, 1998): 189–209. https://doi.org/10.2307/3985379.

Quam-Wickham, Nancy. "'Sacrificed on the Alter of Oil': Los Angeles' Uneasy Relationship with Petroleum." *Progressive Democracy*, February 2015. http://taipd.org/node/262.

Ranganathan, Malini. "Thinking with Flint: Racial Liberalism and the Roots of an American Water Tragedy." *Capitalism Nature Socialism* 27, no. 3 (July 2, 2016): 17–33.

Redeemer Community Partnership. "Update on Activity at Jefferson Drill Site." March 15, 2016; http://makejeffersonbeautiful.weebly.com/news-blog/update-on-activity-at-jefferson-drill-site.

Reyes, Emily Alpert. "Community Group Petitions City to Enclose South L.A. Drilling Site." *Los Angeles Times*, June 9, 2016. http://www.latimes.com/local/lanow/la-me-ln-oil-drilling-building-20160609-snap-story.html.

Reyes, Emily Alpert. "Environmental Advocates Sue L.A., Accusing It of 'Rubber Stamping' Oil Drilling Plans." *Los Angeles Times*, November 6, 2015. http://www.latimes.com/local/lanow/la-me-ln-lawsuit-oil-drilling-20151106-story.html.

Ruffalo, Mark. "Why Is L.A. Toxic?" *Huffington Post* (blog), May 11, 2016. http://www.huffingtonpost.com/mark-ruffalo/why-is-la-toxic_b_9910640.html.

Sabin, Paul. *Crude Politics: The California Oil Market, 1900–1940.* Edited by Philip Rousseau. Berkeley: University of California Press, 2004.

Sadd, James, and Bhavna Shamasunder. "Oil Extraction in Los Angeles: Health, Land Use, and Environmental Justice Consequence." In *Drilling Down: The Community Consequences of Expanded Oil Development in Los Angeles*, 7–14. Los Angeles: Liberty Hill Foundation, 2015. https://www.libertyhill.org/sites/libertyhillfoundation/files/Drilling%20Down%20Report_1.pdf.

Sahagun, Louis. "Chemical Odor, Kids' Nosebleeds, Few Answers in South L.A. Neighborhood." *Los Angeles Times*, September 21, 2013. http://articles.latimes.com/2013/sep/21/local/la-me-0922-oil-20130922.

Schoch, Deborah. "Toasting Industry as Art." *Los Angeles Times*, September 13, 2006. http://articles.latimes.com/2006/sep/13/local/me-islands13.

Sharp, Renee, and Bill Allayaud. *California Regulators: See No Fracking, Speak No Fracking.* Environmental Working Group, February 2012. http://static.ewg.org/reports/2012/fracking/ca_fracking/ca_regulators_see_no_fracking.pdf.

Shonkoff, Seth B. C., and Donald Gautier. "A Case Study of the Petroleum Geological Potential and Potential Public Health Risks Associated with Hydraulic Fracturing and Oil and Gas Development in the Los Angeles Basin." In *Independent Scientific Assessment of Well Stimulation in California: An Examination of Hydraulic Fracturing and Acid Stimulations in the Oil and Gas Industry*. California Council on Science and Technology, 2015. https://ccst.us/publications/2015/vol-III-chapter-4.pdf.

Signal Hill Petroleum. "Signal Hill Petroleum Successfully Conducts 1st Aerial Gravity Gradiometry Survey of Los Angeles Basin." December 2, 2014. http://mb.cision.com/Main/8678/9690258/319785.pdf.

South Coast Air Quality Management District. "Oil and Gas Well Electronic Notification and Reporting." Accessed November 28, 2016. http://www.aqmd.gov/home/rules-compliance/compliance/1148-2.

STAND-L.A. "Acid-Filled Trucks Arrive in South L.A., Residents Hold a Communal Phone-In to Report Odors." STAND-L.A., November 8, 2015. http://www.stand.la/2/post/2015/11/acid-filled-trucks-arrive-in-south-laresidents-hold-a-communal-phone-in-to-report-odors.html.

Stringfellow, William T., Mary Kay Camarillo, Jeremy K. Domen, Whitney L. Sandelin, Charuleka Varadharajan, Preston D. Jordan, Matthew T. Reagan, Heather Cooley, Matthew G. Heberger, and Jens T. Birkholzer. "Identifying Chemicals of Concern in Hydraulic Fracturing Fluids Used for Oil Production." *Environmental Pollution* 220, Part A (January 2017): 413–20. doi:10.1016/j.envpol.2016.09.082.

Sze, Julie, and Jonathan K. London, "Environmental Justice at the Crossroads." *Sociology Compass* 2, no. 4 (July 1, 2008): 1331–354.

Taylor, Alan. "The Urban Oil Fields of Los Angeles." *Atlantic*, August 26, 2014. http://www.theatlantic.com/photo/2014/08/the-urban-oil-fields-of-los-angeles/100799/.

Taylor, Dorceta. "The State of Diversity in Environmental Organizations." July 2014. https://orgs.law.harvard.edu/els/files/2014/02/FullReport_Green2.0_FINALReducedSize.pdf.

U.S. Environmental Protection Agency. "Sector Performance Report." Washington, DC, 2008. https://nepis.epa.gov/Exe/ZyPDF.cgi/P1001IJT.PDF?Dockey=P1001IJT.PDF.

Warner, Barbara, and Jennifer Shapiro. "Fractured, Fragmented Federalism: A Study in Fracking Regulatory Policy." *Publius: The Journal of Federalism* 43, no. 3 (July 1, 2013): 474–96.

White House. "Advancing American Energy." 2015. https://www.whitehouse.gov/energy/securing-american-energy.

Yerkes, R. F., T. H. McCulloh, J. E. Schoellhamer, and J. G. Vedder. "Geology of the Los Angeles Basin, California: An Introduction." U.S. Geological Survey Professional Paper 420-A (January 1, 1965). http://www.osti.gov/scitech/biblio/6736958.

PART THREE

Community Contestation, Expanding Expertise

9

Atomic Bomb Survivors, Medical Experts, and the Endlessness of Radiation Illness

Naoko Wake

The atomic bombs dropped on Hiroshima and Nagasaki in 1945 have unfailingly provoked historical fascination. In particular, recent scholarship has highlighted, often with little connection to local contexts, how the bomb resulted in nation-specific, gendered understandings of Americans as masculine victors and the Japanese as feminine victims in medical and cultural discourses in the nuclear age. In these discourses, the bomb's survivors are often helpless "guinea pigs" at U.S. scientists' disposal or "keloid girls" whose scarred beauty could be retrieved only by America's advanced medical technologies.[1] Much scholarly attention, too, has focused on institutional medicine, such as the genetic research conducted by the Atomic Bomb Casualty Commission (ABCC) or the plastic surgeries performed in the mid-1950s on Japanese women whose faces were disfigured by the bomb.[2] My examination of about one thousand survivors who reside in America today—U.S.-born, U.S. citizens of Japanese ancestry who happened to be in Japan in 1945 in addition to the Japanese who came to America after the war and became citizens at some point—challenges these dualistic, institution-based understandings of survivors, exploring their history outside rigid gender norms, medical categories, and national boundaries. By illuminating multilayered Japanese American experiences of Hiroshima and Nagasaki, I hope to begin to destabilize categorizations such as “feminine victims-as-patients” and “masculine victors-as-doctors.” U.S. survivors’ experiences, too, help us discover lost local settings for understanding the bomb’s human costs, including not only immediate deaths and injuries but also long-standing radiation illness that affects survivors regardless of their nationality.[3] Survivors sought medical care that attended to emotional and social difficulties of living with radiation illness. And yet, they found medical experts and institutions largely driven by national interests, often privileging scientific research interests over patients’ desire to be treated with care and respect. Moreover, Japanese American survivors have been repeatedly excluded from the scope of state-sponsored medicine both in Japan and the United States, showing the historical limit of such medicine as a response to the nuclear attacks.[4]

This is part of my attempt to integrate richly individual experiences into our historical understanding of nuclear destruction, radiation illness, and medicine. Using oral history interviews that I and others have conducted as my main source, I first look at Japanese American survivors’ memories about a range of folk medicines that they offered to each other at each city’s ground zero. At the same time that I highlight how creatively they used resources at hand, including human bones, sea shells, and vegetables to heal their burns, I focus on how a majority of caretakers were women, whose significance has escaped the scholarly attention given to institutional medicine predominantly conducted by male scientists. I also show how U.S. survivors’ determination to care for each other was intertwined with their cross-national life experiences. These experiences led some U.S. survivors to resist the research conducted by the ABCC, some going so far as to question the commission’s right to collect blood samples. Finally, I examine the rise of a U.S. survivors’ movement in the 1970s and 1980s, which attempted to bring medical care to all survivors in America. In this movement, women again took leadership roles. In quiet yet adamant resistance to the medical establishment’s focus on cancer, mutation, and malformation, Japanese Americans pursued psychological and communal support from both the U.S. and Japanese governments. Getting access to regular, locally available health checkups and medical consultations as well as creating community meeting places where they could share their difficult experiences of being injured by their own government were their priorities. These boundary-crossing aspects of U.S. survivors’ approach to medical care—and women’s effort to acquire it—reveal an understanding of injury and illness missing from the dominant scholarly discourse about the bomb. As their oral histories show,

U.S. survivors have been resourceful makers of medicine that suits their needs, not simply helpless patients or subjects of medical science. Shaped by a realization that medicine driven by national interests does not heal, U.S. survivors' history reveals both the inadequacy of such medicine in the nuclear age and the potency of survivors' agency formed by their persistent cross-national ties.

JAPANESE AMERICANS ON THE GROUND

In many ways, Japanese American survivors of Hiroshima and Nagasaki have been "forgotten." They are little known among scholars and the public alike, and their number is smaller than their Japanese or Korean counterparts.[5] But their existence is not surprising if one considers how movement of people across the Pacific predates the Pacific War and continues to this day. In the latter half of the nineteenth century, Japanese immigrants began to arrive in Hawai'i and on the West Coast of the United States. Many came with an intention of working temporarily but ended up staying permanently. At the same time, they remained connected to their families in Japan. It was common among first-generation immigrants to send their children to the old country for a few years of education.[6] Thus, an estimated fifteen thousand Americans of Japanese ancestry were in Japan when the war broke out in 1941, and more than three thousand of them were in Hiroshima, a prefecture that had sent the largest number of immigrants to Hawai'i and America before the war. Beginning in 1947 and into the 1950s and 1960s, many who survived the bomb came back to the United States.[7] They were reunited with their families who had stayed and been sent to Japanese American concentration camps during the war. As Japanese Americans struggled to rebuild their lives after the war, most remained silent about their experiences of being bombed or incarcerated. Silence over cross-national histories seemed the best strategy for surviving the Cold War culture of conformity. Also coming to America during this time were Japanese citizens who had family and friends in America, seeking educational or occupational opportunities rare in the island nation still reeling from the war. There were quite a few who fell into this category, reaching one thousand by the end of the 1960s. They were not U.S. citizens at the time of the bombing, but many of their siblings, cousins, uncles, and aunts were—so it is when people migrate from one country to another while keeping ties to both. Categories that usually separate people—citizenship and national belonging—become blurred. For U.S. survivors, the war, nuclear destruction, and rebuilding happened in the context of familial bonds bridging the Pacific. The nation-bound understanding of the bomb that has shaped the scholarship—Americans as victors and Japanese as victims—does not hold for U.S. survivors. For them, the bombs came down on people with a range of life histories rooted in cross-national connections.

When U.S. survivors came together in the early 1970s as a self-support group called the Committee of Atomic Bomb Survivors in the United States of America (CABS), first in Los Angeles and then in San Francisco, California, these cross-national ties began to shape a unique body of memories. Initially, CABS consisted of a handful of survivors who wanted to talk to each other about their experiences. But the organization in its first year grew to a few hundred members, and by the mid-1970s, CABS counted about a thousand on its roster. U.S. survivors around this time began a campaign to obtain support from the American and Japanese governments, a development I discuss later. We cannot know all that U.S. survivors said in these early years, but we have three oral history collections to guide us. The earliest set of interviews was conducted in 1976–1991 by members of the Friends of Hibakusha (*hibakusha* means "A-bomb survivors" in Japanese), a grassroots support group for U.S. survivors in San Francisco. The second set of interviews was conducted in 2005–2010 by Mexico-based artist Shinpei Takeda, while the third set was conducted by myself in 2011–2015.[8] These oral histories with approximately 150 survivors, their family members, and their supporters vividly show how U.S. survivors remember and why they choose to tell what they tell. There are recognizable patterns in their memories, told independently yet echoing each other. One such pattern is the story of U.S. survivors' cross-national experiences as a key resource for finding, rescuing, and caring for each other immediately after the bomb's explosion. U.S. survivors' memories of these efforts are crucial for understanding what came afterward, that is, their critical stance toward medical practices and institutions established by nation-states.

Hayami Fukino, born in Redondo Beach, California, in 1929, was nine years old when her parents decided in 1939 to retire in Hiroshima after more than thirty years of working as farmers in the United States. Two years later, in 1941, the war broke out, and soon food and supplies became scarce. Labor shortage, too, was severe, and Fukino started to work at a factory as part of the Imperial Japanese Army's student mobilization. On August 6, 1945, a B-29 dropped the bomb, and as Fukino recalled, her "house came down . . . [and] started to burn." Then, something remarkable happened. When she escaped from the house, she took only one thing with her, a blanket. "And that blanket . . . was the one that . . . [her family] took from America." In fact, it carried a special meaning for her family, because her father once used it in the United States to show hospitality when he was hosting an important guest from Japan, a Buddhist monk, in 1929. Her father put it on the back seat of his brand-new American car, making sure that the monk felt comfortable. Now the blanket had another significant mission, this time not in Redondo Beach but at Hiroshima's ground zero. Escaping, Fukino's family had to cross a river at high tide. It was a struggle to climb onto the other shore hedged by a tall stone wall. The family managed to get over it, but there were others struggling in the river. So her father took out the blanket, and as Fukino recalled: "We used the blanket, so we save a lot of people. So that they could get up. That's the only thing I can remember."[9] Her story of the American blanket helping the Japanese insists on cross-national connections despite the terrible destruction caused by the national conflict.

This kind of story of multi-belonging is salient, too, in U.S. survivors' recollections of the care they offered to each other after surviving the blast. In fact, I see another important pattern in these recollections of what we might call a folk medicine that emerged in the immediate aftermath of the nuclear destruction: a significant role that women played in creating and using medicines. Similar to their cross-national ties, U.S. survivors' memories of women playing the role of caretaker shaped their women-centered activism in the 1970s and 1980s. That there were more women than men in Hiroshima and Nagasaki in 1945 is not surprising, given that many men were away on the war front. As most doctors in town were killed or injured, survivors had to rely on their families and friends for care. What is striking, though, is how stories of women as caretakers became pronounced when they came together as a group of U.S. survivors and claimed their rights to medical care. Their concern was not only that there were not enough physicians in the United States familiar with radiation illness. U.S. survivors were also eager to create a communal space where psychological anguish might be alleviated. This was the first nuclear weapon dropped on humans. Virtually no one knew what it was and how it would affect them. Psychological assurance, as much as physical treatment, was critical from the beginning.

U.S. survivors richly recall the care delivered by women who attended to both physical and mental sufferings. The experience of Junji Sarashina, a sixteen-year-old from Hawai'i, is similar to many others' in that he did not know what happened when the "flashing" hit him. Soon, he found himself "somehow on the ground," miraculously unscarred. But when he went to the first-aid station, he found a nurse "standing there . . . covered with blood. And she opened her mouth, and I saw a piece of glass, about an inch square, stuck in there, in her tongue. So she was yelling, and I went over there, pull the piece of glass from her. . . . I was scared, too; I didn't know what I was doing, and she was scared, too." Terribly shaken, Sarashina hurried back home. Then, "of course, when I walked in, my mother cried, 'Oh, Junji!' And she was trying to feed me, but for some reason I could not eat. . . . I had a diarrhea. It wasn't that severe, and she fed me all sorts of things, home remedy; she knew what's good, what's bad."[10] Clearly, it was a huge relief for Sarashina that his mother's knowledge about "home remedy" survived the blast that transformed so much else.

Tokiko Nambu tells another story of special care offered by a woman. Her sister was at school at the time of the bombing, and her mother could not find her for days. It turned out that the sister had been taken to another school that was transformed into a first-aid station. Her face was so severely burned that her mother did not recognize her at first. With no medicine on hand, the mother decided to take an extraordinary measure. She went to a cremation site that had been hurriedly set up and collected human bones left unclaimed. She ground them finely to apply to her daughter's face. Her mother "knew that it was wrong to take these bones, but her daughter was a girl with a burned face. If left untreated, the skin would harden, and she would look pitiful." It seemed that this emergency procedure prevented the skin from getting hardened. Nambu's sister did not become a "keloid girl" because she "got [her] mother's good treatment."[11] As with Sarashina's, Nambu's memory suggests not only physical aid but also emotional relief that women's care imparted. Women were traditionally chief caretakers of minor illnesses and injuries in Japanese households, contributing to a sense of continuity as women rose to the near-impossible task of caring for scars inflicted by the bomb. Perhaps thanks to this sense of continuity, what Nambu's mother first thought to be "wrong" became a "good treatment" in her daughter's memory.

Nambu's memory also adds perspective to the better-known history of the "Hiroshima Maidens." Her sister was fortunate to be free of keloid, which could have made her rely on U.S. technology for cosmetic surgery. As David Serlin and Robert Jacobs have shown, such reliance was precisely what sent the "maidens," twenty-five young Japanese women whose faces were severely scarred by the bomb, to Mount Sinai Hospital in New York in 1955 to receive "corrective" procedures. Considered unmarriageable until they restored their beauty, these women were featured by U.S. media as a trope of U.S. science, technology, and benevolence. As such, Japanese women and, by extension, Japan itself assumed a position dependent on America. Such dependence was furthered by a belief that, unlike their male soldier counterparts, Japanese female civilians were innocent victims of the war.[12] Women were expected to serve as willing healers of the relationship between the former enemies. As Nambu's story suggests, however, female survivors did not simply fall into this role. A proud memory of a mother's remedy stood as a quiet reminder of an alternative healing.

There were other types of "home remedy" U.S. survivors recalled, the benefits of which are accentuated against the lack of institutional medicine during the days and weeks after the bomb. Medical care as they knew it—hospitals, doctors and nurses, medicines—was gone, and women acting as improvisational caretakers filled the void. After the Nagasaki bombing on August 9, 1945, Michiko Benevedes, along with her mother and sister, went to her grandmother, who was living on an island off the city's shore. There, her grandmother concocted a medicine, a mixture of crab shells and other ingredients, most likely wild herbs. Thanks to this mixture, Benevedes thought, she and her sister began to improve, their wounds gradually healing and their fevers subsiding.[13] Mitsuko Okimoto, a Hiroshima survivor, recalled another method of treatment that her mother used for herself. The crucial ingredient was cucumber. Okimoto's mother used to cut both ends of cucumbers and keep them in a jar to ferment. They made a good painkiller. To heal her nuclear burns, she dipped a piece of old kimono fabric into the jar and covered her wounds with it. Sure enough, the pain diminished. The next step, to dry the burns, was similar to Nambu's mother's remedy: to apply ground human bones she collected from a crematory. Looking back on the episode, Okimoto expressed mixed feelings: "Bones have calcium in them, so they are good for you . . . [but] now that I am thinking about it, she must have been doubly irradiated."[14]

As Okimoto's memory suggests, U.S. survivors were not simply critical of the devastation caused by America. To be sure, as with Okimoto, many were irate because they were not told anything about radiation. The U.S. occupational force until 1952 strictly regulated media coverage about the bomb.[15] But U.S. survivors were also critical of the overall lack of medicine for civilians, part of which was caused by Japan's decision to continue the war without resources and by the government's continuing failure to compensate the sufferers. The Japanese government began to aid Japanese survivors first through the law concerning medical care in 1957, which offered government-funded checkups for healthy and ill survivors alike, as well as free treatment for radiation illness at designated medical facilities. In 1968 the law concerning special measures offered monetary aid not only for hospital treatment but also for care provided at home. But the government refused to define such assistance as compensation. They were "measures" to "aid" Japanese survivors living in Japan, implying that the government admitted no wrongdoing. Moreover, as I discuss later, these laws were not applicable to non-Japanese survivors living outside Japan, in effect excluding U.S. survivors like Okimoto from the laws' beneficiaries.[16] Certainly, U.S. survivors' critical stance toward Japanese policies may have been formed by the fact that they came or returned to America after the war. Nevertheless, I argue that U.S. survivors' understanding of the bomb had its historical origins in the prewar years and continued through the war. The following story of Kazuko Aoki exemplifies this, showing how U.S. survivors' experiences remained distinctively cross-national and how these experiences formed their critical stance toward authorities, including medical experts, in both Japan and America.

Aoki was born in 1931 in Hiroshima. Until several years earlier, her parents were in California and successful in the men's clothing business. When the family returned to Japan in the mid-1930s, her father became a respected figure in his neighborhood because of his success overseas. When the war started, he became outspokenly critical of Japan. He insisted "how it is stupid of Japan to fight such a resource-rich country," a view that worried his friends. If careless, Aoki's father would attract the interest of the military police. When the war ended, Aoki's father was indeed asked to come to the military police, but to that of the U.S. Army. Because of his English-speaking ability, he was appointed as a translator, visiting Japanese households with American military police officers to confiscate swords. He was dismayed by American officers' lack of knowledge about Japanese culture. They did not remove shoes when entering a house, and once one of them broke a beautiful flute over his knee, mistaking it as a weapon. Aoki's father scolded the officer, and, interestingly, the officer did not talk back. He was willing to take some lessons from this bicultural man. Such a respectful relationship made it possible for Aoki to obtain medicines not readily available to Japanese civilians. She recalled, "My father told the MPs that his daughter was injured by the bomb . . . They started to come to my house to deliver a big tube of cream and apply it thickly on my rash . . . they also brought a brown bottle of medicines to take orally, which I think was Vitamin C." Thus, her father's cross-cultural experiences, which made him a fearless critic of both Japanese and American authorities, made it possible to treat Aoki. But this did not mean that he was dependent on institutional medicine. Though a survivor himself, Aoki's father refused the medical care and the monetary allowances offered by the Japanese government. He did not want to rely on support made necessary by what he believed to be misguided decisions that Japan, as well as America, had made.

Aoki and others' memories show traces of what Robert G. Lee, historian of Japanese America, has called "the resistant strain that ruptures the continuity of official history," in this case a dualistic history of Japanese as victims and Americans as victors.[17] U.S. survivors do not fit into either category, and some vehemently refused to be defined by these nation-based categories. U.S. survivors' recollections also destabilize the image of "feminine" bomb survivors that persisted after the war. Indeed, this image went beyond the U.S. depiction of the "Hiroshima Maidens." In Japan, the image of "feminine" survivors was eagerly accepted in order to accentuate the country's victimhood and rebirth as a peace-loving nation. For the United States in the Cold War, this same image served for building an amicable, if unequal, relationship with Japan as an ally and for confirming America as the most powerful, scientifically and medically advanced nation in the world.[18] When deprived of medical experts and facilities, however, U.S. survivors managed to rescue each other, oftentimes using their cross-national experiences as a critical resource. They were also cared for and comforted by women who practiced folk medicine. Even when hurt horribly, U.S. survivors were not helpless, "feminine" victims awaiting treatment or study by medical authorities. They effectively functioned outside established medicine, most of which was obliterated by the explosion. When pressed by need, U.S. survivors, like Aoki's father, mobilized their multi-belonging as a resource to elicit assistance from authorities. This is not to suggest that all U.S. survivors responded to the bomb in the same way, nor do I argue that survivors in Japan and Korea do not have comparable experiences. Nevertheless, it is striking that U.S. survivors' cross-national memories persisted, especially as they confronted the national interests-driven understanding of the bomb. Their "resistant strain" uniquely stood out as a counter-memory of the story of victims and victors. Among the chief architects of this official story were medical experts, with whom U.S. survivors developed highly complex relationships.

CONTESTING MEDICINE: FROM THE 1940S TO THE 1960S

The ABCC had a fraught relationship with Japanese survivors, medical researchers, and institutions from the beginning of its operation in 1947. Established as a U.S. facility by the directive of President Harry S. Truman but billed as a binational research institute for the study of radiation effects on humans, it quickly became obvious to Japanese scientists that their U.S. counterparts did not consider them equals. Japan was under the occupation of the Allied Powers until 1952, making it difficult for a collaborative effort to form on an equal basis. Some of the data Japanese scientists had collected before 1947 was either discarded as useless or became integrated into U.S. studies without full acknowledgment of those who had gathered it. As M. Susan Lindee has shown, survivors themselves often regarded the ABCC with distrust. In particular, the ABCC's no-treatment policy was a roadblock to any appreciation that it might have enjoyed otherwise. The commission's aim was to study abnormal changes in somatic cells, genetics, aging, and mortality rates, and it was mandatory that the ABCC medical researchers offered no treatment to survivors. The medical experts' interest in these long-term epidemiological studies, combined with their apparent lack of interest in caring for survivors, made survivors feel that they were treated as guinea pigs.[19] The resentment over this policy, shaped by both the lack of funding and U.S. officials' concern that treatment would be seen as atonement for the use of the bomb, further intensified in Japan after 1954. During this year, the Lucky Dragon Five incident, in which twenty-three Japanese fishermen were irradiated by the U.S. Castle Bravo test in the Bikini Atoll, stirred national debate about the danger of radioactive fallout and gave birth to a series of international movements against atomic and hydrogen bombs. Some Hiroshima and Nagasaki survivors became outspoken participants of these movements.[20]

U.S. survivors' view of the ABCC in some ways echoed that of Japanese survivors. But in other ways, U.S. survivors' uniquely cross-national ties were salient in their critique of the institution. Yoko Monroe, for instance, held a half-admiring, half-critical view of the commission. As with many

Japanese survivors, Monroe recalls how talk of the bomb was considered taboo. But Monroe also knew of a neighbor who worked for the ABCC. She remembers how this neighbor was "such a modern-looking person." She states, "I have never talked to her, . . . but she was working there, probably making good money. So my impression of the United States was shaped by that."[21] Here, Monroe in one sense is complimenting the striking beauty of powerful things; no one in her neighborhood was as put-together as this ABCC employee. Simultaneously, she implies that America is all about money. In a postwar Hiroshima, where poverty was the norm, "good money" associated with the United States implied a critique of American materialism. Joe Ōhori's stance toward the ABCC was also distinctive. As a Canadian national, Ōhori is not part of the U.S. survivors' group. But his story resonates with U.S. survivors', offering a clue to how they might have formed their critical understanding of medical authorities as people of cross-national backgrounds.

Similar to many U.S. survivors born in America, Ōhori was sent before the war from Canada to Japan at the age of nine because of his parents' desire to educate him in Japan. They were worried about Canadian schools, which they thought were entrenched in racism and were not able to offer opportunities for Asian children. Having survived the bomb in Hiroshima, Ōhori began to experience hair loss in November 1945. He first encountered the ABCC soon after its building was erected in front of his high school. Their "truck" began to come to "round up" student survivors. When he learned that they wanted to draw his blood, he was upset. He asked if they "had a permission from the Canadian government to draw my blood . . . I am not a Japanese but a Canadian. If you are to draw blood from a Canadian citizen you should get permission from the Canadian government." This resistance did not work, so Ōhori asked if he could be paid. Of course he could not, and he concluded: "That means that they are testing us so that they can drop a bomb again. The blood work has nothing to do with curing us of radiation illness." He wanted nothing to do with the ABCC. Whenever their "truck" showed up, he hid in his school's bathroom.[22]

Just as Aoki's father used his connections with America to obtain medicine, Ōhori mobilized his Canadian citizenship to protest. Ōhori may even be seen as trying to privilege himself by bringing up his citizenship in an Allied country. This was another way in which the categories of victors and victims overlapped, particularly in relationship to medical experts. By asserting his background, which made him both a victim and a victor, Ōhori was also claiming, if implicitly, a cross-national history that neither U.S. nor Japanese authorities recognized: that the bombs were not singularly dropped on "the Japanese" but on people with a range of cross-national ties. In this light, the medical care for survivors should not be constrained by narrowly defined national interests, as it was at the ABCC. By upholding the no-treatment policy, the commission was falling short of survivors' expectation that medical experts make survivors-as-patients a priority.

U.S. survivors' relationship to medicine raised a host of questions that remain unanswered, continuing to shape their distrust of nation-bound, institution-based medicine. Kazue Kawasaki became a U.S. citizen because of her marriage to a Japanese American in the early 1950s. Many members of both families had migrated to Hawai'i, America, and Brazil since the early twentieth century, and these connections helped bring the pair together after the war. Shortly after August 1945, while she was still in Japan, she fell ill. Her blood cell counts were abnormal, and mysterious spots all over her body persisted. When she began to see her future husband in postwar Hiroshima, Kawasaki thought: "It is the United States that dropped the bomb, so maybe they can cure me if I went to America." But after marriage and migration to the United States, she quickly discovered that American doctors knew no better how to heal her than their Japanese counterparts. In fact, American physicians seemed largely unfamiliar with radiation illnesses.[23] Kazuko Aoki, who came to Hawai'i after marriage, experienced a similar disappointment. U.S. physicians seemed to lag behind Japanese physicians not only in diagnosis but also in sympathy. Continuing to suffer from conditions similar to Kawasaki's, Aoki told her doctor in Hawai'i that her symptoms might be related to the bomb. The physician's response was: "Oh, you think that the bomb is the reason? It's a long time ago!" Disheartened, Aoki thought: "Well, for the doctor, it has been fifteen years [since 1945], and it may be a long time . . . but because of what happened 'a long time ago' I soon developed a tumor in my thyroid." It was not her doctor in Hawai'i who discovered the cancer. Instead, a doctor based in Hiroshima found it during her stay in Japan. She recalled, "A medical checkup in Hiroshima carefully scrutinizes conditions related to radiation . . . it was such a small tumor that it was difficult to spot. But the doctor said yes, there is a tumor." This physician in Hiroshima wrote a letter to her Hawai'ian doctor, and a swift surgery saved her voice. Now Aoki had to laugh: "So I went to see my doctor here [in Hawai'i], to tell him that a tumor is here . . . yes, here, feel a bump? . . . The doctor touched it and went 'Um? Ah, Oh!'"[24]

By reiterating in her 2013 oral history how funny this Hawai'ian physician looked, Aoki in effect turned an episode into a larger question yet to be fully answered: Why can't doctors treat us better? Another implicit question is why no country has encouraged medical experts to truly care for them. To be sure, Japanese doctors seemed better experienced than their American counterparts. In Hiroshima and Nagasaki, where survivors were numerous, physicians were given unprecedented opportunities to gain clinical knowledge and bedside manners suitable for sufferers of disasters. Their skills and experiences clearly made a difference for survivors. And yet, these Japanese doctors were far away, not available to care for day-to-day sickness among American survivors. Moreover, the Japanese laws passed in 1957 and 1968 to assist Japanese survivors, mentioned earlier, were not applicable to American survivors for decades, making it necessary for survivors like Aoki to pay entirely out of pocket for medical services offered in Japan. Until the late 1970s, Japanese medical facilities stipulated in the 1957 law were not available to American survivors free of charge as they were for Japanese survivors. Similarly, U.S. survivors were not entitled to Japanese monetary allowances specified by the 1968 law until the early 2000s.[25] While their families in Japan received monthly deposits in their bank accounts to help pay for health care services, U.S. survivors received no such support for more than forty years. The ABCC's no-treatment policy was reason enough for them to feel mistreated by medical authorities. Living with the policy's long-term consequences proved equally difficult, especially when coupled with the lack of coverage by the Japanese laws. The mistrust often led to a poignant critique of state-sanctioned medicine. When Aoki's ever-outspoken father received a cancer diagnosis, his family was reluctant to put him through surgery. But he told his Hiroshima doctor: "Open me up anywhere you want; it will be helpful for you. Treat me like I am a guinea pig, then take a look at me." By speaking sharply, he claimed his dignity as a patient in the only way that he knew that spoke to medical experts—to offer himself as a study subject.

Sue Carpenter, a Nagasaki survivor married to an American navy officer who came to the United States in 1966, also offers a compelling critique of medicine. She has suffered cataracts, multiple miscarriages, and vertigo in addition to head and leg injuries that took a long time to heal after the bomb. Having supported her for many years, her husband, Lonnie, expressed a bitter sentiment about her doctors in both the United States and Japan:

> Navy doctors [in the United States] can't say it is not a problem, but they can't explain why it is a problem. So you see the government thinking that if you begin to say that [it is a problem], all of these people [here] are a part of what happened in 1945. . . . But her problems are real. I mean, they can't figure out why. In Japan, they keep her in a hospital for two weeks, [conduct] a lot of exams, and if they can't find out why, then it kind of excuses [them]. Make it go away and wait until we go away.[26]

Here again, one sees a legacy of the no-treatment policy, especially as it has affected U.S. survivors in both Japan and America. First implemented by the ABCC, the policy has been based on an association made by U.S. officials both within and outside of the ABCC among treatment, atonement, and apology. In their understanding, the bombs were used because of "Japanese leaders who refused to surrender." The Japanese, not Americans, should be responsible for treating radiation illness. Decades later, survivors in the eyes of the U.S. officials and the general public alike remained "political and historical symbols" of Japanese mistakes, not American flaws.[27] To treat symptoms such as Carpenter's as related to irradiation would reopen the question of whether the United States was justified in dropping the bombs. Moreover, as Carpenter's story shows, Japanese physicians were not always sympathetic to American survivors. Some doctors regarded U.S. survivors as national betrayers who had abandoned bomb-struck Japan after the war.[28] In light of this, Kazuko Aoki, who received a careful examination of her throat from a Hiroshima doctor, might have been fortunate. Thus it is not surprising that, although the majority of U.S. survivors had been receiving at least some monetary aid from the Japanese government by the time most oral history interviews were conducted, they still expressed a strong sense of having been disregarded. In 1945 U.S. survivors were left with little more than each other to devise treatment. Fundamentally, this has not changed many decades later.

CREATING MEDICINE: IN THE 1970S AND BEYOND

Just as they were not passive victims at ground zero, U.S. survivors refused to remain helpless. As Aoki's oral history reveals, she went to Japan to see Japanese doctors. Continuing to struggle with her inexplicable health problems, Carpenter, too, sought medical attention in Japan. Indeed, this is a tactic that some U.S. survivors used when they were relatively young and mobile. In 1978 the Japanese government began issuing a certificate of survivorhood to Korean survivors visiting Japan for medical treatment. Although the certificate became ineffective once Korean survivors left Japan, this change in the issuing policy was a breakthrough for non-Japanese survivors formerly excluded from any benefit of the 1957 and 1968 laws.[29] Some U.S. survivors, too, obtained the certificate around this time, so that they could see Hiroshima or Nagasaki physicians free of charge, at least while they were visiting Japan. American survivors' family and friends in Japan helped with their applications for the certificate by getting application forms and finding witnesses deemed necessary to "prove" one's survivorhood. Many U.S. survivors also stayed with their families while in Japan, reducing the cost of international travel. Here again, U.S. survivors used their cross-national connections as their resource. Until the aforementioned breakthrough of the early 2000s, which made at least a portion of the Japanese monetary allowances stipulated by the 1968 law available to U.S. survivors, they filled the holes in nation-bound medicine at their own expense.[30]

At the same time as seeking treatment in Japan, U.S. survivors began a movement to create a system of medical checkups in the United States, conducted free of charge by Hiroshima and Nagasaki physicians.[31] In 1977 a medical team consisting of Japanese physicians arrived in San Francisco and Los Angeles, where most U.S. survivors lived. Since then, the medical checkups for U.S. survivors by Japanese physicians have been conducted biannually, continuing to this day. Although U.S. survivors initially sought to establish a domestic system similar to these biannual checkups by obtaining legislative support from California, their effort was unsuccessful. Their similar effort throughout the late 1970s and early 1980s to obtain support from the U.S. government also failed because of the persistent view in Congress that U.S. survivors are former enemy nationals and that offering them treatment meant atonement. In discussion of the bills to assist U.S. survivors, their supporters attempted to counter a still-dominant understanding of the Hiroshima and Nagasaki bombing as a "legitimate payback to the sneak attack on Pearl Harbor," as well as a result of Japanese wartime leaders' failure to surrender in a timely way. But these views, which ignored the fact that a majority of American survivors were American citizens or their family members, continued to prevail among U.S. lawmakers. For instance, a California senator proclaimed during the hearing for the state bill that would have offered medical assistance to California survivors: "These people were our enemies!" Another senator also referred to U.S. survivors as "our enemies," asking, "Why should we help these people?"[32] U.S. survivors' failure to attract political support also originated in the difficulty of linking their cause to other societal issues. Asian American organizations such as the Japanese American Citizens League (JACL) offered only lukewarm support for U.S. survivors in the early 1970s, because the organization was initially unsure if U.S. survivors truly represented Japanese American interests. In the era of the civil rights movement, cross-national ties that defined U.S. survivors' histories still raised questions about their belonging to the United States and, by extension, their suitability to claim citizenship rights. Among conservative members of the JACL, for instance, the assumption was that Japanese Americans who had been in the United States and sent to internment camps during the war were unquestionably qualified to make a claim for lost property, rights, and dignity. Despite the gross injustice of internment, these Japanese Americans had remained "loyal" to the United States. The Redress Movement for former internees gained momentum in the 1970s and 1980s, partly because of this reasoning.[33] Americans overseas, in contrast, did not share this wartime sacrifice, raising a question about their right to make any request for compensation after the war. In fact, in the JACL's view, these Japanese Americans' out-of-country status might indicate that they were "disloyal" to America, thus not worthy of the Japanese American community's support.[34] Moreover, in the early 1980s, when the antinuclear movement reached its peak, nuclear disarmament, not justice for survivors, became the primary concern. When a public hearing was held in the Senate in 1980 to consider a nuclear-freeze bill introduced by Senator Edward Kennedy, U.S. survivors were invited as witnesses. But their mission was to help abolish nuclear weapons by telling the horror of ground zero, not to assert U.S. survivors' unique history and their pursuit of medical care.[35] The prevention of future nuclear attacks became more important than caring for the past war's victims.

When the biannual checkup in the United States, conducted by Japanese physicians, became a reality in 1977, it was funded by the city and prefectural governments of Hiroshima and Nagasaki as well as regional medical associations in these prefectures. Then, over the years, the checkups began to receive a formal budget from the Japanese government. This trajectory of U.S. survivors' effort yet again shows their continuing reliance on their cross-national connections. They sought U.S. support first, assuming that this was reasonable given their American citizenship. When this effort failed, they turned their attention to Japan. Simultaneously, U.S. survivors enlisted American physicians and hospitals to host the checkups by Japanese physicians, an effort that required careful legal preparation and sustained collaboration with American medical associations. In this way, U.S. survivors continued to use their dual connections to acquire medical care that at least partly suited their needs. Unlike the care offered at Japanese hospitals, which included treatment, these checkups in America consisted solely of examinations. Still, many U.S. survivors felt that Japanese doctors' approach to radiation illness was distinct from that of U.S. physicians, making it easier for patients to discuss psychological and physical problems. It spoke volumes that these Japanese doctors had abundant experience with hibakusha and were prepared to consider trauma and social isolation as a major problem with which their patients grappled. That these doctors were a self-selected group willing to engage American survivors likely facilitated doctor-patient relationships. Less sympathetic doctors in Japan, such as the one featured in Lonnie Carpenter's story, were unlikely to be in this group of visiting physicians.

U.S. survivors' quest for medical treatment in the 1970s and 1980s is fascinatingly multifaceted, and full consideration of it is beyond the scope of this chapter. On the remaining pages, I focus on how U.S. survivors sought communal and psychological support, again highlighting the role of women in creating a system of care outside institutional medicine. Decades earlier, in 1945, the care that women offered provided a sense of continuity and comfort. Years later, as U.S. survivors came together under very different circumstances, they found that two-thirds of them were women. These women took leadership roles, some official and clearly defined, others unofficial and more subtly played. For instance, Ayako Elliott, the current president of CABS's San Francisco branch, considered her role an extension of what she had started in Japan. When she was growing up in Hiroshima in the 1950s, there were orphanages for children who had lost their parents to the bomb. When Elliott went to the orphanages to donate used school items, she noticed that visitors included U.S. military officers. They were from the Iwakuni military base, thirty miles from Hiroshima. They impressed her as evidence of Americans' commitment to charity. Later, Elliott joined the World Friendship Center in Hiroshima, founded by U.S. peace activist Barbara Reynolds, to assist Korean survivors in Japan. Lacking any recognition by the Japanese government, many Korean survivors resided in the city's poorest neighborhood. Given these experiences, it came naturally to Elliott to assist CABS after she migrated to the United States.[36] The importance of her cross-national experiences is palpable in her narrative and reminiscent of Fukino's story of the blanket that kept her family's ties to Japan in America, America in Japan.

Elliott's leadership has been helped by other female CABS members. Sachiko Matsumoto is a vivid example of how women took care of things effectively, if subtly. Unlike Elliott, Matsumoto never occupied a formal leadership position in CABS. Nonetheless, she played a central role in communicating with its members. One challenge for CABS was to find members. When CABS started, no one knew how many U.S. survivors existed. The only way to find out was to attract them to meetings. In addition to posting notices in newspapers, someone had to write, call, or meet with the survivors. That Matsumoto had been doing all these things became clear because many U.S. survivors mentioned her. Sayoko Utagawa, of Fresno, California, explained how she had obtained a certificate of survivorhood at Matsumoto's urging: "Ms. Sachiko Matsumoto in San Francisco told me that I should get a certificate, and I started to think that, well, maybe I will get it the next time I visit Japan."[37] Lisa Gendernalik's residence in Arizona has made it difficult to keep in touch with others. "But with Mrs. Matsumoto I occasionally talked over the phone. . . . I would call her when I have a question, and she would explain a variety of things for me."[38]

During my interview with Matsumoto, I was struck by the meticulous CABS records she kept. They seemed to appear from every drawer, and she was able to tell stories about each member.[39] My first thought was that she does an incredible amount of supportive work. As I conducted more oral histories, however, I realized how wrong I was. *This* work of keeping in touch *is* the main work. Aiko Tokito had not told her nonsurvivor friends that she is a survivor. But when she was in a waiting room for the biannual medical checkup, she noticed others in the room talking about the bomb. So Tokito started the conversation: Which town are you from? How did *it* happen to you?[40] For Michiko Benevedes, this kind of connection with other survivors was a catalyst for breaking her long-kept silence: "When I joined CABS, I gained courage to write a memoir. I had had my mouth shut for fifty years. But after being in CABS, I gained courage . . . the president of CABS [Kanji Kuramoto] said to me, write it, write it! I told him that I don't like to write, I am not well educated, and I don't have a confidence. But he told me that I can write simply as I remember."[41] Not only did she write a memoir, but Benevedes also served as vice president of CABS's San Francisco branch for ten years. Clearly, a range of connections with others, underpinned by women-driven actions on the ground, allowed U.S. survivors to discover each other. They told stories that they had not told before, an act of sharing that offered relief. They communicated and connected with each other, forging a sense of community. For a group whose concerns have been repeatedly neglected, this was no small accomplishment. Though imperfect, theirs is a system of care that stood outside the medical establishment, a quiet reminder of the limits of nation-driven, institutionally based medical expertise.

CONCLUSION

U.S. survivors' life histories are a rich source of their understanding of the bomb, injury and illness, and medicine. There are distinctive patterns in U.S. survivors' narratives, and they challenge better-known views of survivors as helpless or as well-suited promoters of national interests. Instead, U.S. survivors, particularly women, actively and creatively employed their cross-national connections and knowledge of folk medicine. These in turn allowed them to create means of emotional relief and medical care. U.S. survivors still lack any formal recognition by the U.S. government, and its Japanese counterpart still does not fully apply the laws created for Japanese survivors to non-Japanese survivors. It is not surprising, then, that voices of these neglected survivors are rife with critique of medical expertise. When survivors sought treatment, they were subjected to research that stripped them of their dignity. Struck by radiation illness and the fear of it, survivors confronted state-sanctioned medical professionals in both America and Japan who refused to see them as *their* patients. These experiences have prompted U.S. survivors to critique nation-bound understandings of the bomb and its aftermath, understandings that have allowed medicine to disregard the variety of patients and their histories.

ACKNOWLEDGMENTS

This chapter started as a conference presentation at the American Association for the History of Medicine's annual meeting in 2014, and I am grateful to the audience for giving me the encouragement I needed to develop the presentation into scholarship. I also thank all the participants of the 2015

Contested Expertise, Toxic Environments workshop in Claremont for their engaging, collegial, and useful critiques on this chapter. A special thank-you goes to the organizers-turned-editors, Janet Farrell Brodie, Vivien Hamilton, and Brinda Sarathy.

NOTES

1. Serlin, *Replaceable You*; Shibusawa, *America's Geisha Ally*; Simpson, *An Absent Presence*. Keloids are painful growths of scar tissue common after burns.

2. Lindee, *Suffering Made Real*; Schull, *Effects of Atomic Radiation*.

3. There is one book-length scholarly study of U.S. survivors: Sodei, *Were We the Enemy?*, first published in Japan in 1977–1978. There are a series of articles about psychosocial challenges that U.S. survivors face. See, for instance, Ikeno and Nakao, "Zaibei hibakusha kyōkai bunretsu no yōin bunseki to kongo no enjo no kadai." See also Tamura, "Zaigai hibakusha no konnichi teki kadai," and Tamura, "Zaigai hibakusha enjo no genjō to kadai" for legal studies of U.S. survivors.

4. This kind of exclusion has been prevalent in other radiation-affected communities consisting of socioeconomically marginalized populations. See, for example, Barker, *Bravo for the Marshallese*; and Iverson, *Full Body Burden*; Johnston, ed., *Half-Lives and Half-Truths*.

5. On Japanese survivors, see Minear, ed. and trans., *Hiroshima*; Hein and Selden eds., *Living with the Bomb*; Orr, *The Victim as Hero*; Treat, *Writing Ground Zero*; and Zwigenberg, *Hiroshima*. These studies are either a collection of survivors' writings or a study of survivors in cultural and political representations, indicating a limited range of studies about survivors available in English. This is because of the significant focus that U.S. scholars of the bomb have placed on three fields of study—the making of the bomb (including the history of science), military history, and diplomatic history, none of which has explored survivors' history as its central focus. On Korean survivors, see Ichiba, *Hiroshima wo mochikaetta hitobito*; and Hirano, *Umi no mukō no hibakusha tachi*. Yoneyama, *Hiroshima Traces*, offers a valuable comparison of Korean and Japanese survivors.

6. Azuma, *Between Two Empires*, 136–37; Takahashi, *Nisei/Sansei*, 74–79.

7. Hiroshima Prefecture, *Hiroshima kenjin kaigai hatten shi nenpyō*, 13; Glenn, *Issei, Nisei, War Bride*, 25; Lee, "Introduction," 18–19; Sodei, *Were We the Enemy?*, 11–18. The precise number of Japanese Americans in Japan or Hiroshima in 1941 is unknown because of the lack of reliable statistics.

8. These three sets of oral histories are: eighteen oral histories at the Regional Oral History Office, Bancroft Library, University of California–Berkeley (FOH hereafter); fifty-six oral histories at the Robert Vincent Voice Library, Michigan State University, East Lansing (ST hereafter); and eighty-four oral histories in possession of the author (NW hereafter). I use either real names or pseudonyms for interviewees based on their preferences.

9. Interview with Hayami Fukino, June 6, 2012, NW. For a similar story of bicultural connections, see Interview with K. Y., May 12, 1989, FOH.

10. Interview with Junji Sarashina, June 6, 2012, NW. Many survivors recalled the lack

of medical care for a long time after the bomb. See, for example, interview with Setsuko Thurlow, April 2, 2010, ST.

11. Interview with Tokiko Nambu, June 19, 2012, NW. For a similar story about the use of human bones, see interview with Yasuko Kashihara, June 2, 2012, NW. It is likely that ground bones were used to cover wounds and prevent infection, in effect helping the healing process. Although human bones were not used in ordinary times, Japanese people have long eaten fish bones as part of their diet and used them as a fertilizer. It was also not unusual to keep bones of deceased family members at home for a long time after cremation. Thus, it is plausible that people's familiarity with bones' benefits and their intimate presence in daily life, combined with the lack of resources at ground zero, prompted the therapeutic use human bones. Survivors also used ash, oil, rice flour, baked potatoes, grated pumpkins, and used green tea leaves for similar healing effects. See interview with Nobuko Fujioka, June 14, 2012, NW; interview with Alfred Kaneo Dote, June 25, 2012, NW.

12. Serlin, *Replaceable You*, chapter 2; Jacobs, "Reconstructing the Perpetrator's Soul by Reconstructing the Victim's Body."

13. Interview with Michiko Benevedes, July 21, 2011, NW.

14. Interview with Mitsuko Okimoto, June 6, 2012, NW.

15. Braw, *The Atomic Bomb Suppressed*; Sasamoto, *Beigun senryō-ka no genbaku chōsa*.

16. Tamura, "Zaigai hibakusha no konnichi teki kadai."

17. Lee, "Introduction," 4.

18. On the image of "feminine" survivors in both Japan and the United States, see, for example, Orr, *The Victim as Hero*, 3–7, 55–58; Shibusawa, *America's Geisha Ally*, 232–38, 245–50.

19. Lindee, *Suffering Made Real*, chapter 5.

20. Yamamoto, *Grassroots Pacifism in Post-War Japan*, 15, 166–69; Tachibana, "The Quest for a Peace Culture"; Zwigenberg, *Hiroshima*, 79–82.

21. Interview with Yoko Yamada, June 12, 2012, NW. Scholars have suggested a complex relationship between Hiroshimans and the ABCC. For instance, some American men "had come to Japan in the army, fallen in love, and, unable to wed as soldiers [because of the military's restrictive policy on international marriage], taken civilian employment with the Atomic Bomb Casualty Commission in Hiroshima" (Zieger, *Entangling Alliances*, 183).

22. Interview with Joe Ōhori, April 2, 2010, ST.

23. Interview with Kazue Kawasaki, Nobuko Fujioka, and Wes Aoki, July 22, 2011, NW; interview with Kazue Kawasaki, June 18, 2012, NW.

24. Interview with Kazuko Aoki, June 13, 2013, NW.

25. Hiroshima District Court, *Hanketsu*.

26. Interview with Sue and Lonnie Carpenter, April 27, 2005, ST. Other U.S. survivors expressed the difficulty of telling U.S. physicians about their survivorhood. See, for example, interview with A. G., December 12, 1989, FOH.

27. Lindee, *Suffering Made Real*, 118, 121–22.

28. Interview with Tokie Akihara, July 25, 2011, NW.

29. On the lawsuits by non-Japanese survivors against the Japanese government, see Tamura, "Zaigai hibakusha enjo no genjō to kadai."

30. The difference between the benefits offered by the Japanese government to U.S. survivors and the aid that Japanese survivors receive remains substantial and complex. As of

2016, one of the major differences is the amount of monetary support. American survivors do not receive any subsidies for their insurance premiums, while their Japanese counterparts are eligible for full coverage by the government. Also, Japanese survivors are eligible for coverage for medical services, including prescriptions, without any out-of-pocket expenses, while U.S. survivors receive only up to a set amount of reimbursement per year. See the Ministry of Health, Labour, and Welfare's website, http://www.mhlw.go.jp/stf/seisakunitsuite/bunya/kenkou_iryou/kenkou/genbaku/index.html (accessed on November 8, 2016).

31. The following discussion of U.S. survivors' efforts relies on Sodei, *Were We the Enemy?* and the following primary sources: Itō, *Hazama ni ikite gojūnen*; Kuramoto, *Zaibei gojūnen*; Hiroshima Prefectural Medical Association, *Hiroshima-ken ishikai zaibei genbaku hibakusha kenshinjigyō suishin sanjusshūnen kinenshi*; Committee of Atomic Bomb Survivors in the United States of America (CABS) Papers, November 1971–June 1975, California State Archives, Sacramento; Committee of Atomic Bomb Survivors in the United States of America, *CABS Newsletters*, November 1976–July 2004 (given to the author by Setsuko Kuramoto); *The Paper Crane* (FOH's newsletter), Spring 1983–Winter 1991 (given to the author by Jenni Morozumi); *Payments to Individuals Suffering from Effects of Atomic Bomb Explosions.*

32. Sodei, *Watakushi tachi ha teki datta no ka*, 181, 194–96.

33. Murray, *Historical Memories of the Japanese American Internment and the Struggle for Redress*, chapter 7.

34. See, for example, Japanese American Citizens League's organ *Pacific Citizen*, August 13, 1972 and August 16, 1974, where their reporters failed to acknowledge U.S. citizenship of U.S. survivors, implying instead that they were Japanese citizens. Some U.S. survivors expressed their frustration with JACL's lack of engagement. Interview with Tokie Akihara, July 25, 2011, NW.

35. *CABS Newsletter*, January 21, 1980; interview with Francis Mitsuo Tomosawa, June 7, 2012, NW.

36. Interview with Ayako Elliott, June 11, 2012, NW.

37. Interview with Sayoko Utagawa, June 13, 2012, NW.

38. Interview with Lisa Gendernalik, July 17, 2011, NW. See also Interview with Yamashita Yasuaki, February 15, 2006, ST.

39. Interview with Sachiko Matsumoto, July 23, 2011, NW.

40. Interview with Aiko Tokito, June 19, 2013, NW.

41. Interview with Michiko Benevedes, July 21, 2011, NW.

BIBLIOGRAPHY

Azuma, Eiichiro. *Between Two Empires: Race, History, and Transnationalism in Japanese America*. New York: Oxford University Press, 2005.

Barker, Holly M. *Bravo for the Marshallese: Regaining Control in a Post-Nuclear, Post-Colonial World*. Toronto: Wadsworth, 2004.

Boyer, Paul S. *By the Bomb's Early Light: American Thought and Culture at the Dawn of the Atomic Age*. New York: Pantheon, 1985.

Braw, Monica. *The Atomic Bomb Suppressed*. Armonk, NY: M. E. Sharpe, 1991.

Committee of Atomic Bomb Survivors in the United States of America. *CABS Newsletters.* November 1976–July 2004.

Committee of Atomic Bomb Survivors in the United States of America Papers. November 1971–June 1975. California State Archives, Sacramento.

Glenn, Evelyn Nakano. *Issei, Nisei, War Bride: Three Generations of Japanese American Women in Domestic Service.* Philadelphia: Temple University Press, 1986.

Hein, Laura, and Mark Selden, eds. *Living with the Bomb: American and Japanese Cultural Conflicts in the Nuclear Age.* Armonk, NY: M. E. Sharpe, 1997.

Hirano, Nobuto. *Umi no mukō no hibakusha tachi: zaigai hibakusha mondai rikai no tame ni.* Tokyo: Hachigatsu shokan, 2009.

Hiroshima District Court. *Hanketsu.* Hiroshima, Japan: Hiroshima District Court, June 17, 2015.

Hiroshima Prefectural Medical Association. *Hiroshima-ken ishikai zaibei genbaku hibakusha kenshinjigyō suishin sanjusshūnen kinenshi.* Hiroshima, Japan: Hiroshima Prefectural Medical Association, 2007.

Hiroshima Prefecture. *Hiroshima kenjin kaigai hatten shi nenpyō.* Hiroshima, Japan: Hiroshima Prefecture, 1964.

Ichiba, Junko. *Hiroshima wo mochikaetta hitobito: Kankoku no Hiroshima ha naze umareta no ka.* Tokyo: Gaifū-sha, 2000.

Ikeno, Satoshi, and Nakao Kayoko. "Zaibei hibakusha kyōkai bunretsu no yōin bunseki to kongo no enjo no kadai." *Ningen fukushigaku kenkyū* 6, no. 1 (2013): 47–68.

Itō, Chikako. *Hazama ni ikite gojūnen: zaibei hibakusha no ayumi.* Walnut, CA: Committee of Atomic Bomb Survivors in the United States of America, 1996.

Iverson, Kristen. *Full Body Burden: Growing Up in the Nuclear Shadow of Rocky Flats.* New York: Crown Publishers, 2012.

Jacobs, Robert A. "Reconstructing the Perpetrator's Soul by Reconstructing the Victim's Body: The Portrayal of the 'Hiroshima Maidens' by the Mainstream Media in the United States." *Intersections: Gender and Sexuality in Asia and the Pacific* 24 (June 2010). Accessed August 26, 2015. http://intersections.anu.edu.au/issue24/jacobs.htm.

Johnston, Barbara Rose, ed. *Half-Lives and Half-Truths: Confronting the Radioactive Legacies of the Cold War.* Santa Fe, NM: A School of Advanced Research Resident Scholar Book, 2007.

Kuramoto, Kanji. *Zaibei gojūnen: watashi to America no hibakusha.* Tokyo: Nihon tosho kankō-kai, 1999.

Lindee, M. Susan. *Suffering Made Real: American Science and the Survivors at Hiroshima.* Chicago: University of Chicago Press, 1994.

Lee, Robert G. "Introduction." In *Dear Miye: Letters from Japan, 1939–1946,* edited by Mary Kimoto Tomita, 1–22. Palo Alto, CA: Stanford University Press, 1995.

Minear, Richard H., ed. and trans. *Hiroshima: Three Witnesses.* Princeton, NJ: Princeton University Press, 1990.

Murray, Alice Yang. *Historical Memories of the Japanese American Internment and the Struggle for Redress.* Palo Alto, CA: Stanford University Press, 2007.

Oral Histories of American Survivors. Regional Oral History Office, Bancroft Library, University of California–Berkeley.

Oral Histories of North- and South American Survivors. Robert Vincent Voice Library, Michigan State University, East Lansing.

Oral Histories of US Survivors, Family Members, and Supporters. Conducted by Naoko Wake and in her possession in East Lansing, MI.

Orr, James J. *The Victim as Hero: Ideologies of Peace and National Identity in Postwar Japan*. Honolulu: University of Hawai'i Press, 2001.

Pacific Citizen. August 1972–August 1974.

Payments to Individuals Suffering from Effects of Atomic Bomb Explosions: Hearing before the Subcommittee on Administrative Law and Governmental Relations of the Committee on the Judiciary. March 31 and June 8, 1978. Washington, DC: Government Printing Office, Serial No. 43.

Sasamoto, Yukuo. *Beigun senryō-ka no genbaku chōsa: genbaku kagaikoku ni natta nihon*. Tokyo: Shinkan-sha, 1995.

Schull, William. *Effects of Atomic Radiation: A Half-Century of Studies from Hiroshima and Nagasaki*. New York: Wiley, 1995.

Serlin, David. *Replaceable You: Engineering the Body in Postwar America*. Chicago: University of Chicago Press, 2004.

Shibusawa, Naoko. *America's Geisha Ally*. Cambridge, MA: Harvard University Press, 2006.

Simpson, Caroline Chung. *An Absent Presence: Japanese Americans in Postwar American Culture, 1945–1960*. Durham, NC: Duke University Press, 2002.

Sodei, Rinjiro. *Watakushi tachi wa teki datta no ka*. Tokyo, Japan: Ushio shuppan-sha, 1978.

Sodei, Rinjiro. *Were We the Enemy? American Survivors of Hiroshima*. Boulder, CO: Westview Press, 1998.

Tachibana, Seiitsu. "The Quest for a Peace Culture: The A-bomb Survivors' Long Struggle and the New Development for Redressing Foreign Victims of Japan's War." *Diplomatic History* 19, no. 2 (Spring 1995): 329–46.

Takahashi, Jere. *Nisei/Sansei: Shifting Japanese American Identities and Politics*. Philadelphia: Temple University Press, 1997.

Tamura, Kazuyuki. "Zaigai hibakusha enjo no genjō to kadai." *Chingin to shakai hoshō* 1390 (March 2005): 4–21.

Tamura, Kazuyuki. "Zaigai hibakusha no konnichi teki kadai." In *Shakai hoshō-hou, fukushi to rōdō-hou no shintenkai*, edited by Araki Seiji and Kuwahara Yōko, 585–98. Tokyo: Shinzan-sha, 2010.

Treat, John Whittier. *Writing Ground Zero: Japanese Literature and the Atomic Bomb*. Chicago: University of Chicago Press, 1995.

Yamamoto, Mari. *Grassroots Pacifism in Post-War Japan: The Rebirth of a Nation*. London: Routledge, 2004.

Yoneyama, Lisa. *Hiroshima Traces: Time Space, and the Dialectics of Memory*. Berkeley: University of California Press, 1999.

Zieger, Susan. *Entangling Alliances: Foreign War Brides and American Soldiers in the Twentieth Century*. New York: New York University Press, 2010.

Zwigenberg, Ran. *Hiroshima: The Origin of Global Memory Culture*. Cambridge: Cambridge University Press, 2014.

10

On Sovereignty, Deficits, and Dump Fires

RISK GOVERNANCE IN AN ARCTIC "DUMPCANO"

Alexander Zahara

On May 20, 2014, the local dump in the Arctic community of Iqaluit—Canada's northernmost and smallest capital city (population: seven thousand), located in Nunavut territory—spontaneously caught fire for the fourth time in less than a year.[1] In Canada, as elsewhere, landfill fires are a relatively common occurrence—the by-product of metabolically active waste materials and chemical oxidation.[2] However, unlike most major Canadian and American urban centers, which contain sophisticated technology for the detection, suppression, and disguising of landfill fires, Nunavut dump sites lack this infrastructure. Additionally, because Iqaluit is only accessible by airplane (and sealift during the summer), wastes accumulate but rarely, if ever, leave.

Within twenty-four hours of igniting, the Iqaluit dump fire spread throughout the city's four-story, football field–size active waste disposal site. While the city's previous dump fires had been restricted to a small corner of the dump and were able to be extinguished, initial attempts to douse the fire forced the suspension of the city's water supply. Despite being located on a peninsula only two kilometers away from the city center (figure 10.1), use of seawater to extinguish the fire was not recommended due to concerns over increased contaminant concentrations.[3]

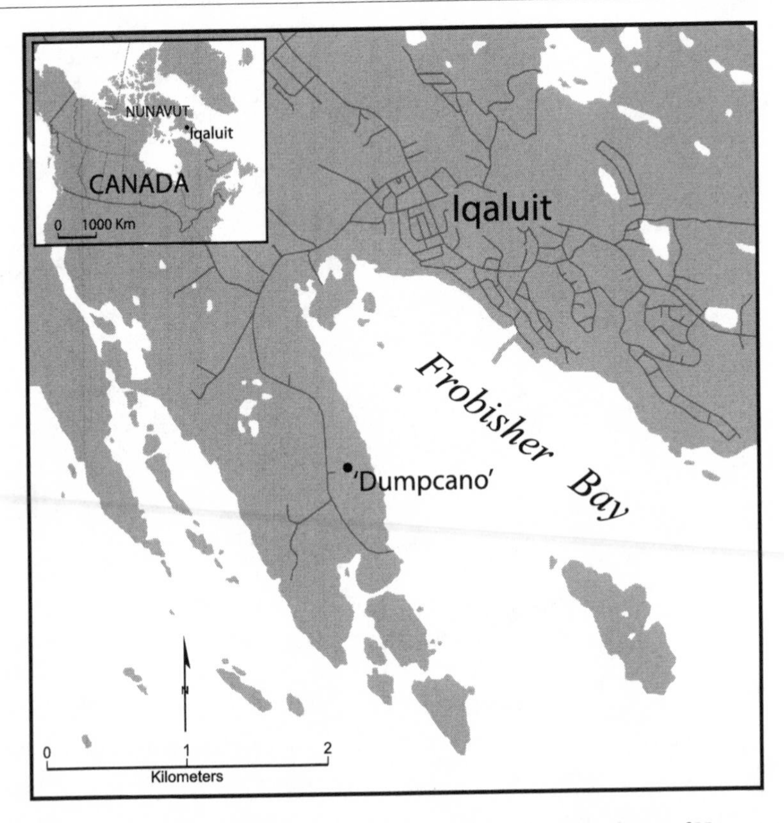

Figure 10.1. Map showing the locations of (1) Iqaluit, the capital city of Nunavut Territory in Canada, and (2) the "dumpcano," which was situated approximately two kilometers from Iqaluit's city center, adjacent to Frobisher Bay. Created by Alex Zahara.

Out of necessity, then, the city abandoned its efforts to put out the fire and shifted their strategy from extinguishment to containment. Fire crews dug additional firebreaks, including one between the smoldering section of the dump and a smaller, older pile that made up "the most hazardous part of the landfill."[4] Hereafter, residents of the majority-Inuit community spent nearly four months heeding public health warnings and breathing in contaminant-laden dump smoke (figure 10.2).[5] In July, when contaminant concentrations were revealed to be above the standards set by many southern Canadian provinces, publics responded to risk management frames not through complacency or ambivalence, as risk theorist Brian Wynne suggests, but by protests aimed at long-standing issues of inequity and misunderstandings of what "healthy living" means to northern communities.[6]

May	June	July	August	September
20th: Iqaluit dump ignites. 23rd: 'Dumpcano' Twitter account created; People with breathing issues warned not to go outside; City announces plan to let fire burn.	10th: Air quality monitoring begins. 11th: Council votes to put out fire. 12th: Dump fire working groups begin meeting. 30th: Landfill fire expert provides Council with Draft Extinguishment Plan.	4th: 'Taima' image published online. 17th: Air quality exceeds Ontario standards, pregnant women warned not to go outside. 20th: Iqalummiut for Action (IFA) group forms, begins lobbying levels of government. 23rd: City Council approves scaled down extinguishment plan.	1st: City denied funding to put out fire. 12th: IFA requests public meeting about dump fire. 15th: Deputy Mayor asks DND for assistance with fire, is denied. 21st-29th: Operation Nanook held in Iqaluit 25th: Public dump fire meeting held; Prime Minister Stephen Harper visits Iqaluit	1st: Dump fire extinguishment begins. 16th: Iqaluit Dump fire extinguished

Figure 10.2. Time line of Iqaluit dump fire over the summer of 2014. Created by Alex Zahara.

In this chapter, I highlight the ways in which government practices of risk management—what are based primarily on what *is known or knowable* through Western scientific knowledge and analysis—often serve to decontextualize and therefore speak past local understandings of risk. In examining the 2014 Iqaluit dump fire event, I suggest that efforts to protect human health and the environment that do not listen to the concerns of impacted publics may themselves cause harm, particularly when they ignore historical inequities and local understandings of risk and well-being. To demonstrate this, my argument is twofold. First, I showcase that risk management practices privileged a "molecularization" of the dump fire issue, where well-being is understood almost exclusively in terms of calculable levels of contaminant exposure, and wherein government decision-making practices were primarily delimited to scientific analysis, knowledge, and frames.[7] This was done, I argue, because territorial and federal government risk management protocols are governed through a deficit model framework, where public controversies associated with risk management are framed as the outcome of a public "deficit" in scientific and institutional knowledge and/or trust. Second, I showcase how community resistance to the dump fire was an effort to foreground risk management protocols within Iqaluit's historical and cultural context, and in ways that promote Inuit governance, knowledge, and relations. As Iqaluit residents (known locally as Iqalummiut) show, doing so involves contending with Arctic Canada's situated geographies of settler colonialism, Iqaluit's military history, and often contested North-South relations that inform governance practices throughout the territory.

I begin this chapter by providing an overview of the contemporary social science literature that examines efforts to integrate Indigenous knowledge into Euro-Canadian governance practices throughout Canada's north. In doing so, I discuss how risk management practices that overlook public concerns about contaminant issues themselves ignore Indigenous knowledge and community-specific ways of knowing and being. I then turn to the controversy surrounding the Iqaluit "dumpcano" (a local nickname given to the dump fire, a combination of the words "dump" and "volcano") to examine how matters of public concern were governed through a deficit model risk management framework. The technical configuration of the dump fire led to risk management strategies (e.g., directing children and Elders to stay indoors during the short Arctic summer) that, to many Iqalummiut, were considered an affront to community health and well-being.[8] I then finish the chapter by showcasing how and why the dump fire controversy culminated with Iqalummiut formally requesting the federal government to abandon a yearly Arctic military and sovereignty exercise, Operation Nanook, in favor of tackling a "real emergency" that impacts the lives of northerners.[9] Doing so, I argue, was a strategic decision aimed at repoliticizing the dump fire within the region's history of colonialism and long-standing issues of inequity. As an effort to place community understandings of health and well-being at the forefront of risk management decision making, I suggest that the "dumpcano" protests should be viewed within the context of Inuit self-determination and efforts to place community-specific understandings and values within territorial governing practices.

THE DEFICIT MODEL AND INDIGENOUS GOVERNANCE IN NUNAVUT

In Nunavut Territory, where Inuit self-governance formed the basis of the Nunavut Land Claims Agreement (NLCA) in 1999, how and whether community understandings are included within institutionalized governance practices is particularly relevant. Nunavut (translated as "Our Land" from Inuktitut) is Canada's largest aboriginal land claim, encompassing over 1.5 million square kilometers and nearly 25,000 Inuit beneficiaries. The NLCA was negotiated with the federal government over a period of nearly two decades, from 1976 to 1994, directly following what is considered to be Nunavut's major colonial period. From the 1940s to the 1960s, Inuit experienced radical changes as they were forced to move off the land into government-created communities.[10] Within a single generation, Inuit shifted from a mixed economy of subsistence hunting and trade (of animals such as Arctic hare, polar bear, caribou, and seal, and metals such as copper) to one predominated by wage labor and community living. Children were placed in residential schools, where they were taught in Euro-Canadian school curricula, and prevented from speaking Inuktitut and engaging in Inuit cultural and spiritual practices.[11] As a consequence of these Canadian government policies (which have since been recognized as cultural genocide), when Inuit negotiated the NLCA, they highlighted the importance of including and supporting *Inuit Qaujimajatuqangit* (Inuit knowledge, both traditional and contemporary) within territorial policy and practice.[12] Through the NLCA, Inuit negotiated for the creation of a number of comanagement boards, including those involved in overseeing environmental and human health issues, including land use and development, and water and waste management.[13]

Despite Inuit success in negotiating the NLCA, many have criticized how Inuit knowledge is included within territorial governing practices, noting the government's propensity to have biases toward settler governing practices and their conceptualizations of nature and health.[14] Additionally, as Emilie Cameron points out, when discussing Indigenous knowledge, both administrators and scholars alike actively seek out forms of knowledge that adhere to settler understandings of Indigeneity—as "local" or "traditional" rather than fluid, current, and adaptable. This in turn "delimit[s] the ways in which northern Indigenous perspectives, concerns, and critiques can be heard and can be effective."[15] Accordingly, Inuit concerns about management issues are often made legible within government institutions "in a manner compatible with Western science and logic," transforming into what Mario Blaser refers to as "informational inputs"—bits of information that do not push up against or otherwise question existing government practices and infrastructure.[16] Specifically, institutionalized comanagement frameworks often favor "a naturalized discourse that specifically excludes political and ethical considerations" from what constitutes Indigenous knowledge.[17] In this way, Indigenous knowledge may be used to identify community sources of contaminant exposure (popular fishing spots, berry-picking sites, etc.) but rarely as a way of exploring causes or solutions to public health concerns.

Risk management, among other forms of governance in Nunavut, exists within Canada's settler colonial context, wherein the inclusion of Inuit knowledge into government management practices is frequently called into question. Here, the federal government departments charged with advising the Government of Nunavut (GN) in matters of public health operate within what scholars of science and technology studies refer to as a "deficit model" framework. The deficit model is so-called because it frames the problem of a controversy as existing within an "ignorant and uninformed public" that is deficient in scientific or institutional knowledge and/or trust.[18] As a result, governments often address public health controversies as "risk communication problems," wherein solutions lie in educating the public about the "correct" understanding of a given issue.[19] Solutions, then, are often predetermined, without taking public concerns seriously. Health Canada, for example, distinguishes between "actual factors that affect people's level of risk" and public perceptions, which they term "risk beliefs"—understandings of risk that stem from psychosocial factors such as anxiety, fear of future dangers, or perceptual cues such as colored water and odors.[20] While public participation in risk management is encouraged as a form of "best practice" within Indigenous communities, this is done explicitly to "increase trust and understanding" of a given issue.[21] The very notion that Indigenous communities might understand but disagree with government frames is not considered. Instead, efforts are directed toward scientific education, or what is termed "capacity building" or "empowerment" by public health officials.[22] As a consequence, governments perpetuate the notion that publics are only concerned about whether risks have been correctly accounted for and mitigated, and not, as Brian Wynne suggests, "about upstream (usually unaccountable) driving human visions, interests and purposes."[23] Deficit-style risk management frameworks, then, which are devoid of listening and acting on community concerns, often "miss the music," as Wynne puts it—a critique that has been applied to both Inuit and non-Inuit encounters with Western governance systems.[24]

As I show in this chapter, understanding that sustaining power relations is implicit to deficit model frameworks is particularly important given ongoing efforts by Inuit to include Inuit knowledge and governance within territorial management practices. In what follows, I showcase how government risk management practices served to reinforce scientific understandings of harm but were responded to by protests aimed at rearticulating and historicizing Iqaluit's waste issues.

RISK MANAGEMENT AND THE IQALUIT DUMP FIRE

As noted earlier, Iqaluit City Council's initial decision to keep the dump fire burning coincided with public health advisories from the GN. As the summer progressed, new information about Iqaluit's contaminant concentrations changed how the dump fire was managed by government officials and experienced by community members. Beginning in mid-June, air-monitoring equipment was put in place to allow for daily and hourly analysis of contaminants—a process that took approximately thirty days for results to be analyzed in southern Canada and returned to Nunavut officials. Shortly after monitoring began, a press conference was held at an Iqaluit City Council meeting, where Environment Canada officials simultaneously informed Iqaluit residents that they were in "little immediate danger" of health effects from dump smoke, while acknowledging that concentrations of dioxins and furans—considered to be the most harmful contaminants—were as of yet unknown.[25] Iqaluit residents were forced to sieve through the territorial government's messaging while dealing with very real physiological responses to the dump smoke, including sore throats, nausea, and headaches. Due to pressure from constituents, and also due to their own concerns regarding dump smoke, by mid-June the Iqaluit City Council reversed its decision, instructing the fire chief to "put the fire out in a way that he sees fit."[26]

Based on the preliminary contaminant analysis that was taken by Environment Canada, public health officials considered the dump fire to be of relatively low health risk to Iqaluit residents. Yet despite reassurances by the GN and Health Canada, by June 10—less than three weeks after the dump fire began—around ten people had already visited the local hospital for dump smoke–related ailments.[27] When city councilors reversed their decision and voted in favor of extinguishing the fire, they did so in light of concerns over residents' quality of life. As one Iqaluit councilor explained to a local newspaper, "Council's been getting a backlash [about the fire]. . . . Schools are being closed, kids told to stay inside. Summer's here. People want to get outside and enjoy the pristine air."[28] Council's directive to extinguish the fire prompted the formation of at least two dump fire working groups, comprised of "relevant" GN, federal, and city department staff.[29] The working groups met throughout the summer to discuss issues such as the dump fire extinguishment plan, air-quality monitoring, and public messaging. Crucially, the groups did not include nonexperts. Other than the council's final decision-making authority, citizens were essentially excluded from decision-making practices.

On July 17, nearly a month and a half after the fire began, test results revealed that concentrations of dioxins and furans had been above the standard set for the southern Canadian province of Ontario since the beginning of the summer.[30] At this time, the GN released a public health advisory notifying all pregnant women and women of childbearing age to avoid going outdoors and breathing in dump smoke, which was now known to cause "a possible risk" of decreased fertility in male offspring.[31] From a public health perspective, the discovery of high levels of dioxins proved the success of risk management decisions; the government's efforts to make the dioxin concentrations knowable had been realized, and public health messaging did not change since the dump fire began ("shelter in place" was still the advice being given).

For the GN, the major issue of the dump fire was becoming one of risk communication. As one public health official stated: "You know, there's pregnant women who've talked to their doctors so I've made a point of keeping the physicians—my colleagues—informed. Um, so I've found that decreases [anxiety]. But I've also found that people don't read the [GN's] Q and As. They don't read the public health advisories, they don't particularly listen—that they hear somebody say something about how awful it is and repeat it. So it makes it very difficult to get the messaging across."[32] As the chief medical officer of health explained at a city council meeting, the main reason for increased public concern was that the messaging was being "interpreted broadly" by community members. In her framing of the issue, if women of childbearing age simply understood that they would only be impacted if they *became pregnant* during the dump fire, then the controversy would subside.[33] For her, public concerns were about levels of contamination that the GN had already determined were "not a public health emergency."[34] That the dump fire was not a public health emergency made little sense to many Iqaluit residents given that pregnant women, children, and those with breathing disorders were being told not to go outside. Shawn Inuksuk and Julie Alivaktuk, a pregnant Inuk couple, were quoted in a local newspaper as saying: "The right to clean air is the most basic of human rights. . . . We feel strongly about it. We want people to know what's going on."[35] Significantly, the GN's risk management framing assumed a misinformed public, thus drawing boundaries around the ways in which public concerns about risks were listened to and addressed.

Throughout the summer, the issue of the "dumpcano" continued to be governed by what Wynne refers to as "correctional idioms of communication" aimed at refocusing the issue toward what is scientifically defined and knowable.[36] The dioxin standard, as the chief medical officer of health explained during a public meeting, was based on animal tests and multiplied by a safety factor to account for differences in human physiology.[37] The public health framing also required a rational decontextualization of the dump fire: the GN remarked that levels of dioxins and furans were significantly lower in Iqaluit than in the nearby Northwest Territories, where forest fires had reduced air quality considerably.[38] Other public health messaging reminded citizens that contaminant concentrations were lower than those received from smoking.[39] From a rational scientific perspective, then, the public had little to worry about. As the chief medical officer of health stated at a public meeting: "I have two daughters in this community and I'm not worried about my future grandchildren."[40] Similarly, a municipal government official charged with managing the fire distinguished between public and expert forms of risk evaluation, stating: "We will be able to see it [contaminant levels]. Quantify it. . . . These are facts, but the exposure—or, basically, the [risks of] exposure to our health—are perceptions."[41] In turn, public understandings of the dump fire were rendered technical, and power was maintained within scientific knowledge frames and networks. While this may have addressed certain public concerns, the government's propositional risk management framework failed to address how the dump fire was understood through Iqaluit's particular "civic epistemology"—or, the "historically and politically situated, culturally specific, public knowledge-ways" of the community.[42] As one long-term Iqaluit resident explained of the GN's messaging: "I don't think any of our questions have been answered except that you won't die by breathing it."[43]

ISSUES OF PUBLIC MEANING: PUBLIC CONTESTATION AND ARCTIC SOVEREIGNTY

A photo taken of Alivaktuk, the pregnant Inuk woman and activist quoted earlier, came to symbolize how the dump fire was experienced by many Iqalummiut (figure 10.3). The photo features a pregnant Alivaktuk standing in front of the dump fire wearing a surgical mask; on the palm of her outstretched hand is the word *Taima* ("enough"), written in Inuktitut syllabics. The image is an example of what Tuck and Yang refer to as a "refusal," as it foregrounds Inuit agency and efforts to make public the histories of colonialism and racialized inequities that contributed to this and many other

Figure 10.3. "Taima" [Enough]. Photo courtesy of Shawn Innuksuk.

Nunavut dump fires.[44] The photo, which obtained hundreds of shares on social media, became the Facebook profile photo of Iqalummiut for Action—Stop the Dump Fires (IFA)—a local protest group that formed halfway through the summer.

Crucially, the Iqaluit "dumpcano" occurred at a moment when Inuit communities throughout the territory began actively engaging in public protest as a way of asserting Inuit values, knowledge, and concerns. As has been noted elsewhere, Inuit have historically engaged in "covert" rather than "overt" strategies of resistance—what is often mistaken for agreeability or complacency.[45] More recently, however, Inuit have begun engaging in highly visible (and confrontational) expressions of resistance alongside Canada's larger Indigenous movement. The Idle No More movement, for example, began in December 2012 with the purpose of bringing attention to the fact that Canada's environmental policy regularly ignores Indigenous sovereignty and land rights. The movement started in southern Canada but quickly spread to locations worldwide, including the Arctic. Inuk activist and artist Laakkuluk Williamson Bathory has explained the importance of Idle No More within a Nunavut context: "While Inuit have been successful in negotiating historic land claim agreements and self-government in both

Inuit regions (Canada and Greenland), there is still a deeply embedded struggle to actually implement the agreements and have a lived experience of sovereignty. . . . The southern swell of Idle No More arrived in the North at a time when many Inuit and Inuit organizations were frustrated in this incongruity [*sic*] of post-colonial Arctic society. . . . Rising up is not simply about protesting federal government bills, but also a call from within to make personal and individual politicized choices."[46] Within the last five years, Inuit have initiated several well-publicized social media campaigns aimed at raising public awareness about issues directly affecting Inuit. Among the most widely publicized of these campaigns is the social media–driven #sealfie campaign (a play off the term "selfie"), where Inuit and other northerners take photos of themselves hunting or wearing seal. The purpose of the movement is to showcase the importance of seal hunting for Inuit livelihoods, including how nonsubsistence seal hunting is important for Inuit participation in the global economy.

To this end, it is perhaps not surprising that Iqaluit residents took to social media in order to help raise the profile of the Iqaluit dump fire. As is typical with municipal landfill fires, the Iqaluit dump fire was what landfill engineers classify as "deep-seated," meaning it originated deep within the dump and is characterized by a smoldering appearance.[47] Describing this type of landfill fire to local media outlets, the Iqaluit fire chief compared the dump to what he referred to as a "smouldering volcano."[48] Local social media users took to Twitter, combining the words "dump" and "volcano," creating the hashtag "#dumpcano" to use when discussing the fire online. Shortly afterward, an anonymous Iqaluit resident created the anthropomorphizing Twitter account @yfbdumpcano, which began tweeting satirical posts about the dump fire and the government's response toward it. The term "dumpcano" caught on and contributed to the dump fire receiving both national and international media attention.[49]

Throughout the summer, a major source of frustration within the Iqaluit community (and one that was increasingly targeted by activists) was the prolonged process of approving the dump fire extinguishment plan, which required the city to secure upward of $7 million. For the city of Iqaluit, whose tax base is only seven thousand people, this expenditure would require "putting a freeze on hiring, delaying capital building plans and halting purchases."[50] Amid many infrastructural issues, Iqaluit has a massive housing shortage, an aging sewage treatment and collection system, and several recreation projects under way. Community members noted that choosing one issue at the expense of others was not a viable option. Moreover, many community members pointed out that waste management was inextricable from other social issues associated with colonialism. A journalist that covered the dump fire extensively explained: "We're washing our berries now. They're warning that mother's sons might have fertility issues. That is not good enough. . . . Imagine a town with the population of Nunavut, 35,000 or 36,000 people—and I've heard this compared to Orillia, Ontario. Imagine if Orillia, Ontario, had forty-five suicides in one year. Orillia, Ontario, would be transformed. . . . But it's like, 'That's Nunavut. It's way up there. People are going to kill themselves.' And it's like—it doesn't get the press. It doesn't get the public's attention that it needs. It doesn't get the money it needs."[51]

Accordingly, when the dump fire extinguishment plan was downgraded to a projected $2.2 million plan—one that was more amenable to the city's budget—community members expressed concern. The local protest group, IFA, began a letter-writing campaign, requesting assistance from territorial and federal politicians for funding to help the city extinguish the dump fire. By early August, both the GN and the Government of Canada denied city council's request for federal and territorial government funds. With higher levels of government refusing to provide additional funding, many locals felt that Iqaluit had been left alone to deal with the situation. As one Iqaluit city councilor noted during a council meeting: "There is no Leona; there's no Taptuna trying to help us out."[52]

After the GN's public health advisory about harmful levels of dioxins and furans was released in July, several formal and informal community protests began throughout the city. Near the end of July, the Qaggiavuut Society, a local arts and culture group, began holding a weekly "Art in the Park" event, where artists engaged in throat singing, drum dancing, acting, and storytelling to channel frustrations and anxieties about the fire into public displays of activism. Additionally, IFA members began attending all council meetings, of which at least two were over capacity.[53] By mid-August, more than five hundred people had liked IFA's Facebook page, which was by then being used as a central hub to post dump-related videos and pictures, council meeting minutes, and letter templates for lobbying territorial and federal government officials. In July, an anonymous videographer from Iqaluit posted a video online about the fire that went viral, receiving over 22,000 views.[54] Again, the purpose of this activism was to call attention to the fact that the dump had been burning for over three months and that the community was receiving little help from outside sources.

The final month of the Iqaluit dump fire coincided with two significant and overlapping events. First, in late August (three months after the dump fire started), then Canadian prime minister, Stephen Harper, visited Iqaluit as part of his annual "Northern Tour" of Canada's Arctic. The second and more highly publicized event was the Canadian Armed Forces' annual military exercise, Operation Nanook, which is considered by the Canadian federal government to be the "largest sovereignty operation in the Canadian North."[55] The exercise was held in Iqaluit August 21–29, bringing hundreds of military personnel (and their waste) into the community. When the dump fire protests began receiving widespread media attention, southern Canadian journalists began linking the prime minister's visit to the dump fire controversy. A column printed in the *Ottawa Citizen* stated: "When a problem is distant, it's easy to ignore. . . . Since taking power, Prime Minister Stephen Harper has gone on eight Arctic tours, the last couple to promote 'Canada's Northern Strategy.' . . . Now, a situation has come up that threatens our environmental heritage and will almost certainly set back social and economic development, and the federal government is silent."[56] In this way, IFA and other Iqalummiut were successful in reframing the "dumpcano" as a justice and governance issue.

As with many governance controversies in Nunavut territory, the dump fire came to be folded into what Inuk scholar Jackie Price refers to as "arctic sovereignty debates."[57] For the Government of Canada, a key priority for the Arctic has been to assert geopolitical dominance through increased research, military activity, and resource extraction—what has been explicitly referred to by the government as "exercising our Northern sovereignty."[58] Many Inuit have critiqued the federal government's particular approach to northern sovereignty, stating that it contrasts with Inuit knowledge systems and reifies a European (and colonial) notion of "rule [of people, of place, and of land] that is not only heritable, but beyond question."[59] By contrast, Inuit writers and thinkers have articulated that Arctic sovereignty "begins at home"[60] and is based on "the self-maintained right [for Inuit] to define themselves, mind and soul" through the ability to engage in land and community practices.[61] According to *A Circumpolar Inuit Declaration on Sovereignty in the Arctic*, doing so necessitates the creation of "healthy arctic communities . . . [and] standards of living for Inuit that meet national and international norms and minimums."[62] The "dumpcano," as evidence of standards of living quite literally not meeting national norms and minimums, highlighted the disconnect between how sovereignty is understood and experienced by northerners and the federal government. Public reactions to the dump fire, then, were meant to point out "when the emperor has no clothes," as Sheila Jasanoff puts it, or to call attention to the fact that Inuit and other northerners are routinely exposed to levels of contamination and standards of living deemed unacceptable by southern Canadian communities.[63]

After the GN and the Canadian federal government denied Iqaluit assistance to put out the fire, IFA began publicly lobbying Canada's Department of National Defence (DND). On August 12, representatives from IFA spoke to the Iqaluit City Council, asking the city to formally request DND's assistance in extinguishing the dump fire. The motion passed, and Iqaluit's deputy mayor, Mary Wilman, submitted a formal request to Operation Nanook, asking for "material," "human," and "logistical" resources to help extinguish the fire—a request that was denied later that month.[64] In a statement echoing the stance of IFA, Councilor Joanasie Akumalik framed the DND's sovereignty exercise within the context of Iqaluit's dump fire and a history of "chronic underfunding" from higher levels of government:[65]

> I am somewhat confused by the message from the Government of Nunavut and the Government of Canada. I don't think we should let them off the hook. They never fund the City very well in terms of infrastructure because Iqaluit is always growing. And both have not fully enforced their own regulations over the years. I'm also confused about the message that the [GN's] Chief Medical Officer is sending. She has indicated that the dump fire is not a medical emergency but the Government of Nunavut has released a health advisory that states the dump fire smoke could have an impact on vulnerable people. . . . There is a risk from dioxins. So that's why I'm confused why both governments are not taking responsibility. Cause we'll be hosting the Prime Minister and the military for Operation Nanook, which in 2012 cost $16.5M. This coming exercise will be a mock exercise to save a fake cruise ship. I think we should ask them to tackle a real problem and put out the dump fire.[66]

In this way, both community activists and certain city councilors became part of Iqaluit's "problem" public, challenging Canadian performances of Arctic sovereignty and calling into question the federal government's priorities and responsibilities.[67]

That the community of Iqaluit would channel its frustrations toward the Canadian military is perhaps not surprising given the community's history of military contamination. At least seven abandoned military dumps are located throughout Iqaluit, including the West 40 site where the "dumpcano" was situated.[68] All of these dumps originated in the 1940s–1960s, when the U.S. and Canadian militaries abandoned these wastes. While reports have called for these sites to be remediated since at least the 1970s, other than the removal of PCB-contaminated materials by the federal government in the early 1990s, these sites remain largely the same.[69] Community members, including elected city officials, have been concerned about these waste sites for decades. In 1995 Joe Kunuk, then mayor of Iqaluit, spoke directly to a Canadian parliamentary committee, asking them to remediate Iqaluit's contaminated sites. His comment echoes many of the concerns expressed by Iqalummiut today:

> In the past our elders were concerned about activities that created the old military dumping sites. . . . As late as last week I heard from all the elders that they were told not to worry because the government knew what it was doing. . . . Now we are being told, through studies and reports, that these sites have unacceptable levels of toxins and chemicals. . . . People should not be frightened of the food they eat and the water they drink. We want to be free to move on the land without worrying about the sites that have been contaminated by the military or other outsiders. . . . With our climate, location and scarce financial resources, we are forced to deal with an environment that is not acceptable to our residents and that we're sure would not be tolerated in the south.[70]

Here, Kunuk not only attributes responsibility for the abandoned waste sites to the Canadian military but links contamination temporally through the region's history of colonization and military settlement. He also articulates one of the many ways in which contaminant issues impact Inuit health and well-being: by preventing people from being able to safely go out on the land, or otherwise eat or harvest food obtained from their own communities. Similar concerns have been raised by other Indigenous groups about public health practices of risk avoidance, particularly as they impact relationships with the land. This can be seen, for example, in Elizabeth Hoover's analysis of fish health advisories in the Mohawk community of Akwesasne.[71] Due to contamination by a nearby Superfund site, Akwesasne Mohawks have been warned against consuming local fish—an important food source whose consumption is understood as necessary for healthy living. In these instances, public health is "maintained" but only through institutionalized practices that negate Indigenous sovereignty and land relations.

Given the military's lack of financial assistance to extinguish the fire, many Iqalummiut described Operation Nanook as doing little to help the lives of northerners whose health was being impacted by the dump fire and its corresponding public health measures. One respondent referred to the exercise as "Stephen Harper coming up to go to sleep on a boat for a night," while another described it as "protect[ing] Arctic sovereignty by burning fossil fuels, or whatever it is that they're doing."[72] In a statement that implicates Iqaluit's waste as part of a larger social and environmental justice issue, the GN official charged with overseeing Nunavut's dump sites noted: "I hope that the dump is still burning when [Stephen] Harper comes up here for his annual 'summer vacation' with the military. I really hope it's burning. And maybe . . . maybe they'll get it. That we have a serious infrastructure deficit up here and that building more garbage dumps is not the answer."[73] To this end, the targeting of DND was meant to rearticulate (and thus reframe) the "dumpcano" issue as one not devoid of history or responsibility. To those involved with IFA, requesting assistance from the Canadian military was not meant to be a practical or even expected solution to the dump fire. Indeed, considering it as such would be a "really simplistic way of looking at it," as one IFA member noted.[74] Rather, it was meant to highlight how various modes of governance ignore "well-being" as it is understood and experienced by those living in northern Canada.

The targeting of Operation Nanook by the IFA was particularly strategic in that it redirected the cause of Iqaluit's waste issues to one of governance and of conflicting understandings of Arctic stewardship. One long-term non-Inuit resident and IFA member explained how northerners and the federal government have different understandings of sovereignty:

> Soldiers and military—whatever you want to say—can *defend* your sovereignty, but they can't create sovereignty. It takes women and children, communities, clam diggers, and fisherman, and berry pickers to create sovereignty. . . . That doesn't play into the photo ops of jets flying overhead. And when we look at the amount of money that we invest in the berry pickers versus that amount that we invest in the jets- that's the discussion around how effectively are we actually addressing- or do we need to address, or how do we address- any kind of sovereignty issues. . . . But that's how northerners think about it. I mean, people use the land- not "use it or lose it" in having guys march around on it- but snow machines. People going on snow machines. People going hunting. Young people having a transfer of hunting skills. Family spending a summer

> camping. Those are the ways that you genuinely know about [the land]. You own the land because you know about it. And you know about it because you live on it and you use it. And you hear stories about it and how history and geography and all those things come together. Family history, natural history, geography- they're all wound together in the way that northerners use the land. That's what really established northern sovereignty.[75]

The focus of IFA, then, was not one of downstream waste management per se, but one of ethics and responsibility. In this way, an investment in community infrastructure and activities that support community health and well-being, and that encourage active participation in, and creation of, the Arctic's natural social environment—what is, for many, the foundation of Inuit governing practices[76]—is, for many northerners, what was being affronted by the dump fire: children and Elders were unable to go outside, berry-picking became a source of anxiety, and community celebrations were postponed. As Iqaluit City Councilor Kenny Bell explained to a local newspaper: "What worries me most is the quality of life, and not just the potential of getting sick or being sick."[77] Discussions of risk management, then, which purposefully ignored public framings of the issue, further reinforced a governance structure for which local understandings of community health and well-being were neglected.

In the end, DND did not provide additional resources to help the community douse the dump fire. Rather, DND officials privately conducted additional testing for dioxins and furans, the results of which were not shared publicly.[78] Indoor air quality was assessed in the locations where military personnel were sleeping, and soil samples were preemptively taken near military headquarters—tests that had not, and were never, conducted elsewhere in the community. While levels of contamination were, by and large, considered to be below allowable standards for human exposure, both pregnant women and asthmatics were not deployed to Iqaluit. In this way, not only were southern Canadian military personnel more informed than Iqalummiut of potential hazards located in their own community, but these risks were deemed hazardous enough to warrant halting the deployment of vulnerable southern Canadians.[79] That military officials who defend sovereignty should have greater protection measures than the community members who create it is perhaps implicit to governance structures wherein sovereignty is defined as land ownership and constituted through military operations.

CONCLUSIONS

On August 25, two weeks prior to the dump fire's eventual extinguishment, nearly sixty Iqalummiut packed into a local hall to learn of the city's recently approved dump fire extinguishment plan. The public meeting was held at the request of IFA on behalf of community members who were concerned about the dump fire and its extinguishment. Despite the dump fire's infamy, this end-of-August public meeting was the first official dialogue between experts and the public.

After spending the first hour of the meeting outlining the expected dump fire extinguishment process, the Iqaluit fire chief turned the floor over to the public to ask questions and voice their concerns. Over the next two hours, community members posed a variety of questions, ranging from logistical constraints ("Do we have enough firefighters?") to contaminant monitoring ("Is there a way to monitor the amount of pollutants in the fish, the sea, and for the berries?"), to concerns over communication ("Have they prepared and informed the Elders who have health problems? Have they informed these people about what to do?"), to issues of equity and air-quality monitoring: "In my area, I think we are the most populated. We're the ones [who] get the smoke probably first, and I don't think the smoke coming to us is the same smoke that's coming up to [the air-quality monitors]. I'm not sure—I could be wrong. But I'm worried . . . We have overcrowding. There's lots of children; there's lots of Elders. There's a lot of people. When you say [people] are at risk, a lot of them are in my area."[80] For those attending the meeting, concerns about contaminant exposure were not just a technical issue but embedded in community histories of contaminant exposure, relationships with colonial government officials, and contrasting ideas, values, and understandings of what it means to live together in healthy Nunavut communities.

As I have showcased throughout this chapter, concerns about the dump fire were framed and responded to differently by community members and government officials. While "deficit model" risk management practices privileged a scientific understanding of harm, the Iqaluit public took to protest, repoliticizing the issue on their own terms. In doing so, Iqaluit residents clarified to a wider audience that the "dumpcano" was not an isolated incident but one that both stemmed from and reinforced institutionalized barriers to healthy living. These structural barriers include a lack of resources provided to Nunavut communities, the inability to engage in community practices without fear of contamination, and risk management practices that stifled avenues for meaningful public engagement. That community concerns about berry-picking were addressed by converting these concerns into "informational inputs"—adding "wash the berries before you eat them" to public health warnings—was evidence of this disconnect.[81] Rather, understandings of risk for many Iqalummiut are fundamentally embedded in community histories of contamination and Inuit conceptualizations of health and livelihood. For governance practices to help Nunavut residents secure and maintain their health, well-being, and future, risk management practices must incorporate local understandings of justice, sovereignty, and well-being—not just as points of data but as central to the very framework.

ACKNOWLEDGMENTS

I thank the Inuit of Nunavut, whose lands this research was conducted on. I thank Myra J. Hird, Arn Keeling, Donny Persaud, Ignace Schoot, the editors of this collection, and an anonymous reviewer for providing feedback that greatly improved this chapter. Thanks also to David Mercer, Jess Melvin, and Amanda Degray for help with map making. Research was funded by an Ontario Graduate Scholarship, Social Science and Humanities Research Council of Canada (SSHRC) CGS-M, and SSHRC Insight Grant (#43502013–0560). All fieldwork was approved by the Iqaluit City Council and licensed through the Nunavut Research Institute (#0102114N-M).

NOTES

1. According to federal government documents, the exact origins, depth, and diversity of wastes in the West 40 site are unknown; though, as is evidenced by abandoned military equipment found in the dump, the site is assumed to date back to the city's origins as a World War II military base. The dump became the city's active waste disposal site in 1995 (a previous dump had reached capacity), and its use was intended to be temporary; however, due to limited financially feasible alternatives and the high costs associated with infrastructure development in the Arctic, the site remains in use. For a brief history of the dump, see Public Works Canada, *Literature Review on Abandoned and Waste Disposal Sites in Iqaluit Area, Northwest Territories*, 6–9.

2. In the early 2000s alone, the United States had approximately 8,300 landfill fires annually. Federal Emergency Management Agency, *Landfill Fires*, 20–21. In many Nunavut communities, living with smoke from landfill fires is common.

3. Doing so was thought to increase contaminant concentrations throughout the community. Sperling, "Iqaluit Landfill Fire Control." Moreover, the ocean was covered in sea ice until early July.

4. Iqaluit fire chief Luc Grandmaison, quoted in Varga, "No End in Sight for Iqaluit Dump Fire, Officials Say."

5. This chapter draws from three months of ethnographic research, including participant observation and twenty-seven semistructured interviews with community members, government officials, and others involved in managing the dump fire. Fieldwork was conducted from June 2, 2014, to September 5, 2014. See Zahara, "The Governance of Waste in Iqaluit, Nunavut" (master's thesis, Queen's University, 2015).

6. Wynne, "May the Sheep Safely Graze?," 47.

7. Liboiron, "On Solidarity and Molecules (#MakeMuskratRight)."

8. "Iqalummiut" is the local term used to describe those living in Iqaluit.

9. Interview with Iqalummiut for Action member, August 9, 2014; interview with a local Iqaluit journalist, September 3, 2014.

10. Qikiqtani Inuit Association, *QTC Final Report*, 22–26.

11. Tester and Irniq, "*Inuit Qaujimajatuqangit*," 52. Truth and Reconciliation Commission of Canada, *Honouring the Truth, Reconciling for the Future*, 82.

12. Tester and Irniq, "*Inuit Qaujimajatuqangit*," 58.

13. Members of the boards are elected by Inuit governments and the territory. Of course, Indigenous governing practices also take place outside of government bureaucracy; see Todd, "Fish Pluralities," 218; Watts, "Indigenous Place-Thought and Agency amongst Humans and Non-Humans," 25–28.

14. Jackie Price, "Tukisivallialiqtakka: The Things I Have Now Begun to Understand: Inuit Governance, Nunavut and the Kitchen Consultation Model" (master's thesis, University of Victoria, 2007), 36.

15. Cameron, "Securing Indigenous Politics," 104.

16. Tester and Irniq, "*Inuit Qaujimajatuqangit*," 50; Blaser, "The Threat of the YRMO," 15.

17. Nadasdy, "The Anti-Politics of TEK," 216; emphasis original.

18. Irwin, "Risk and Public Communication," 2.

19. Wynne, "Public Engagement as a Means of Restoring Public Trust in Science," 216.

20. Health Canada, *Addressing Psychosocial Factors through Capacity Building*, 5.

21. Health Canada, *A Guide to Involving Aboriginal Peoples in Contaminated Site Management*, 10.

22. Health Canada, *Addressing Psychosocial Factors through Capacity Building*, 1; Canadian Council of Academics, *Health Product Risk Communication*, 39.

23. Wynne, "Public Engagement as a Means of Restoring Public Trust in Science," 217.

24. Wynne, "Public Engagement as a Means of Restoring Public Trust in Science," 211. Cameron, "Securing Indigenous Politics," 110; Jasanoff, "Breaking the Waves in Science Studies," 398.

25. Environment Canada, paraphrased in Varga, "Environment Canada."

26. City councilor Kenny Bell quoted in Peter Varga, "City Council Orders Iqaluit Fire Department to Extinguish Dump Fire."

27. Varga, "Environment Canada."

28. Iqaluit city councilor Terry Dobbin, quoted in Varga, "City Council Orders Iqaluit Fire Department to Extinguish Dump Fire."

29. I include "relevant" in scare quotes here to showcase how siloed risk management processes were within scientific frames. Specifically, the Dump Fire Working Group included representatives of the GN's Departments of Environment, Health, Executive and Intergovernmental Affairs; the director of Protection Services; representatives of Emergency Response and Recovery for Aboriginal Affairs and Northern Development Canada; and manager of the Iqaluit International Airport. City of Iqaluit representatives were from the Department of Engineering and the Fire Department. The Air Quality Monitoring Working Group was comprised of GN, Health Canada, Environment Canada, AANDC, and, later on, Department of National Defence officials. Note that representatives from GN's Department of Culture and Heritage, local or national Inuit organizations, were not included in either group.

30. The Government of Nunavut does not have standards set for dioxins and furans, and therefore went by the most up-to-date standard, which was the province of Ontario's. The standards used in Ontario are based on recommendations of the World Health Organization.

31. For an overview of contaminant exposure and reproductive health issues in Indigenous communities, see Hoover et al., "Indigenous Peoples of North America."

32. Interview with GN Health official in Iqaluit, August 8, 2014.

33. Field notes, August 1, 2015.

34. Maureen Baikie, quoted in Varga, "Iqaluit Dump Fire Smoke Not a Public Health Emergency, GN Says."

35. Quoted in Murphy, "Pregnant Nunavut Mom Worried about Dump Smoke Toxins."

36. Wynne, "Risk as Globalizing Democratic Discourse?," 66.

37. Field notes, August 25, 2014.

38. Field notes, August 25, 2014.

39. Interview with GN Health official in Iqaluit, August 8, 2014.

40. Field notes, August 1, 2014.

41. Interview with government official in Iqaluit, August 27, 2014.

42. Jasanoff, *Designs on Nature*, 251.

43. Interview with long-term Qallunaac resident in Iqaluit, August 27, 2014.

44. Tuck and Yang, "R-Words: Refusing Research." See also Zahara, "Refusal as Research Method in Discard Studies."

45. Tester and Irniq, "*Inuit Qaujimajatuqangit*," 49.

46. Williamson Bathory, "Naamaleqaaq!"

47. Environment Agency, *Review and Investigation of Deep-Seated Fires within Landfill Sites*, 4.

48. Iqaluit fire chief Luc Grandmaison, quoted in Varga, "City Can't Douse Iqaluit's Latest Massive Dump Fire."

49. Stories referring to the "dumpcano" appeared in major Canadian newspapers such as the *National Post* and *Globe and Mail*, and were circulated by the Canadian press to local newspapers across the country. Others appeared in *Vice News* and the *Huffington Post*.

50. City of Iqaluit chief administrative officer John Hussey, quoted in Murphy, "Landfill Expert Says $3.5-Million Dunking Best Solution for Iqaluit Dump Fire."

51. Interview with Iqaluit journalist, September 3, 2014.

52. City councilor Joanasie Akumalik, field notes, August 1, 2014. He is referring to minister of environment and Nunavut MP Leona Aglukkaq and Nunavut premier Peter Taptuna.

53. Field notes, August 1, 2014; August 12, 2014.

54. "Iqaluit's Dump Fire."

55. National Defence and the Canadian Armed Forces, "Operation Nanook."

56. Gordon, "Dumpcano and Canada's Northern Hypocrisy."

57. Price, "But You're Inuk, Right?"

58. Canada's Northern Strategy, "Exercising our Arctic Sovereignty."

59. Qitsualik, "Innummarik," 23.

60. Simon, "Inuit and the Canadian Arctic," 250.

61. Qitsualik, "Innummarik," 31.

62. Inuit Circumpolar Council, *A Circumpolar Declaration on Sovereignty in the Arctic*, 1.

63. Jasanoff, "Breaking the Waves in Science Studies," 298.

64. Deputy Mayor Mary Wilman, email to Ed Zebedee, August 15, 2014.

65. LeTourneau, "Planning Commission Slams Ottawa."

66. City councilor Joanasie Akumalik, field notes, August 12, 2014.

67. See Latour, "Turning around Politics," 816, for a discussion of how publics become a "problem."

68. Hird and Zahara, "The Arctic Wastes."

69. Government of Northwest Territories, *Report*, 1; Environmental Services Group, *Environmental Study of a Military Installation and Six Waste Disposal Sites at Iqaluit, NWT*, vol. 1.

70. Joe Kunuk, evidence, Parliamentary Hearings on Canadian Environmental Protection Act, May 9, 1995. http://www.parl.gc.ca/content/hoc/archives/committee/351/sust/evidence/121_95-05-09/sust121_blk-e.html (accessed June 25, 2015).

71. Hoover, "Cultural and Health Implications of Fish Advisories in a Native American Community," 10.

72. Interview with long-term resident of Iqaluit, August 1, 2014; interview with long-term resident and member of IFA, August 25, 2014.

73. Interview with GN government official, July 28, 2014.

74. Interview with IFA member, August 9, 2014.

75. Interview with IFA member, August 25, 2014.

76. Price, "Tukisivallialiqtakka," 36. Qitsualik, "Innummarik," 28–30.

77. Varga, "City Council Orders Iqaluit Fire Department to Extinguish Dump Fire."

78. This information was made available through the completed access to information request, #A-2014-00932.

79. The exact wording of the internal memo reads "as a precautionary approach, those with severe lung disease (asthma) as well as pregnant women should not be part of the deployment to Iqaluit."

80. Field notes, August 25, 2014.

81. Blaser, "The Threat of the YRMO," 12; Government of Nunavut, *Public Health Advisory*.

BIBLIOGRAPHY

Blaser, Mario. "The Threat of the YRMO: The Political Ontology of a Sustainable Hunting Program." *American Anthropologist* 111, no. 1 (2009): 10–20.

Canada's Northern Strategy. "Exercising Our Arctic Sovereignty." Accessed June 20, 2015, http://www.northernstrategy.gc.ca/sov/index-eng.asp.

Cameron, Emilie. "Securing Indigenous Politics: A Critique of the Vulnerability and Adaptation Approach to the Human Dimensions of Climate Change in the Canadian Arctic." *Global Environmental Change* 22, no. 1 (2012): 103–114.

Canadian Council of Academics. *Health Product Risk Communication: Is the Message Getting Through?* Ottawa: Canadian Council of Academics, 2015.

Environment Agency. *Review and Investigation of Deep-Seated Fires within Landfill Sites.* Bristol: Environment Agency, 2007.

Environmental Services Group. *Environmental Study of a Military Installation and Six Waste Disposal Sites at Iqaluit, NWT.* Vol. 1. Ottawa: Indian and Northern Affairs Canada and Environment Canada, 1995.

Federal Emergency Management Agency. *Landfill Fires: Their Magnitude, Characteristics, and Mitigation.* Arlington, VA: United States Fire Administration, 2002.

Gordon, James. "Dumpcano and Canada's Northern Hypocrisy." *Ottawa Citizen,* August 7, 2014. http://ottawacitizen.com/news/national/gordon-dumpcano-and-canadas-northern-hypocrisy.

Government of Northwest Territories. *Report: Environment: Frobisher Bay.* Yellowknife: Northwest Territories, 1972.

Government of Nunavut. *Public Health Advisory: Iqaluit Dump Fire Smoke Public Health Update.* Iqaluit: Department of Health, Government of Nunavut, 2014.

Health Canada. *Addressing Psychosocial Factors through Capacity Building: A Guide for Managers of Contaminated Sites.* Ottawa: Queen's Printer, 2005.

Health Canada. *A Guide to Involving Aboriginal Peoples in Contaminated Site Management.* Ottawa: Queen's Printer, 2005.

Hird, Myra J., and Alexander Zahara. "The Arctic Wastes." In *Anthropocene Feminism,* edited by R. Grusin, 121–45. Minneapolis: University of Minnesota Press, 2017.

Hoover, Elizabeth. "Cultural and Health Implications of Fish Advisories in a Native American Community." *Ecological Processes* 2, no. 4 (2013): 1–12.

Hoover, Elizabeth, Katsi Cook, Ron Plain, Kathy Sanchez, Vi Waghiyi, Pamela Miller, Renee Dufault, Caitlin Sislin, and David O. Carpenter. "Indigenous Peoples of North America: Environmental Exposures and Reproductive Justice." *Environmental Health Perspectives* 120, no. 12 (2012): 1645–649.

Inuit Circumpolar Council. *A Circumpolar Declaration on Sovereignty in the Arctic.* Ottawa: Inuit Circumpolar Council, on Behalf of Inuit in Greenland, Canada, Alaska, and Chukotka, 2009.

"Iqaluit's Dump Fire." Posted by *Carbon Capture,* July 7, 2014. YouTube video, 1:49. https://www.youtube.com/watch?v=a1GTSlAYQ5c.

Irwin, Alan. "Risk and Public Communication: Third-Order Thinking about Scientific Culture." In *Handbook of Public Communication of Science and Technology,* edited by Massimiano Bucchi and Brian Trench, 199–212. New York: Routledge, 2008.

Jasanoff, Sheila. "Breaking the Waves in Science Studies: Comment on H. M. Collins and Robert Evans, 'The Third Wave of Science Studies.'" *Social Studies of Science* 33, no. 3 (2003): 389–400.

Jasanoff, Sheila. *Designs on Nature: Science and Democracy in Europe and the United States.* Princeton, NJ: Princeton University Press, 2007.

Latour, Bruno. "Turning around Politics: A Note on Gerard de Vries' Paper." *Social Studies of Science* 37, no. 5 (2007): 811–20.

LeTourneau, Michele. "Planning Commission Slams Ottawa." *Northern News Services Online*, June 23, 2014. https://www.nnsl.com/archives/2014-06/jun23_14slam.html.

Liboiron, Max. "On Solidarity and Molecules (#MakeMuskratRight)." *Discard Studies* (blog), October 20, 2016. https://discardstudies.com/2016/10/20/on-solidarity-and-molecules-makemuskratright/.

Murphy, David. "Landfill Expert Says $3.5-Mllion Dunking Best Solution for Iqaluit Dump Fire." *Nunatsiaq News*, July 1, 2014. http://www.nunatsiaqonline.ca/stories/article/65674landfill_expert_says_3.5-million_dunking_best_solution_for_iqaluit_dum/.

Murphy, David. "Pregnant Nunavut Mom Worried about Dump Smoke Toxins." *Nunatsiaq News*, July 4, 2014. http://www.nunatsiaqonline.ca/stories/article/65674pregnant_nunavut_mom_worried_about_dump_smoke_toxins/.

Nadasdy, Paul. "The Anti-Politics of TEK: The Institutionalization of Co-management Discourse and Practice." *Anthopologica* 47, no. 2 (2005): 215–32.

National Defence and the Canadian Armed Forces. "Operation Nanook." Accessed May 12, 2015. http://www.forces.gc.ca/en/operations-canada-north-america-recurring/op-nanook.page.

Price, Jackie. "But You're Inuk, Right?" Paper presented at Indigenous Governance speaker series, Victoria, BC. February 12, 2013. YouTube video, 1:31:44. https://www.youtube.com/watch?v=W36cGxXpjWw.

Price, Jackie. "Tukisivallialiqtakka: The Things I Have Now Begun to Understand: Inuit Governance, Nunavut and the Kitchen Consultation Model." Master's thesis, University of Victoria, 2007.

Public Works Canada. *Literature Review on Abandoned and Waste Disposal Sites in Iqaluit Area, Northwest Territories*. Edmonton: Environmental Services, Pacific Western Region, 1992.

Qitsualik, Rachel. "Innummarik: Self-sovereignty in Classic Inuit Thought." In *Nilliajut: Inuit Perspectives on Security, Patriotism and Sovereignty*, edited by Scot Nickols, Karen Kelley, Carrie Grable, Martin Lougheed, and James Kuptana, 23–34. Ottawa: Inuit Tapiriit Kanatami, 2013.

Qikiqtani Inuit Association. *QTC Final Report: Achieving Saimaqatiqiigniq*. Iqaluit: Inhabit Media, 2013.

Simon, Mary. "Inuit and the Canadian Arctic: Sovereignty Begins at Home." *Journal of Canadian Studies* 43, no. 2 (2009): 250–60.

Sperling, Tony. "Iqaluit Landfill Fire Control." Presentation to Iqaluit City Council, June 30, 2014.

Tester, Frank, and Peter Irniq. "*Inuit Qaujimajatuqangit*: Social History, Politics and the Practice of Resistance." *Arctic* 61, no. 5 (2007): 48–61.

Todd, Zoe. "Fish Pluralities: Human-Animal Relations and Sites of Engagement in Paulatuuq, Arctic Canada." *Études/Inuit/Studies* 38, nos. 1–2 (2014): 217–38.

Truth and Reconciliation Commission of Canada. *Honouring the Truth, Reconciling for the Future.* Ottawa: Truth and Reconciliation Commission of Canada, 2015.

Tuck, Eve, and K. Wayne Yang. "R-Words: Refusing Research." In *Humanizing Research: Decolonizing Qualitative Inquiry with Youth and Communities,* edited by Django Paris and Maisha T. Winn, 223–48. Thousand Oaks, CA: SAGE, 2014.

Varga, Peter. "City Can't Douse Iqaluit's Latest Massive Dump Fire." *Nunatsiaq News,* May 29, 2014. http://www.nunatsiaqonline.ca/stories/article/65674city_cant_douse_iqaluits_latest_massive_dump_fire/.

Varga, Peter. "City Council Orders Iqaluit Fire Department to Extinguish Dump Fire." *Nunatsiaq News,* June 12, 2014. http://www.nunatsiaqonline.ca/stories/article/65674city_council_orders_iqaluit_fire_department_to_extinguish_dump_fire/.

Varga, Peter. "Environment Canada: Iqaluit Dump Smoke Tests Reveal Little Immediate Danger." *Nunatsiaq News,* June 11, 2014. http://www.nunatsiaqonline.ca/stories/article/65674environment_canada_iqaluit_dump_smoke_tests_reveal_little_immediate_da.

Varga, Peter. "Iqaluit Dump Fire Smoke Not a Public Health Emergency, GN Says." *Nunatsiaq News,* August 4, 2014. http://www.nunatsiaqonline.ca/stories/article/65674iqaluit_dump_fire_smoke_not_a_public_health_emergency_gn_says/.

Varga, Peter. "No End in Sight for Iqaluit Dump Fire, Officials Say." *Nunatsiaq News,* May 26, 2014. http://www.nunatsiaqonline.ca/stories/article/65674no_end_in_sight_for_iqaluit_dump_fire/.

Watts, Vanessa. "Indigenous Place-Thought and Agency amongst Humans and Non-Humans (First Woman and Sky Woman Go on a European World Tour!)." *Decolonization: Indigeneity, Education and Society* 2, no. 1 (2013): 20–34.

Williamson Bathory, Laakkuluk. "Naamaleqaaq! Idle No More in the Arctic." *Northern Public Affairs* 1, no. 3 (Spring 2013): 39–41.

Wynne, Brian. "May the Sheep Safely Graze? A Reflexive View of Expert-Lay Knowledge Divide." In *Risk, Environment and Modernity: Towards a New Ecology,* edited by Scott Lash, Bronislaw Szerzynski, and Brian Wynne, 44–83. London: SAGE, 1996.

Wynne, Brian. "Public Engagement as a Means of Restoring Public Trust in Science—Hitting the Notes but Missing the Music?" *Community Genetics* 9, no. 3 (2006): 211–20.

Wynne, Brian. "Risk as Globalizing Democratic Discourse? Framing Subjects and Citizens." In *Science and Citizens: Globalization and the Challenge of Engagement,* edited by Melissa Leach, Ian Scoones, and Brian Wynne, 66–82. Portland: Ringgold, 2005.

Zahara, Alexander. "The Governance of Waste in Iqaluit, Nunavut." Master's thesis, Queen's University, 2015.

Zahara, Alexander. "Refusal as Research Method in Discard Studies." *Discard Studies* (blog). March 21, 2016. https://discardstudies.com/2016/03/21/refusal-as-research-method-in-discard-studies/.

Epilogue

Containment

DISCUSSING NUCLEAR WASTE WITH PETER GALISON

Interview by Vivien Hamilton and Brinda Sarathy

The rich and nuanced stories in this volume deepen our understanding of the multiple ways in which toxic environments have been created and continue to be sustained. You may feel, as we do, overwhelmed by all that remains debated, unknown, and uncertain. However, simply acknowledging that uncertainty seems to be an important first step in confronting our toxic present. As so many of our cases have shown, the creation and regulation of contaminated places has historically favored economic growth and military-industrial interests. Uncertainties around toxicity have either been ignored or used to justify action rather than caution. But in recognizing that pattern, we can imagine acting otherwise in the future: we might move ahead with greater caution, anticipating and planning for toxic substances rather than reacting to an environmental or health crisis already under way.

The histories in this collection also help us to imagine a more active partnership between citizens and scientific experts. These cases show how an easy reliance on expert opinion is complicated when experts from different disciplines, or even within the same discipline, disagree about the nature and safety of toxic spaces. Even when there has been clear consensus, expert knowledge has often been suppressed due to structures of military or industrial secrecy. This prompts us to have a more critical and deliberate political conversation about what is known and by whom, and to do a better job of acknowledging the voices of those harmed by toxic exposures so that we can incorporate that knowledge into policies and decision making around toxic substances.[1]

As we begin to envision ways to minimize and better regulate toxics moving forward, we also need to confront the reality of the toxic landscapes we already inhabit. This volume has shown how the invisible nature of toxic agents can lead to appearances of normalcy or innocuousness, which contribute to a lack of urgency around risks of exposure. Whether it is urban drilling sites that have been artfully masked by landscape beatification schemes, the silent seep of chemicals into underground aquifers, or the sheer insensibility of radiation, the imperceptibility of toxicity has obscured a clear path to action.

This collection is one attempt to make the all-pervasive reach of toxic spaces more visible, but it is clear that there is no silver bullet for the problem of toxic contamination. The very nature of nuclear and industrial waste, and the entanglement of petrochemical products in almost every facet of life, renders the issue of toxic pollution a "wicked problem." Such problems are large scale and constitute "long-term policy dilemmas in which multiple and compounding risks and uncertainties combine with sharply divergent public values to generate contentious political stalemates."[2] Yet, stalemates or not, it is incumbent on society to do something about toxic contamination. As historian of science Peter Galison states in the interview that follows: "There is no way of avoiding these questions. We have to figure out what's the best among alternatives, none of which are perfect."

We have chosen to end this collection with a consideration of one of the most urgent challenges currently confronting us: the problem of how to store nuclear waste safely and communicate the dangerous nature of that waste to civilizations that will exist millennia from now. In the remaining pages, we engage in conversation with Peter Galison, whose recent film with Robb Moss, *Containment* (2015), explores the fraught problem of long-term nuclear waste storage. The film takes us to radioactive wilderness in Fukushima, Japan, to nuclear weapons plants and a waste storage facility deep underground in the United States, and to a distant future in which generations will need to be warned about the presence of this almost timeless danger. We encourage readers to engage with this interview as a companion to the film and also to this book. Our conversation with Galison connects with many of the themes covered in this volume (scientific expertise, toxic contamination, and political contestation) with a focus on the most terrifying example of toxicity we have considered yet—that of nuclear waste.

SPACE AND TIME

VIVIEN: One of the big questions that emerges from the case studies in our book is one of responsibility. Who feels responsible for managing toxic spaces and for communicating risk? Who do we feel responsible to protect? Whose danger gets acknowledged?

PETER: Hard questions of responsibility arise when we look at how to dispose of uranium, plutonium, and all of the by-products of fission up and down the periodic table. What is impressive to me is that such a concrete issue of waste throws into question fundamental aspects of how we think about the politics of space and time.

For instance, if you ask the people who are living right near the Carlsbad, New Mexico, salt mine (the Waste Isolation Pilot Plant, WIPP) that is being used as a deep ecological repository for nuclear waste, many favor it—the work has launched a significant economic boom. As former Carlsbad, New Mexico, mayor Bob Forrest says in the documentary, it provides a thousand jobs that, with benefits, pay $100,000 apiece in a part of the world that is not filled with high-paying jobs. So, if you think of families of four, in a town of 25,000, you have got four thousand people, or 17 percent of the city directly affected. Then there are all those businesses that support or are supported by the effort: in total, the waste industry is a big part of the local economy. The WIPP is a powerful part of the economic life of Carlsbad.

But what about the rest of the state? The northern part of New Mexico frequently votes Democrat; the southeast is distinctly Republican. There are differences in people's relationship to the federal government as well as different patterns of ethnic and political composition. In the south, it is not uncommon to hear people talking about how great it has been to have the WIPP, and in the north people have often been quite opposed—I am generalizing to be sure—but that is an overall feeling. This split goes back decades. So, then, the question is, who should decide? Is this an issue for the proximate neighbors of the plant in Carlsbad and a few surrounding towns? Does New Mexico as a whole get to make the decision? What about the states around New Mexico? The waste is transported through other states on its way to WIPP. Do citizens in those other states have a vote? What about the country of Mexico? Mexico is closer to the WIPP site than the WIPP site is to the biggest city in New Mexico—Albuquerque. In all the research I have done over the past decade, I never saw a single document that raised the idea that Mexico might have a view about this. Energy policy has rarely been able to transcend national boundaries, even if nuclear accidents do. Chernobyl's plume did not stop at the Ukrainian border.

In addition to spatial questions, there are the temporal ones. As we make decisions today, we can ask: To which generation are we responsible? Are we only obligated to people living now? To the generation of our children? To that of our grandchildren? Five generations? A hundred generations? Four hundred generations? In order to open the WIPP site in New Mexico, Congress had the Environmental Protection Agency (EPA) determine the period during which people should be adequately warned against inadvertent intrusion into the waste. Taking into account the half-lives of the radioactive materials (e.g., plutonium-339 has a half-life of 24,100 years), balancing that against the age of our civilization since writing, the EPA set the era of immediate responsibility at ten thousand years. After protest from various quarters, and reckoning with some of the much-longer half-lives involved, the National Academy of Sciences urged the Yucca Mountain site in Nevada to follow a *million*-year period in protecting and warning the future. A million years from now, we may not even be the us of our species self. I mean that literally: "we," as homo sapiens, emerged from homo erectus around 200,000 to 300,000 years ago; we homo sapiens were surely not in existence in our anatomically modern form a million years ago. There is no reason to expect that "we" (in that sense) would be around a million years from now. Does this mean giving up on the far future altogether?

I do not see a conflict between the future and the present: you cannot safeguard the future without taking care of the present. Indeed, to think about the future can reinforce our care in taking action now; to care for the present is to make a first and needed step toward safeguarding the far future. However, it is sobering to think how difficult it is to contemplate hundreds or thousands of years down the line. We may be ill-equipped to address the future, but we have created a world with nuclear materials that makes thinking about that future mandatory.

BRINDA: In the documentary, it was striking to see so much thought and resources put into conveying dangers to future generations that we will not even know. By contrast, there was little warning or explanation for communities in the present. I am thinking of the scene in Burke County near the Savannah River Site, a vast nuclear complex owned by the Department of Energy. There is a "no trespassing" sign but not much else in the way of explanations about danger.

PETER: That is true. The young woman driving the boat on the Savannah River says, "It says 'no fishing,' but it doesn't say *why* no fishing." And she adds, "People think it's a territory thing, not a radioactive fish thing." Burke County (Georgia) is right across the river, just fifty feet from the Savannah River Site (SRS is in South Carolina). An SRS advisory urges people to eat just one meal a week of mudfish or largemouth bass from the river because of the cesium-137, Strontium-90, and mercury.[3] On the Georgia side, they report that there are no consumption guides for the cesium and strontium issued.[4] I read an article once warning people against eating more than twenty-five kilos (about fifty-five pounds) of fish a year caught near the weapons facility. The article had an asterisk, and in the footnote it said that, in fact, many African Americans in this area eat more than that. It turns out that many people in Burke County rely on fishing for food; a lot of their protein is from fish from the Savannah River. It is a matter of pride for some in the county that they are not on a public handout program—and they have been fishing for generations. These are old communities that go back to the time of slavery. These people are more than an asterisk. It is indeed important that contemporary fish consumption warnings be clearer and more widely distributed as well as consistently presented across state lines.

So, no doubt the present is important, urgent, a matter of health and justice. That said, I do not think that the resources spent on warning the future are pointless or a distraction. By forcing us to think about future dangers and communication with the future, we can find better ways to assess contemporary dangers, mitigation, and warnings. I do not see it as a choice that *either* you warn the future *or* you warn the present. We must understand the reality that subsistence fishing in this area is an integral part of some residents' food supply. To be effective, warnings must intersect with the lives people lead.

COMMUNICATING RISK

VIVIEN: How should we most effectively communicate risk and danger when we are thinking about many, many generations in the future? In the film, you raise the possibility that myths and stories containing some kind of archetypal character might be the most effective. And yet if we are thinking about the SRS and what we have just been talking about, it seems so important to situate any communication in the particular local experiences of a community. It seems that those two modes of communication are very different from each other. How do we balance talking to a particular culturally, temporally situated group with trying to talk to an imagined future where we have no idea, culturally or socially, what it is going to look like?

PETER: Thinking about the culture surrounding the messenger, message, and recipient is crucial. We do not do a good job with that. If you go on YouTube, you can see people cheerfully, proudly breaking into the old atomic airplane research station (Georgia Nuclear Aircraft Laboratory), where much radioactive material lies buried; people have cut the chain-link fence, squeezed underground, clambered onto the structures for the frisson of penetrating the secret and the dangerous. Once something is no longer in use, it is hard for companies or governments to find the motivation to guard them with the utmost vigilance. Almost instantly, abandoned sites get transformed for reasons ranging from the recreational to the economic. As Adriana Petryna has documented in her work, local scavengers regularly go into Chernobyl to dig up radioactively hot copper piping and sell it on the black market.[5] No mystery here: they are poor, and they need the money. People break into other places as a kind of dark tourism, curiosity, or adventure.

As long as the WIPP site is in use, no one is going to bust into it. Private security guards provided with trucks and automatic weapons guard the facility day and night. I was filming there with Robb Moss one day after we had detailed discussions with the head of the WIPP site about filming after the plant was closed for the evening. I said, "I want to make sure that nobody thinks we're illegally there." The plant director said, "No, no, don't worry, we'll take care of it." We set up the camera, we started to shoot, and a militarized vehicle pulled up with armed guards not at all happy to see us. I'm glad the site is well guarded. That is now. A very different situation will surely exist thirty years after the site is closed.

Judging by our experience with ancient sites, the more types of warning we establish, the better our chance of being understood now and in the future: inscribed text, multiple languages, easily understood images—even stories. Questions of preventing exposure obviously are particularly salient in the nuclear case, but there are similar issues with myriad other substances, from e-waste and mining slag to chemical effluents. Thinking about communication is crucial, and if we could learn something about the difficulty of warning the distant future, perhaps it could help us in the here and now.

SECRECY, EXPERTISE, COMMUNITY

BRINDA: Could you speak a little bit about the nature of secrecy while you were making the film? It was interesting to hear one of the interviewees saying, "There are problems with nuclear storage, but I can't talk about that." It seems that secrecy poses limits to communication and sharing specific information.

PETER: That comment was from Allison Macfarlane, the former chair of the Nuclear Regulatory Commission (NRC), where she served from 2012 to 2014. She started to talk about things that could go wrong, and she said that she could not speak about certain classified dangers, dangers presumably having to do with a potential terrorist attack at some of the nuclear power sites. The problem occurs in any discussion about risky technologies: How can we discuss dangers—to prepare for them—without also giving an instruction manual to people who would want to do us harm? It is clear, for instance, that any cutoff of water to the waste fuel pools would be a very bad thing. In Fukushima, you recall, the plant lost outside power because of the earthquake, and then the tsunami flooded and destroyed the backup diesel-powered pumps. Without cooling, the stored fuel rods overheated, and the pools started to boil off. The then prime minister of Japan, Naoto Kan, says in *Containment* that if this stored hot fuel had caught fire, the airborne contaminants could have been a threat to greater Tokyo. According to Naoto Kan, some fifty million people might have had to be evacuated, a calamity that would have endangered the very existence of the modern state of Japan. No one has any idea how to evacuate so many people in one urgent go. It has never been done: the vast evacuation in Britain at the outset of World War II saw some three million people moved to the countryside. Because nuclear disasters are thinkable at such a scale, secrecy matters, not just for military nuclear capacities but for the civilian sector as well.

I think there are ways that secrecy can actually make things more dangerous. Secrecy is complicated. We do not want to see on the web detailed instructions for how to make nerve gas. I am really happy that that is not on the web (or I hope it isn't). At the same time, there are cases where secrecy can cover up bad or shoddy practices and lead to things being much more dangerous than they would have been otherwise.

VIVIEN: One of the things we have grappled with in this book is understanding the role that scientific experts have played in making decisions about safe levels of exposure for different toxic agents. It feels as though there is a continuum between things that are deliberately kept secret and issues that come into play when you have scientific experts with a special kind of knowledge. This makes some of the assessment of risk inaccessible simply because of a particular specialized vocabulary. Communities then react to scientific pronouncements with trust or suspicion, without necessarily being able to follow or understand all the decisions that are being made by experts.

PETER: Some people living in a technical bubble simply do not think about communicating with people beyond that world. Such outward address seems unimportant, not their job, not interesting, or not rewarding. That is one problem. Some companies keep things secret because they want to protect proprietary information, save money, resist public criticism, and negotiate from a position of strength with employees, unions, and regulators. Sometimes secrecy is just a prosaic cover over cost savings: with the Bhopal disaster in India, one of the significant problems was that the company that ran the plant had cut back on safety procedures to save money and speed up production.

Then there is the secrecy around the specifics of nuclear power and nuclear weapons, which have been shrouded since the beginning of the nuclear age in the 1940s; that has gotten us into trouble over many decades. Just think back on the U.S. Atomic Energy Commission's experiments.

Pregnant women, prisoners, soldiers, and "downwinders" were exposed to some very high doses of radioisotopes. It was only after 1994, under Department of Energy chair Hazel O'Leary in the Clinton administration, that much of this information (some 1.6 million pages of documentation) was released. Unsurprisingly, that decades-long period of secrecy and misrepresentation exacerbated mistrust toward things atomic and made it harder for the Department of Energy to persuade people even when their science and intentions would have benefited the civilian population. All over the world, nuclear technologies and secrecy have been intertwined.

In some ways, we live in a more open world now. In other ways, secrecy has *increased* as more and more of the nuclear establishment has become corporate and the big weapons labs and other labs have shifted from government to private control. You might think, "Oh, well, that's better. The government is the most secret entity." I do not think that is true. There is no Freedom of Information Act for corporations. My big worry, actually, in many of these cases is about corporate secrecy more than it is about government secrecy. If you go to a private chemical plant and say, "I'd like to see documents surrounding your founding and safety records," they will simply say "no." That is the end of the discussion. If you go to a government plant that does the same things, they often say, "Well, here are the things we can release, and here are the things we cannot. You can then apply for those documents through the Freedom of Information Act, and if that fails, you can move it up the chain to the appeals process. In the event of an appeal, the request goes all the way up to the highest-level group that deliberates about what can and cannot be disclosed. Even then, at that ultimate level of adjudication, they release a significant fraction of the materials. There is no such analog process in the private sphere, and I think this kind of secrecy is often not discussed. The more restrictive element is often now lodged in the private sphere—this was indeed my experience with Robb Moss in making *Containment*.

BRINDA: And in addition, in the corporate sector, research is often done by corporate scientists to deem whether something is indeed safe or not. For example, how do we actually regulate practices like hydraulic fracking, which may lead to contamination?

PETER: In Josh Fox's film *Gasland* (2010), he talks to the companies that inject the liquids into the ground, which often contaminate the groundwater, but those processes are proprietary. What these companies do is held back but not for some national security reason. No, the secrecy is about protecting their particular mix of chemicals used under pressure to release the gas from the matrix of rocks. So they say, "No, we absolutely are not going to tell you what is in it." It has been a long-running legal battle to get companies to disclose the chemical mix being used, to standardize their practices, and to conform to EPA standards for keeping the water supply safe.

In disposing of waste, both government and the private sector need to consult the community. The tendency toward secrecy has, over and over again, led to confrontation. I was reading recently about an experimental borehole proposed by the Battelle Memorial Institute, three-mile-deep holes that might be prototypes for a means of burying nuclear waste—an alternative to the deep-mine repository, like the one in Carlsbad. A first recent attempt was in North Dakota. The local rural community found out about it in February 2016 and then only by reading the local newspapers. They were furious. No one had told them what these holes were for or what was going to happen if the tests were successful, or even that this was preliminary to the burial of nuclear waste.[6]

Several hundred people came to a meeting (not a trivial fraction of the community), and they blocked it. Then the project leaders moved their experiment to South Dakota, where local rural resistance stopped it again. This reluctance to be open about nuclear matters runs long and deep—and the scientific-technical planning was for naught. There was nothing inherently secret about digging a three-mile-deep hole in the ground, and the company did not even plan to put nuclear materials in it. However, they were not up front with the community, and two communities, infuriated, struck back. This cycle repeats itself over and over with toxic materials. When corporations or government agencies do not communicate with the community, when they pull wool over the eyes of the public, when they think they will reduce conflict by squelching disclosure, they often discover that secrecy makes things much worse.

VIVIEN: I wonder about what is really being communicated and what the community is hearing. At the WIPP site, people were told that science had proven that nuclear storage was going to be safe because salt beds at the site had not dissolved over many millions of years, and so the nuclear waste was going to be perfectly stable. Did local residents consciously embrace risk in order to get the economic benefit from the waste site? Or was it more of a blind trust in the idea that there are scientific experts out there who have deemed something to be safe?

PETER: Think about manned spaceflight. You cannot make launching rockets perfectly safe. You are dealing with extremely high temperatures, and you have people sitting on top of a vast tank of exceedingly explosive material. We do not *have* to send manned missions into space; if we do, we need to understand the risks. We must, however, deal with nuclear waste because it is already among us, left over from almost seventy-five years of Cold War weapons and a half century of nuclear power production. Bad planning, national security pressure, and economic demands join forces: together they have left us with a vast legacy of waste.

We cannot leave the nuclear waste untouched in its unstable pools, tanks, and canisters, but there are no perfect solutions. The worst of all the solutions is to leave the 80 percent or so of fuel rods that now sit in cooling pools packed to the gills with both new, hot fuel and older cool fuel. Many of these pools are high aboveground, high in the air to facilitate transfer from the reactors. If a storage pool loses the cooling water, the hot fuel can catch fire and ignite the older, cold fuel. Among other measures, it seems clear that we must minimize the used reactor rods contained in these pools—that is, take out the old, cold rods that have been there three, four, or five years or more, reducing the consequences of a loss-of-cooling event. Then the question is, What are you going to do with that waste? It is not a choice between the one true, perfect solution and other bad solutions. It is a choice among imperfect solutions, each with risks, some more risky than others. When I talk about transparency, I mean not pretending that these things are risk-free, but instead I am urging that we talk soberly and realistically about the alternatives.

Containment judges that, ultimately, the waste is safer underground than it is on the surface. That does not mean that putting it underground is going to be absolutely safe. Indeed, one of the buried underground drums had the wrong mix of chemicals in it, leading to a fire and a leak. Former chair of the NRC Allison Macfarlane is clear that there is "no magic," as she puts it—the waste will not go away. Her point is reinforced by the accident: the underground is not perfect.[7] Still, had that same accident occurred aboveground it would have been far worse. A crucial part of communication is giving people an understanding of the real risks and the tools they need to assess realistic alternatives.

(IN)VISIBILITY AND MATERIALIZATION

BRINDA: This raises the larger question of what is visible and what is not. Nuclear weapons were out of sight of the public for so long, but in this book we are also talking about everyday waste and thinking critically about the processes of production. The United States is always talking about growth, naturalizing economic growth, and waste is the invisible outcome of that. There is justifiably so much fear around the issue of nuclear waste, but toxic waste is also the invisible by-product of consumer society.

PETER: Radioactivity is not apparent to us through sight, taste, or touch. It is not like a river turned to green suds or the carcass of a freighter being disassembled in Bangladesh. We use instruments to try to make radioactivity visible. Then there is the waste itself. As soon as something is dubbed waste, we want to avert our eyes. We do not want to look at sewage and sludge. We react with disdain, even disgust. There is a remark attributed to various famous physicists that put it directly: "No one will win a Nobel Prize for solving the nuclear waste problem." If you are a physicist, chemist, or metallurgist, prestige goes to new theories, novel instruments, and innovative experiments, not to sludge treatment.

A large part of the work I do both in print and film is designed to make things concrete. I want to see abstractions through their concrete manifestations: seeing the size of a million-gallon tank, or peering into those tanks; looking at the x-rays of waste material as it is being brought to WIPP; and seeing the trucks, staging areas, cooling pools, salt mines, the SRS, and Fukushima accident video records. Understanding means recognizing that these things have, as Bertolt Brecht might have said, a name and an address. Because waste is often hidden—or set in sparsely populated or impoverished counties—it is crucial to make the problem visible. As long as these issues remain cloudy abstractions, they remain both out of sight and out of mind.

It is worth underscoring that waste is often invisible not because of physics or our reluctance to think about the discarded, but because it is put into communities that are racially or economically marginalized. Most Americans probably do not know that a big part of the uranium for the American nuclear arsenal came from the Navajo Nation. Indeed, the Navajo Nation has had a thousand uranium mines on it, and those mines operated with poor safety oversight. Accidents keep happening even after the mines are closed. Houses have had to be torn down, and there are always worries about groundwater contamination. A particularly bad accident occurred on July 16, 1979, when United Nuclear Corporation's Church Rock uranium mill (northeast of Gallup, New Mexico) had a catastrophic failure of its retaining pool for radioactive sludge. Some thousand tons of solid radioactive detritus and ninety-three million gallons of acidic sludge broke its pool and ran into the Puerco River. When that pool broke, the sludge came in a flash flood down the hills—children were playing in it, and animals drank from it. It was a disaster, but the general public, away from what is now a Superfund site, is largely unfamiliar with the incident. More generally, many people may not even be aware of the crucial part the Navajo Nation played in providing uranium for the Cold War weapons complex. We need to make things visible in order to confront their materiality and their location, and to make it both understandable and politically addressable.

Brinda: There are often harmful yet innocuous-looking landscapes in highly populated areas. Superfund sites are a good example of this. You might see injection wells in the ground not far from residential communities but have no meaningful understanding of what you are seeing or the history of a particular site. We live a mile away from freeways where populations are exposed to fine particulate matter that is known to be carcinogenic. All this is visible but also invisible because it is normalized. How, then, can we mobilize against such ubiquitous toxicity?

Peter: To me, an essential first step is materializing things. If you do not have an idea of where toxicities are or what they are like, it is hard to address them. We barely grasp what we cannot picture. That is why when I look at the history of physics I am interested in laboratories, equipment, and the procedures of image making. Learning from the materialization of things often tells us about how to understand the abstract. And if we do not have a sense of where toxic things are, what they are like, and whom they affect, it is hard to mobilize any kind of action.

There are different strategies for materializing (or perhaps rematerializing) toxic environments. We can write with the kind of evocative, human, spatial specificity we need. I think many of the authors in this volume are doing just that. It is crucial, I think, that other forms of visibility—film, photography, exhibits—complement the written word. There are also other ways of making things visible: for example, new forms of maps that show the distribution of plastics or organic compounds or nuclear materials. These maps let people know more about the dimensions and shapes of things such as the radon prong that goes up through Pennsylvania, which has affected lots of people. This kind of information can alert people about installing the right kind of equipment so that the presence of radioactive dust can be known.

VIVIEN: I understand what you are saying about the need to make something that is invisible visible by emphasizing its materiality. Yet, as the case studies in this book have shown, some of the effects of toxicity on the body are complex and difficult to establish epidemiologically. When we reduce toxic effects on the body to material tumors, for example, are we perhaps ignoring more holistic understandings of mental and emotional health and well-being?

PETER: By materialization, I do not just mean of the contaminants. I mean we have to make visible the real lives of people who live in and among toxic environments. When we look, really look, it is clear that lives are not run only on doses and dose-response curves. Mr. Sasaki, the older man in *Containment* who goes back to his house in Fukushima every other day, shows us a great deal. He would like to move back permanently but cannot: put aside the cesium-137, and there are no stores, no nurses, no doctors, no hospitals; he cannot bring back family since there are neither jobs nor schools.

Second, there is, as your question suggests, fear. Too often we read or hear experts say that fear is not real, that their calculations suggest that people could move back to the areas around Chernobyl or Fukushima or some Pacific islands without a large increase in cancer rates. Here the various themes we have been discussing intersect. Secrecy and the concomitant minimization of risk by authorities in many countries have left people without faith in nuclear reassurance. Couple that with a casual dismissal of trauma and anxiety, and you have a recipe for never understanding the lives people actually live. You can tell people that their children would only suffer a 1 percent chance of contracting cancer due to exposure near the Fukushima plant, but is it then really a surprise that people say: "Thanks, but I'll live elsewhere." In recent years, at long last, PTSD has gotten better traction, at least for returning soldiers, as a "real" disorder. We can only hope that our societal stance toward acute anxiety, depression, and other difficulties will come to be treated with the seriousness with which we treat cardiac arrest or staph infections.

Third, there are challenging technical questions. What dose did people get? This is often *very* hard to determine. Unless you are near the explosion of an atomic bomb, where simple geometry can tell you how much gamma radiation you received, it is more or less impossible to calculate a dose of radiation accurately, even if you know a worker was in a certain plant over a period of, say, ten years. Exposure could be entirely different on one side of a hallway than on the other and depend on whether the worker was laboring where radioactive dust was present, or whether the worker had a mask or a breathing apparatus, or whether he or she was typically standing behind a protective vent, next to a milling machine, or outside a glove box. Exposure depends on a great range of details, and, historically, there was very little monitoring, so people do not really know what the doses were. I am speaking here about radioactive doses, but analogous questions arise with myriad industrial chemicals, including asbestos, heavy metals, organic compounds, and other contaminants.

Then there is the question of the relationship between the dose and resulting cancer. That, too, requires statistics, and even the most basic data are debated. At very high levels of radiological exposure, you get immediately sick and can die. Acute radiation poisoning is tragically clear. At low levels, it is still hugely debated. There are even people who say low levels of radiation are good for you. That claim is widely deprecated as bad science, but it is a view that one hears among some pronuclear activists. So, there are confounded levels of uncertainty: if you develop cancer, you can never prove in a particular case what the origin of that cancer was. That is, even if you know your exposure, and even if you know the statistical correlation between exposure and cancer rates, you still cannot know that *your particular* eye cancer, liver cancer, breast cancer, or testicular cancer came from a specific source.

There are many kinds of anxieties that are associated with living in a contaminated area. What does such acute anxiety about exposure do to people as they worry about their future health, or the future health of their children? No doubt that living in the midst of a radioactive area can be very psychologically damaging, even if these psychological impacts are hard to prove. In Adriana Petryna's *Life Exposed* (2002), the work I referenced earlier, she talks about how people struggled to try to make a claim for psychological damage beyond the physiological damage to activate compensations or housing or schooling or whatever it was that they needed in that all-too-turbulent period of Ukrainian history. Very few paths led to help for people with often debilitating anxiety from having to live in a region of constant radioactivity.

Sometimes people will say, "Well, why did people leave Fukushima? It's really damaging to them to be nuclear refugees." But could you really have stayed there? During the accident sequence itself, no one knew how bad it would get; no one knew if fuel pool 4 would collapse; no one knew whether the reactor meltdowns would result in a far worse loss of containment. It was uncertain for months, by which time the problem of reinhabitation was truly acute, not to mention the ordinary, nonnuclear damage to houses from moisture, mold, and fauna. When experts and politicians ask with surprise why people do not head back, they ignore the trauma that those 150,000 people have undergone.

NUCLEAR LESSONS

PETER: There is a remark at the beginning of Kant's *First Critique* that states that there are some problems we face that we cannot avoid and cannot solve. Some nuclear issues are like that. We cannot turn away and say we do not have to deal with this. You could say we do not have to go to Mars. We could choose not go to Mars or decide not go to Mars for fifty or a hundred or five hundred years. That is a choice. But we cannot leave the waste in these big swimming pools. It is not an option. People say, "Just bury it and don't mark it." Well, oil companies cannot wait to dig underneath the WIPP site because of the staggering amount of oil and gas there. You *cannot not* mark such sites; ethically, it would be wrong and people recognize that. So then you mark the site, but the various marking techniques each present great challenges. Do you make a nuclear Rosetta Stone, do you use pictures, do you use cartoons, do you bury samples, do you make stories up, or do you try to do a little of columns A, B, C, and D?

I do not think there is any way of avoiding these questions. We have to figure out what the best among the alternatives is, none of which are perfect. The larger lesson, it seems to me, is that we need to think about the consequences of producing the waste as part of the production process itself. If you want to make aluminum or you want to make electronics, you should have to bear responsibility for the cost of that socially, politically, financially, and environmentally. We have to think of waste not as something that we can avert our eyes from but as part of the process. That is a big change because people offload those things: "Oh, the mountaintop's unstable in West Virginia because we strip mined it, but that's a problem for the government, for some other generation." We want to get that coal out now, and we have immediate needs. Or we want to make iPhones cheaply in China, and we will deal with the acid effluents some other time. But it is always more expensive to deal with it once it is decontained. It will require a real political and economic change to think about waste as part of the process of production itself. Nobody wants to do that.

BRINDA: Do you see links between concerns about climate change, which is a slow-moving crisis of planetary proportions, and concerns about nuclear catastrophe? Is there hope in the face of such monumental problems?

PETER: In a way, as big an issue as nuclear waste is, it is a model, maybe, for the even bigger issue of global warming, where we have done a terrible job of imagining the future. Even though the water is bubbling up in Miami through the waste grates on the ground, you can still have a state government banning discussion or use of the phrase "global warming." If thinking about nuclear waste gets us to reason beyond the political periods of election cycles or the economic periods of fiscal quarters, that would be a great thing.

Is *Containment* pessimistic? I think the fact that in this one solitary instance it actually came to pass that several dozen people with backing from the government produced some thought about how to mark nuclear waste for the distant future is actually a very optimistic turn of events. Mostly, we think in super-short time frames that are not even remotely close to a human lifetime. True, each of the schemes for marking seem to pale before the vast expanse of ten thousand years. Yet there is something very moving, inspiring even, about the government moving beyond the demands of a fiscal quarter to put the ethics of ten millennia on its agenda.

If the ten-thousand-year ambition inspires us to think a little bit more about lead and mercury and organic compounds, if it prompted questioning about the long-term effects of fracking fluids in the water supply and plastics in the ocean, that would be good. To think a little bit beyond the immediate: to think up to parents and grandparents or great-grandparents and down to children and grandchildren and great-grandchildren. Maybe we think that far, but, relative to the long half-life of plutonium, that is a pretty narrow horizon. Thinking spatially and thinking temporally about these issues is really important. As difficult as it is to get at, the nuclear is still a finite thing. There are four hundred nuclear power plants in the world. There are maybe fifteen thousand nuclear weapons. It is not an infinite task, and so grappling with this, and the mining from the beginning of the cycle to the burial, is something that we could at least get our heads around. Perhaps this view beyond our immediate temporal horizon could inform these other issues of prevalent waste, contamination, and, at the largest scale, global warming.

NOTES

1. For a careful consideration of what it might mean to include citizens in environmental policy decisions, see Bäckstrand, "Civic Science for Sustainability."

2. Balint, *Wicked Environmental Problems*. Book description, https://islandpress.org/book/wicked-environmental-problems.

3. South Carolina Department of Health and Environmental Control, "Eating Fish from the Savannah River."

4. Georgia Department of Natural Resources, "Guidelines for Eating Fish from Georgia Waters," 40.

5. Petryna, *Life Exposed*.

6. North Dakota residents of Spink County first heard about the borehole project from local newspapers, greatly worsening the controversy. See, for example, Vossen, "Protests Spur Rethink on Deep Borehole Test for Nuclear Waste."

7. U.S. Department of Energy, "What Happened at WIPP in February 2014."

BIBLIOGRAPHY

Bäckstrand, Karen. "Civic Science for Sustainability: Reframing the Role of Experts, Policy Making and Citizens in Environmental Governance." In *The Postcolonial Science and Technology Studies Reader*, edited by Sandra Harding, 439–58. Durham, NC: Duke University Press, 2011.

Balint, Peter J. *Wicked Environmental Problems: Managing Uncertainty and Conflict.* Washington, DC: Island Press, 2011.

Georgia Department of Natural Resources. "Guidelines for Eating Fish from Georgia Waters." 2013. https://epd.georgia.gov/sites/epd.georgia.gov/files/related_files/site_page/GADNR_FishConsumptionGuidelines_Y2013.pdf.

Petryna, Adriana. *Life Exposed: Biological Citizens after Chernobyl.* Princeton, NJ: Princeton University Press, 2013.

South Carolina Department of Health and Environmental Control. "Eating Fish from the Savannah River." Accessed December 20, 2016. http://www.srs.gov/general/pubs/fish100101final.pdf.

U.S. Department of Energy. "What Happened at WIPP in February 2014." Accessed December 20, 2016. http://www.wipp.energy.gov/wipprecovery/accident_desc.html.

Vossen, Paul. "Protests Spur Rethink on Deep Borehole Test for Nuclear Waste." *Science* 27 (September 2016). http://www.sciencemag.org/news/2016/09/protests-spur-rethink-deep-borehole-test-nuclear-waste.

Contributors

Janet Farrell Brodie has a PhD in history from University of Chicago and a BA from University of California–Berkeley. She is a professor in the History Department at Claremont Graduate University. Her recent scholarship focuses on secrecy in the nuclear era.

Lindsey Dillon is an assistant professor of sociology at University of California–Santa Cruz. Her research examines the intersection of race and urban environmental issues in U.S. cities, focusing on San Francisco. Her work combines political ecology, feminist geography, critical race studies, and science and technology studies. She holds a PhD in geography from University of California–Berkeley.

Peter Galison is the Pellegrino University Professor in history of science and physics at Harvard University. In 1997 Galison was awarded a John D. and Catherine T. MacArthur Foundation Fellowship, in 1998 the Pfizer Award in history of science (for *Image and Logic*), in 1999 the Max Planck and Humboldt Stiftung Prize, and in 2018 the Abraham Pais Prize for history of physics. His books include *How Experiments End* (1987), *Einstein's Clocks, Poincaré's Maps* (2003), and *Objectivity* (with Lorraine Daston, 2007), and his films include *Ultimate Weapon: The H-bomb Dilemma* (with Pamela

Hogan); *Secrecy* (with Robb Moss, Sundance, 2008), on national security secrecy; and *Containment* (Full Frame, 2015), on guarding radioactive materials for ten thousand years. Galison has collaborated with South African artist William Kentridge on a multiscreen installation titled "The Refusal of Time" (2012) and is now at work on a film about black holes and the limits of knowledge.

Vivien Hamilton is an associate professor of history of science and director of the Hixon-Riggs Program for Responsive Science and Engineering at Harvey Mudd College. Her work examines the history of medical technologies, focusing on questions of authority, expertise, and cross-disciplinary collaboration. She holds a PhD in history of science from the University of Toronto and is currently completing a book examining the role of physics in the early history of radiology.

James G. Lewis is the historian at the Forest History Society, executive producer of the Emmy Award–winning documentary *America's First Forest: Carl Schenck and the Asheville Experiment* (2016), and author of *The Forest Service and the Greatest Good: A Centennial History* (2005).

Char Miller is the W. M. Keck Professor of environmental analysis at Pomona College and author of *Not So Golden State: Sustainability vs. the California Dream* (2016) and *America's Great National Forests, Wilderness, and Grasslands* (2016).

William Palmer is a doctoral student in the History Department at Claremont Graduate University. His dissertation, in progress, is titled "The Burden of the Cold War Body."

Brinda Sarathy is a professor of environmental analysis and director of the Robert Redford Conservancy for Southern California Sustainability at Pitzer College. She holds a PhD in environmental science, policy, and management from the University of California–Berkeley. In addition to the books *Pineros: Latino Labour and the Changing Face of Forestry in the Pacific Northwest* (2012) and *Partnerships for Empowerment: Participatory Research for Community-Based Natural Resource Management* (2008), Sarathy has published articles in *Journal of Forestry, Society and Natural Resources, Policy Sciences, Race, Gender and Class,* and *Local Environment.*

Bhavna Shamasunder is an assistant professor in the Urban and Environmental Policy Department at Occidental College in Los Angeles. She teaches and conducts interdisciplinary research on environmental health and justice, often in partnership with disproportionately impacted communities. She has published in scientific and social scientific journals, including the *American Journal of Obstetrics and Gynecology, International Journal of Environmental Research and Public Health,* and *Environmental Justice.* She is currently working on a book manuscript on urban oil drilling.

Sarah Stanford-McIntyre received her PhD from William & Mary in 2017. She is the 2017–2018 Bernard Majewski Fellow and an instructor of history at the University of Wyoming. Her current book manuscript examines efforts to control desert ecology and manage workplace discontent within a quickly globalizing, and increasingly multiethnic, Texas oil industry.

Naoko Wake is an associate professor of history at Michigan State University. Her field of specialization is the history of medicine, gender, and sexuality in the United States and the Pacific Rim, and she is author of *Private Practices: Harry Stack Sullivan, the Science of Homosexuality, and American Liberalism* (2011) and coauthor (with Shinpei Takeda) of *Hiroshima/Nagasaki Beyond the Ocean* [*Umi o koeta Hiroshima Nagasaki*] (2014). She is currently completing her second monograph, titled *Bombing Americans: Gender and Trans-Pacific Remembering after World War II,* which explores the history of Japanese American and Korean American survivors of the 1945 atomic bombings of Hiroshima and Nagasaki, with a focus on their cross-national and gendered memory, identity, and activism.

Alexander Zahara is a PhD candidate in geography at Memorial University and cross-disciplinary researcher examining the intersections of pollution and colonialism in Northern Canada. Alex is a contributing editor of the *Discard Studies* blog, and his work has appeared in journals such as *Environmental Toxicology and Chemistry, Environmental Humanities,* and the edited collection *Anthropocene Feminism.* His research interests include action-oriented research and feminist, queer, and anti-colonial science studies.

Index

Note: Page numbers in *italics* indicate figures and tables.